Medicine and Biomedical Sciences in Modern History

The aim of this series is to illuminate the development and impact of medicine and the biomedical sciences in the modern era. The series was founded by the late Professor John Pickstone, and its ambitions reflect his commitment to the integrated study of medicine, science and technology in their contexts. He repeatedly commented that it was a pity that the foundation discipline of the field, for which he popularized the acronym 'HSTM' (History of Science, Technology and Medicine) had been the history of science rather than the history of medicine. His point was that historians of science had too often focused just on scientific ideas and institutions, while historians of medicine always had to consider the understanding, management and meanings of diseases in their socio-economic, cultural, technological and political contexts. In the event, most of the books in the series dealt with medicine and the biomedical sciences, and the changed series title reflects this. However, as the new editors we share Professor Pickstone's enthusiasm for the integrated study of medicine, science and technology, encouraging studies on biomedical science, translational medicine, clinical practice, disease histories, medical technologies, medical specialisms and health policies.

The books in this series will present medicine and biomedical science as crucial features of modern culture, analysing their economic, social and political aspects, while not neglecting their expert content and context. Our authors investigate the uses and consequences of technical knowledge, and how it shaped, and was shaped by, particular economic, social and political structures. In re-launching the Series, we hope to build on its strengths but extend its geographical range beyond Western Europe and North America.

Medicine and Biomedical Sciences in Modern History is intended to supply analysis and stimulate debate. All books are based on searching historical study of topics which are important, not least because they cut across conventional academic boundaries. They should appeal not just to historians, nor just to medical practitioners, scientists and engineers, but to all who are interested in the place of medicine and biomedical sciences in modern history.

More information about this series at
http://www.palgrave.com/gp/series/15183

Stuart Anderson

Pharmacy and Professionalization in the British Empire, 1780–1970

palgrave
macmillan

Stuart Anderson
Centre for History in Public Health
London School of Hygiene &
Tropical Medicine
London, UK

Medicine and Biomedical Sciences in Modern History
ISBN 978-3-030-78979-4 ISBN 978-3-030-78980-0 (eBook)
https://doi.org/10.1007/978-3-030-78980-0

Cover photo: Singapore Dispensary, c.1890. Credit: Chronicle/Alamy Stock Photo

This Palgrave Macmillan imprint is published by the registered company Springer Nature Switzerland AG
The registered company address is: Gewerbestrasse 11, 6330 Cham, Switzerland

For Liz

Preface

Pharmacy—the occupation concerned with the making, supply, and use of medicines—has received surprisingly little attention from scholars. Whilst sociologists have examined its professionalization, medical historians have explored the evolution of the apothecary, and pharmaceutical historians have described its people and practices, such studies have rarely been brought together to present a broader picture. Most relate to a single country, with few providing a basis for international comparison. Studies of colonial pharmacy are even fewer. Yet a better understanding of the development and nature of pharmacy requires an awareness of its professionalization across time and space.

Those who venture into colonial history are quickly confronted by many questions, not least those posed by Alan Lester in his 2006 (4, p. 124) article in *History Compass*: how to write about the many, diverse places that constituted the British Empire in the same text? And how to conceive of both the differences and the connections between Britain and its various colonies? If these questions have been considerable challenges for imperial historians, for others the obstacles are huge indeed. But Lester's questions suggest that many different histories can, and indeed should, be written.

This is especially true of fields that have been relatively neglected by historians such as pharmacy. There is much to be said about pharmacy and empire, and many different histories to be told. In their contribution to discussions about the future of the history of pharmacy, Anna Greenwood and Hilary Ingram note that 'there is still much work required to better situate pharmacy history within colonial, post-colonial, and international frameworks, and to show the multifarious ways pharmaceutical trades, fashions and cultures moved around, against and beyond, colonial and commonwealth networks' (*Pharmacy in History*, 61 [2019], p. 129).

The limited interest shown in the subject to date by medical and imperial historians is disappointing but understandable. For medical historians, pharmacy is often of peripheral interest to the professionalization of medicine. Studies of early practitioners of both medicine and pharmacy, such as the ship's surgeon and the apothecary, invariably focus on their medical rather than pharmaceutical activities, and on their relevance to the history of medicine rather than that of pharmacy. For imperial historians, pharmacy is usually a marginal and easily ignored topic of limited intrinsic interest.

One of the factors contributing to this limited interest is the lack of a broad overview of the subject. The boundary between medicine and pharmacy, and how (and indeed if) pharmacy became a separate and autonomous profession is often unclear, and there is a need for greater clarity. In Britain, that process occurred in the late nineteenth century at a time when the Empire was expanding rapidly. This therefore raises interesting questions about the relationship between a pharmacy that is professionalizing in Britain, and its regulation and development in the colonies. How were the two linked? In short, how British was pharmacy in the British Empire?

This book, then, is a pharmaceutical history of empire. It is a top-down narrative history that draws on research in the history of pharmacy, that has often been carried out some years ago in former British colonies, usually in isolation from work done elsewhere and in other fields. It embraces a large geographical area, but focusses narrowly on processes of professionalization, and on commonalities and differences between colonies. It makes no claim to in-depth historical analysis, nor does it seek to provide important new insights into professionalization, pharmacy, or colonialism.

Its aim is a simple one: to bring together material relating to the history of pharmacy across the countries of the former British Empire that is often difficult to source, in an organized and accessible form. Its purpose is to provide a framework which will hopefully be of use to others wishing to carry out more detailed studies. Its shortcomings and limitations are readily acknowledged; there is an over-reliance on a limited range of secondary sources, and on contemporary professional and trade journals. A variety of primary sources have been accessed, principally official documents, circulars, and letters, but little use has been made of social historical sources such as letters and diaries.

It is hoped that the book will be of interest to medical and imperial historians as well as to historians of pharmacy. Whist I have cited some important scholarship in imperial and medical history, I have not attempted to engage extensively with it. Neither have I tackled issues of gender, exploitation, consumerism, and imperial hegemony in the book, although I hope it might stimulate and provide a possible framework for such studies in the future. And whilst the book has little to say about pharmacy in hospitals and other settings, or about the wholesaling, manufacturing, or marketing of pharmaceuticals, I hope it suggests ways in which pharmaceutical history might better engage with broader historical and sociological literatures.

It is hoped too that the book might help pharmacy students—along with pharmacists new and old—across the English-speaking world, to gain a better understanding of the different ways in which pharmacy evolved as a profession and what factors have helped shaped it. Despite its limitations, I hope this volume stimulates interest in the history of pharmacy and empire and provides a basis for further studies.

London, UK — Stuart Anderson

Acknowledgements

This book has been ten years in the making. Its origins lie in conversations with pharmaceutical historians from former British colonies about the lasting impact of British pharmacy on education and practice in their countries, and about the role played by British authorities in shaping it. Since then, I have received many suggestions and a great deal of information and advice from friends and colleagues across the world. Some have visited archives on my behalf, and others have scanned and sent me documents. I am extremely grateful to them all, some of whom sadly are no longer with us.

Others have read through drafts of individual chapters and offered helpful suggestions and corrections. Any remaining errors are mine. Those to whom special thanks are due include John Bachynsky, Greg Bond, John Joseph Borg, Tony Cartwright, Patrick Chiu, John Crellin, John Ferguson, Ellen Grizzle, Greg Haines, Andrew Hersom, Greg Higby, Peter Homan, Briony Hudson, Augustina Koduah, Geoff Miller, Harkishan Singh, Thomas Paraidathathu, Ray Pogir, Halil Tekiner, Nina Thune, and Nick Wood. Thanks also to Lee Williams for help in producing the maps.

I am grateful to the staff at several archives in both Britain and overseas for their help and support. Particular thanks are due to those at the British Library, The National Archives, The Wellcome Library, the Royal College of Physicians, and the Archives of the Society of Apothecaries in London. Staff at the Royal Pharmaceutical Society in London, especially

Jane Trodd and Karen Horn from the Library and Catherine Walker from the Museum, are due a special mention, as is Sophie Clapp, Curator of the Walgreens Boots Alliance Archive in Nottingham.

I am extremely grateful to the peer reviewers for many helpful comments and suggestions, and to the Series Editors for their encouragement and support. Thanks are also due to the team at Palgrave, particularly Lucy Kidwell, Redhu Ruthroyoni, and John Justin Thomyyar, for efficiently seeing through the editing and production processes for the book. I would also like to thank my colleagues at the Centre for History in Public Health at the London School of Hygiene and Tropical Medicine, particularly Virginia Berridge, Martin Gorsky, John Manton, Susanne McGregor, and Alex Mold, for their continuing support and encouragement.

My greatest debt, however, is to my wife Liz who accompanied me on many trips and who has been unfailing in her support throughout the process of research and writing. The book is dedicated to her.

Stuart Anderson

Contents

About the Author

Stuart Anderson is Professor Emeritus of the History of Pharmacy at the Centre for History in Public Health at the London School of Hygiene & Tropical Medicine (LSHTM), UK. He was previously Associate and later Acting Dean of Education at LSHTM until 2015. He has been researching and writing about the history of pharmacy for over 30 years, and his output has included over 30 chapters in edited volumes and over 60 articles in peer-reviewed journals.

Stuart edited *Making Medicines: A Brief History of Pharmacy and Pharmaceuticals*, published in 2005, and is now the editor of the international peer-reviewed journal *Pharmaceutical Historian*. He is a former chair of the Society for the Social History of Medicine, a past president of the British Society for the History of Pharmacy, and a Fellow of the Royal Pharmaceutical Society. He was President of the International Academy for the History of Pharmacy between 2009 and 2017.

Abbreviations

BACD	British Association of Chemists and Druggists
BMA	British Medical Association
BMJ	British Medical Journal
CAP	Caribbean Association of Pharmacists
CDTA	Chemists and Druggists' Trade Association
CMO	Chief Medical Officer
CPA	Commonwealth Pharmaceutical Association
FIP	International Pharmaceutical Federation
FMS	Federal Malay States
GCPDU	Gold Coast Pharmacist and Druggist Union
GMC	General Medical Council
HKGG	Hong Kong Government Gazette
IMS	Indian Medical Service
KC	King's Counsel
MP	Member of Parliament
NHI	National Health Insurance
PSA	Pharmaceutical Society of Australia
PSGB	Pharmaceutical Society of Great Britain
PSGC	Pharmaceutical Society of the Gold Coast
PSGH	Pharmaceutical Society of Ghana
PSHK	Pharmaceutical Society of Hong Kong
PSI	Pharmaceutical Society of India
PSN	Pharmaceutical Society of Nigeria
PSNSW	Pharmaceutical Society of New South Wales
PSNZ	Pharmaceutical Society of New Zealand
PST	Pharmaceutical Society of Tasmania

PSV	Pharmaceutical Society of Victoria
PSWA	Pharmaceutical Society of Western Australia
QPS	Queensland Pharmaceutical Society
RCP	Royal College of Physicians
RPU	Retail Pharmacists Union
SAPS	South Australian Pharmaceutical Society
TCM	Traditional Chinese Medicine
TNA	The National Archives, London
USCD	United Society of Chemists and Druggists

List of Figures

List of Tables

CHAPTER 1

Pharmacy and the British Empire

Soon after the Pharmaceutical Society of Great Britain (PSGB) was founded in 1841 its relationship with the Empire was made explicit. In 1844 Jonathan Pereira, a physician who had been appointed its first professor of materia medica, proposed that a committee be formed that could "be made the means of declaring to the remotest part of our colonial possessions the wants of the mother country, and conversely, of making known to England the capabilities of the different portions of the British Empire".[1] The new society was as much concerned with medicines and their sources as it was with developing the profession of pharmacy. It actively pursued both objectives across the Empire. Over the following decades colonial authorities sought the PSGB's advice in formulating legislation concerning the regulation of pharmacy and the control of the sale of poisons. In 1881 a draft of a proposed Sale of Drugs and Poisons Ordinance for Jamaica sent to the PSGB for comment was one of many such exchanges.[2] By the 1890s a clear policy of 'imperializing' the pharmacopoeia (the list of officially approved medicines) had been agreed, and when the fifth edition of the *British Pharmacopoeia* (BP) was published by the General Medical Council in 1914 it was described as being "suitable for the whole Empire".[3] Its preface listed the 'Divisions of Empire' where it was to apply (Table 1.1).[4]

S. Anderson, *Pharmacy and Professionalization in the British Empire, 1780–1970*, Medicine and Biomedical Sciences in Modern History,
https://doi.org/10.1007/978-3-030-78980-0_1

Table 1.1 Divisions of the British Empire referred to in the *British Pharmacopoeia 1914*

Division	*Colonies*
Indian Division	Ajmer-Merwara, The Andamans, Assam, Bengal, Bihar and Orissa, Bombay, Baluchistan, Burma, The Central Provinces and Berar, Coorg, Delhi, Madras, The North-West Frontier Province, the Punjab, United Provinces of Agra and Oudh
African Division	Basutoland, Bechuanaland Protectorate, Gambia, Gold Coast, Nigeria, Northern Rhodesia, Southern Rhodesia, Sierra Leone, Swaziland, The Union of South Africa (provinces of Cape of Good Hope, Natal, Orange Free State, Transvaal) Saint Helena
Australasian Division	New South Wales, Queensland, South Australia, Tasmania, Victoria, Western Australia, Northern Territory of Australia, Federal Capital Territory; forming the Commonwealth of Australia. New Zealand, Fiji Islands, Papua, Western Pacific
Eastern Division	Ceylon, Hong Kong, Labuan, Mauritius, Seychelles, Straits Settlements, Weihaiwei
Mediterranean Division	Cyprus, Gibraltar, Malta
North American Division	Alberta, British Columbia, Manitoba, New Brunswick, North-West Territories, Nova Scotia, Ontario, Prince Edward Island, Quebec, Saskatchewan, Yukon; forming the Dominion of Canada. Newfoundland
West Indian Division	Bahama Islands, Barbados, Bermuda Islands, British Guiana, British Honduras, Jamaica and Turks and Caicos Islands, Leeward Islands (Antigua, Dominica, Montserrat, Saint Christopher and Nevis, Virgin Islands), Trinidad and Tobago, Windward Islands (Grenada, Saint Lucia, Saint Vincent)
South Atlantic	The Falkland Islands

Source *British Pharmacopoeia,* London, 1914, p. xxii

The professionalization of pharmacy in Britain in the second half of the nineteenth century coincided with a rapid expansion of the Empire, which reached its peak around 1921 (Fig. 1.1). But for over one hundred years substantial numbers of pharmacy practitioners left Britain to begin new lives as settlers or colonists. They took with them a 'British model' of pharmacy which usually formed the basis of education, training, and practice in the colonies, potentially opening the door to the free movement of pharmacists between colonies. In 1871 Ontario recognized the qualifications of pharmacists trained in Quebec, and by 1881 Australian pharmacists in Victoria had sought the reciprocal recognition of pharmacy qualifications between itself and Britain. It took over thirty years for agreement to be reached. In 1913 a statement on pharmaceutical qualifications and registration in the British Empire, including its terms of reciprocity agreement, was published by the PSGB (Chapter 12).[5]

These events raise a host of questions in the history of pharmacy relating to the relationship between the centre and the periphery—between Britain and the colonies—and its nature. These questions relate to the medicines used and the people who made and sold them, to the interaction between western and indigenous pharmacy, and to the people who used, rejected, or were denied access to them. Many varied histories need to be written, but to date few have, as pharmacy has received little attention from medical and colonial researchers, whilst pharmaceutical historians have largely ignored the imperial context of pharmacy.

This book aims to stimulate future studies in this field and to provide a framework for them. It examines how pharmacy developed in different parts of the Empire, and considers to what extent the professional framework of pharmacy—its institutions and legislation, its arrangements for education and registration—were based on the British model. Was this model simply transplanted into the colonies? Did professionalization follow the same pattern in the colonies as it did in Britain? How did the emergence of pharmacy as a profession distinct from medicine differ between colonies? How much influence did the PSGB have? And how uniform was the experience between colonies?

Britain had an Empire long before the mid-nineteenth century, and the nature of the relationship between pharmacy and medicine differed considerably between colonies. In the West Indies pharmacy was inextricably linked to slavery; in Malta pharmacy and medicine had been separate and distinct professions for centuries before the British arrived; whilst in India pharmacy remained undeveloped throughout British rule. The central theme of this book is the professionalization of pharmacy in the colonies—the process by which it emerged as a separate and autonomous profession. At its heart is a simple question; how British was pharmacy in the British Empire?

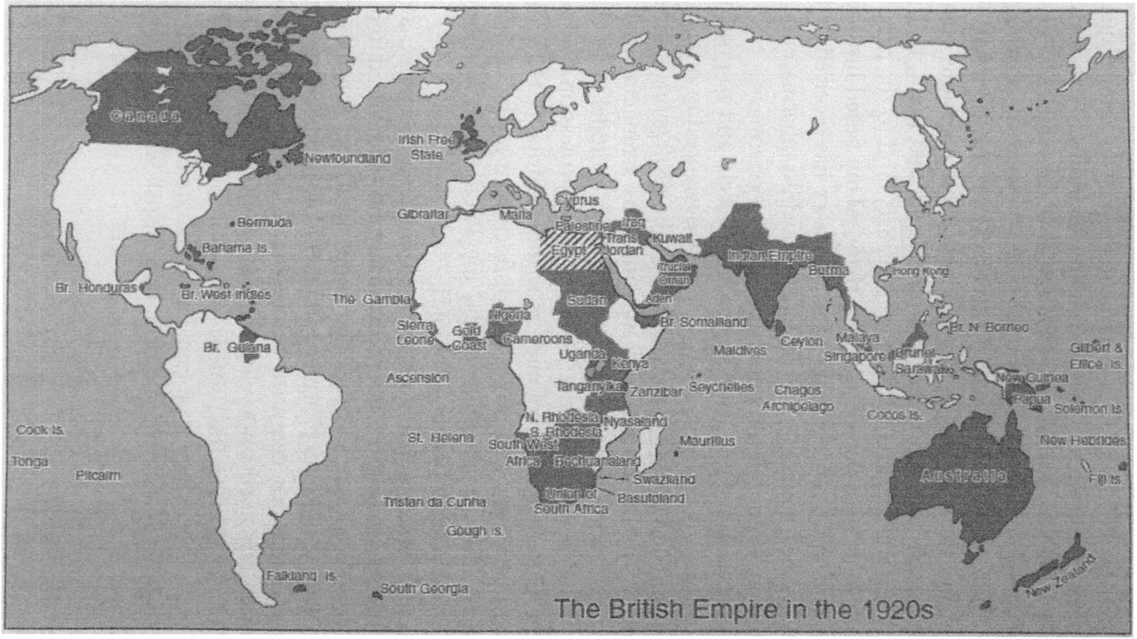

Fig. 1.1 Map of British Empire, 1920s (Ashley Jackson, *The British Empire: A Very Short Introduction*. Reproduced with permission of the Licensor through PLSclear)

The Development of an Empire

The British Empire developed in a number of distinct phases.[6] The 'first British Empire'—roughly spanning the period 1583 to 1783—was an Atlantic empire, based on north America and the West Indies. It was a commercial empire founded on extensive British settlement overseas. It began with early incursions into Atlantic space and continued until the end of the American Revolutionary War.[7] The 'second British Empire' was characterized by British expansion into India, south east Asia, and Australia, and extended roughly from 1783 to around 1860. A third British Empire is linked to the 'scramble for Africa' that occurred from the mid-1870s.[8] The British Empire reached its greatest extent, in terms of both population and area, in 1921, prior to the loss of the Republic of Ireland. Thereafter followed a period of gradual decolonization, with many former British colonies gaining independence during the 1950s and 1960s.

Empire formation was a dynamic process, with territories being gained and lost as a result of wars, treaties, occupations, requests for protection, or other reasons. Borders moved, territories changed their names, and alliances were made, and the status of the territories of Empire was highly variable; some were colonies, some were protectorates, others were dominions. Those claimed in the name of the king and controlled directly by the Crown became Crown Colonies, the model used as the main method for creating and governing colonies in the nineteenth century. Many colonies were part of the British Empire one year but not another. In this book I focus on the British Empire as it was in 1914.

The first British Empire was accompanied by emigration on a massive scale. During the seventeenth century around 1 million people migrated, about 70 per cent of whom were English.[9] Between 1646 and 1670 some 400,000 English and Irish people settled in America or the West Indies. Another 217,000 emigrated from the British Isles to the thirteen American colonies between 1700 and 1775.[10] Virtually the entire colonial medical profession was Scottish trained, and more than 150 Scottish doctors emigrated to America during the eighteenth century. Further mass British emigration occurred as new regions of empire opened up in India, Australia, and Africa. People from diverse backgrounds took with them very different skills and expectations. Alongside countless unskilled labourers were those with agricultural experience, skills in setting up plantations and in running merchant houses.[11] There was also a steady stream of professional men and artisans, including teachers and accountants, weavers and smiths, bakers, and carpenters. And amongst them were doctors, apothecaries, chemists, and druggists, along with others with few if any qualifications claiming to be so.

The reasons people undertook such life-changing journeys were many and varied. Early colonists were often settlers needing to escape religious persecution in their own land. Others moved because of the lure of a better life and increased prosperity. Later, people were transported as convicts, and in due course colonial governments began to encourage settlement in their countries with assisted migration schemes. Some went to make their fortune, eventually returning to Britain to live in comfortable retirement. Each wave of migration usually included individuals engaged in pharmacy in one form or another. They went out for the same reasons as everyone else, whether to make their fortune in India, through transportation to a penal colony in Australia, or through assisted migration to New Zealand.

They came from all parts of Great Britain, and migration continued well into the twentieth century. Between 1853 and 1920 there were over 8 million outward passenger journeys from England alone. By the early twentieth century Irish colonists constituted a significant proportion of settler populations. In 1911 there were around 14,600 Irish natives in South Africa, 12,200 in India, 1000 in the Maltese islands, 400 in Ceylon, 250 in the Straits Settlements, and 160 in the Federated Malay States.[12] People with pharmaceutical skills or aspirations went to the colonies to set up businesses. Their backgrounds were highly varied; some were originally employees of trading companies such as the English East India Company; others went out with dreams of making their fortune; still others went out as prisoners or as migrants hoping for a better life. By the nineteenth century, substantial numbers went out as assisted migrants, to Australia, New Zealand, and South Africa.

Empires have been the subject of extensive scholarship, and the imperial literature is vast. The field now has an extensive theoretical foundation, with much of it relating to post-colonial analysis and reflection. Imperial literature emphasizes the importance of political, economic, and social context in understanding the British Empire. In India, for example, Susan Bayly suggests that in order to understand the complexity of Empire it is necessary to examine the 'interactions between Britons and their Asian subjects in the realms of religion, law, science, education, and modernising social activism'.[13] She examines what she calls 'invisible empire'; the context in terms of the forces shaping local cultural encounters; religion, particularly differences in worship of people of the East or West; and the intersection between culture and politics, including issues of power and authority.

Interaction occurred at the boundaries between British pharmacy and indigenous pharmacy, with each adopting remedies from the other's culture; each was changed by the other. By British pharmacy I mean pharmacy as it was practised in Britain at the time, within the framework of 'western medicine', although this term is contested. In relation to the French in Vietnam, Laurence Monnais notes that the medical system emanating from Europe is variously described as 'modern', 'scientific', 'western', or 'European'. She uses the more neutral term 'biomedicine', to avoid making claims about the 'geographical origins, epistemological universality or temporal status' of the relationship between medicine and biological sciences.[14] Yet in the early nineteenth century British pharmacy still reflected many of the 'pre-biomedical' principles of Galen and Paracelsus, with extensive use of purgatives and emetics. In many ways it was little more biomedical than the indigenous pharmacy it encountered. In this book I use 'western' in favour of 'biomedical' to better reflect contemporary usage.

Medicine and the British Empire

In recent years medicine in empires has been the subject of extensive research by social and medical historians who have done much to establish the field, although rarely with any mention of pharmacy. David Arnold warned against establishing too rigid a barrier between colonial and metropolitan medicine, and the same might be said of colonial and metropolitan pharmacy, although pharmacy is not mentioned.[15] In 1997 Shula Marks asked the question 'what is colonial about colonial medicine?'[16] in the aftermath of a conference on medicine and the colonies. She noted that there was a large degree of overlap between colonial and metropolitan medicine, and was left wondering what if anything was specifically colonial about it.[17] Waltraud Ernst subsequently reflected on the theories and methodologies that characterized history of colonial medicine at the time.[18] She was critical of some of the approaches used and called for fresh ones: 'The seemingly irreconcilable tension and at times unhelpful hostility between proponents of Fanonian and

Foucaultian paradigms on the one hand, and archival, data-focused historians of medicine on the other, needs to be overcome, lest researchers continue to be caught up in either ideologically fraught and conceptually misleading east-versus-west bifurcations or narrowly framed local case studies'.[19] Many medical historians have since heeded her words.

The secondary literature relating to medicine and empire covers extensive spatial and temporal domains. In 1988 Roy MacLeod and Milton Lewis's edited volume *Disease, Medicine and Empire* opened up new perspectives.[20] Since then India has received extensive attention, including Poonam Bala's 1991 work on medicine in Bengal,[21] and Anil Kumar's 1998 review of British medical policy during the Raj.[22] Scholarly interest now extends to colonialization across the globe.[23] Many studies have involved international comparison, analysis, and reflection; in *Psychiatry and Empire,* for example, Sloan Mahone and Megan Vaughan bring together scholars in the history of medicine and colonialism to explore questions of race, gender, and power relations in former colonial states across Africa, Asia, the Caribbean, and the Pacific.[24] In *Crossing Colonial Historiographies* Anne Digby, Waltraud Ernst and Projit Mukharji use a transnational perspective to explore histories of colonial and indigenous medicines.[25] Other important contributions have been *Medicine in an Age of Commerce and Empire*[26] by Mark Harrison, and Pratik Chakrabarti's *Medicine and Empire.*[27]

These works often explore the drugs being used in the colonies, and references to apothecaries are not uncommon, but these usually focus on their medical role rather than their pharmaceutical activities. Nevertheless, several medical historians have made important contributions concerning the impact of British pharmacy in the countries of the empire, although these usually involve detailed studies of one part of one country or explore a single issue across a whole country. Johanne Collin, for example, has considered the relative impact of French and British influence on pharmacy in Quebec[28]; and Dan Malleck suggests that in the 1870s the limit of doctors' influence over the education and licensing of pharmacists in Canada caused them to change their view of their own role, which at the turn of the century was limited to little more than scrutinizing and restricting the patent medicines trade.[29]

Pharmacy and the Medical Profession

Medical practice has always encompassed a broad spectrum of activity, from the highly qualified physicians to the totally unqualified quacks, from the orthodox to the unorthodox. As Bill Bynum and Roy Porter pointed out, 'the frontiers between orthodox and unorthodox medicine have been flexible'.[30] The distinction between them was itself socially constructed. For them, pharmacy operated towards one end of the spectrum; the tussles between the apothecaries and druggists at the end of the eighteenth century were 'towards the lower end of orthodox medicine' and only a short distance from the outer fringe.[31] Irvine Loudon noted that when historians examine the process of medical reform 'the part played by irregular practice, and the importance of the druggist – if recognized at all – tends to be in the background'.[32] And in his chapter, Sidney Holloway suggested that 'in the first half of the nineteenth century, the rank-and-file chemist and druggist was scarcely a part of the system of orthodox professional medicine' at all. He was the product of an older, deeply rooted practice, the tradition of family self-medication. The rise of the chemist and druggist was, he noted, an aspect of the adaptation of folk medicine to industrial, urban society.[33]

In Britain, the transformation of pharmacy—from being an integral part of medicine, to being a branch of medicine, and then to becoming an autonomous profession—took place over several centuries. The apothecaries were originally a specialist group in the Grocers Company until the Society of Apothecaries was founded in 1617. As they fought occupational boundary battles with the physicians, new groups of largely unqualified individuals—the chemists and the druggists—emerged, and by the mid-eighteenth century they were a significant group (Chapter 2). But they were as much part of the medical profession as apothecaries. In 1747 Robert Campbell described the profession of physic in all its branches, i.e. 'the physician, surgeon, chymist, druggist and apothecary'.[34] By 1828 Justice Park noted that the object of the 1815 Apothecaries Act had been to keep the business of the apothecary distinct from the other branches of the profession. He saw 'four degrees in the medical profession; physicians, surgeons, apothecaries, and chymists and druggists... Each is protected in his own branch, and neither must interfere with the province of the other'.[35]

The 'elite' physicians did not engage in pharmacy; it was the job of the apothecaries to dispense their prescriptions. But other medical practitioners such as ships' surgeons did. It was the court ruling to allow apothecaries to practise as general medical practitioners (Chapter 2) that set pharmacy in Britain on a different course to that in Europe. In Europe, the apothecary quickly evolved into the pharmacist; in Britain the apothecary or surgeon-apothecary evolved into a hybrid medical and pharmaceutical practitioner, creating space for the emergence of chemists and druggists who often had little or no training. By mid-nineteenth century a variety of such practitioners had found their way to the colonies.

When the PSGB was founded in 1841, its leaders were in no doubt that they were part of the medical profession. Jacob Bell referred in 1842 to 'the position which pharmacy occupies or ought to occupy as a branch of the medical profession'.[36] Four of the nine Bills introduced for the regulation of the medical profession in the years between 1840 and 1850 included chemists and druggists as part of the profession. In fact, the status of the leading chemists and druggists within their local communities was far higher than that of the general practitioners. It was the 1852 Pharmacy Act which effectively made the exclusion of pharmacists from the medical profession inevitable, as it provided the legal basis for the registration of pharmaceutical chemists; and it was the 1858 Medical Act that drew an inalienable line between those practitioners who were medically qualified and those who were not. It established the legal and institutional boundaries of the medical profession[37]; pharmaceutical chemists had their own register and were no longer a branch of medicine, but completely outside it.

Without the 1852 Act, the 1859 Medical Register might well have been published in four sections rather than three, one of which would have contained the names of the pharmaceutical chemists.[38] Yet to suppose that by 1858 pharmacy in Britain had become an autonomous 'profession' separate from medicine would be inaccurate. The PSGB represented only a small proportion of those engaged in pharmacy. Alternative bodies emerged which presented it with an existential threat. Doctors made renewed attempts to take control of pharmacy. The absence of a curriculum meant that educational standards were variable, and 'crammer schools' emerged to prepare people for the PSGB's examinations by second-guessing questions. Most of these issues were only resolved by the early twentieth century. The professionalization of pharmacy across the Empire was thus taking place at the same time as it was in Britain rather than following it, and in some respects it was ahead. An understanding of pharmacy in the Empire requires an awareness of the situation in Britain at the time, and this is described in Chapter 2.

Pharmacy in the Colonies

Despite a common heritage the experience of pharmacy in the colonies was often very different, even between colonies in the same country. Events were determined by changing local circumstances, actions taken by colonial authorities, by British institutions such as the PSGB, as well as by settlers and colonists, particularly doctors. The pivotal event which led to the foundation of the PSGB was an attempt by the medical profession to demonstrate that chemists and druggists—by counter-prescribing and attending to minor injuries—were practising medicine, and therefore needed to be regulated by the doctors. In many parts of the Empire doctors used the same arguments to bring chemists and druggists under their control. When this failed, other strategies were used; in Australia doctors claimed that chemists could not be trusted to regulate the sale of poisons without statutory controls, and that they needed to be licensed by the medical profession. But the object of the group set up to oppose the threat from the doctors was not to establish a new profession but to safeguard the business of the chemist and druggist.[39] They were as much concerned with protecting the 'drug trade' as raising the standing of pharmacy.

In many of the colonies the early certification of chemists and druggists was undertaken by a local Medical Board, usually consisting entirely of doctors. With the passage of pharmacy legislation registers of chemists and druggists would be maintained by pharmacists themselves. The legal status of a pharmaceutical society would determine whether the society itself would determine registration qualifications, or whether a separate and independent Pharmacy Board needed to be established. The PSGB took on this responsibility itself, under the authority of its own Council.[40] This was to have long-term consequences both for itself and for pharmacy in the Empire. Whilst many colonies followed the British example, others did not, sometimes resulting in different arrangements in colonies in the same country. In Australia, most states followed Victoria in having a separate Pharmacy Board, but Western Australia followed the British model. Later, South Africa followed Victoria in opting for a Pharmacy Board.

Some of these were still subject to medical influence, having doctors as members or as the chair.

The speed at which pharmacy training and education developed and the extent to which clear boundaries were established between pharmacy and medicine, between pharmacy and the 'unqualified', and between pharmacists and subordinate occupations, varied greatly in the different colonies; progress was achieved more quickly in some than in others, and occasionally more quickly than in Britain. The request from Victoria in 1881 for reciprocity in the recognition of pharmaceutical qualifications caught the PSGB by surprise. They took refuge in the wording of legislation, which enabled them to defer the matter indefinitely, only finally implementing the necessary byelaws in 1913. The early Pharmacy Acts related solely to the regulation of those practising in Great Britain. It took the passage of a new Poisons and Pharmacy Act in 1908 to provide for the registration as pharmaceutical chemists or chemists and druggists—under the Pharmacy Acts of 1852 and 1868—without examination, of 'any persons holding colonial diplomas' who produced evidence that they had sufficient skill and knowledge to be so registered. The intercolonial recognition of pharmacy qualifications remained a bone of contention for many years.[41]

The wording of legislation was to play a significant part in the development of pharmacy not only in Britain but also in many colonies. A legal case in 1880 gave rise to company chemists (Chapter 2). Transplantation of wording from British Acts into colonial legislation often left the door open to company chemists elsewhere in the Empire. This often caused long-running disputes between pharmaceutical bodies, companies, and legislatures. But if some interpretations of the law could not have been anticipated, some omissions from legislation could have been. The PSGB was the Pharmaceutical Society of 'Great Britain', not of 'Great Britain and Ireland'; the register of pharmaceutical chemists and chemists and druggists applied to Great Britain but not Ireland, where a similar act was eventually passed in 1875.[42] In some colonies, Irish pharmacists found themselves ineligible for registration, and amendments to legislation had to be made.

Chemists, Druggists, and Pharmaceutical Chemists

One of the challenges in the history of pharmacy is the multitude of terms used to describe practitioners involved in the making and supply of medicines, either as a full-time occupation or as part of a broader occupation as a medical or other practitioner. The dictionary defines 'pharmacy' as 'the art or practice of preparing, compounding, and dispensing drugs, especially for medicinal purposes', or alternatively as 'a drugstore, a chemist's shop, a dispensary'.[43] It thus has a very wide range of meanings; it can mean an occupation, a place, a science, an art, or a technology. It is also an essential component of medicine, which the dictionary defines as 'the art or science of preserving health and curing or alleviating disease, especially as distinguished from surgery and obstetrics'.

In Britain, the term 'druggist' was originally applied to those mainly engaged in the wholesale trade in medicines of plant origin, whilst 'chemist' usually referred to retailers involved in the sale and supply of medicines; the two roles were often combined (Chapter 2). The titles 'pharmaceutical chemist' and 'pharmaceutist' became restricted to qualified practitioners following passage of the 1852 Pharmacy Act, and the 1868 Pharmacy and Poisons Act added 'chemist', 'druggist', 'chemist and druggist', 'pharmacist', and 'dispensing chemist or druggist' to the list.[44] Later, only those whose names appeared on the PSGB register could use certain titles; they were not interchangeable; 'pharmaceutical chemist' became a higher-level qualification granted to those passing the PSGB's Major examination. In the colonies those with a British qualification were included on registers as of right; but colonial registers also included others who might have lesser or even no qualifications, who would nevertheless be entitled to use one of the British restricted titles.

If the boundaries between medical and pharmaceutical practitioners were often indistinct, so too were those between tiers in the pharmaceutical hierarchy. In some colonies the title 'pharmaceutical chemist' was applied to all those having the minimum qualification (typically the British 'chemist and druggist' certificate); in others it was reserved for those who had passed the PSGB's Major examination. The issue is further complicated by the plethora of titles used to describe support staff engaged in the making and supply of medicines. These included 'compounder', 'dispenser', 'dresser', 'assistant apothecary', and 'sub-assistant surgeon'. In some colonies dispensers evolved into pharmacists as educational standards rose; in others they became a subordinate group as pharmacy technicians. In the chapters that follow I use a variety of terms where this can be done without creating unnecessary confusion. Retail chemists played a dominant role in the colonies, although clearly some were engaged in wholesaling and manufacturing. 'Chemist' by itself refers to a retail chemist; non-pharmaceutical chemists are described more fully, such as analytical chemists.

Professionalizing Occupations

Autonomy from medicine, the conflict between dispensing doctors and prescribing pharmacists, the tension between pharmacy as trade and pharmacy as profession, the focus on product rather than patient, and the level of training and education needed to fulfil the role; these issues are at the heart of the story of pharmacy and empire, and what links them all is professionalization. Professions have been the subject of extensive and frequently contentious interpretations in sociological and historical circles for many years. The questions they address are fundamental ones; Why do professions form? What is their relationship with the state and other professions? How can the power, authority, and reach of professions be understood? Yet the answers invariably produce more confusion and complexity. The term 'professionalization' is a hotly contested one, with sociologists tending to a more fixed or standard characterization, and historians usually insisting on a relative and changing one. Such arguments are of little help when considering processes of professionalization over time and space.

It is not the purpose of this book to engage with these debates. Rather, it is to look at change over time, and what it meant to be in a 'profession' and to be 'professional' in very different countries in relation to a single occupation. It is nevertheless necessary to consider definitions that have been proposed for both a 'profession' and 'professionalization', to examine theories put forward to explain them, and to review previous studies of pharmacy professionalization. Many definitions of a profession have been proposed over the years, but a widely accepted one, proposed by Julia Evans, is that 'a profession is a service occupation that is based on extensive technical knowledge, and assists clients in dealing with risk in daily life'.[45] Alan Rushton defines professionalization as 'the process over time that results in 'professional' status for an accomplished occupational in society'.[46] Yet pharmacy fits uncomfortably in most such definitions because of its link with a commodity, medicines. The professionalization of pharmacy went through distinct phases; expertise in the making of medicines evolved from being an essential part of the role of the doctor, to becoming the field of practice of a distinct group of individuals within the medical profession, and eventually to that group establishing itself as an autonomous occupation. It is a process that occurred at very different times in different places.

The professionalization of pharmacy has received considerable attention from both sociologists and historians, and a variety of approaches have been used. Sociological theories are usually divided into two main schools of thought; the trait or functionalist approach, and professionalization.[47] Trait theorists seek to identify attributes which distinguish professions from other forms of occupational grouping; typically, they adopt medicine and law as ideal types, and compare the development of other professions to them. Hughes, for example, argues that for an occupation to be considered a profession, its members must be licensed for practice by the State.[48] But trait theories have significant limitations; Eliot Friedson, for example, argues that 'there is no single, truly explanatory trait or characteristic...that can join together all occupations called professions, beyond the actual fact of coming to be called professions'.[49]

Professionalization theorists argue that a profession can be distinguished from other occupations because it has managed to attain a degree of autonomy as a collectivity to organize its own work. Professionalization is, therefore, according to Paul Bennell, 'an historical process whereby individuals occupying certain positions within the social division of labour, [and] actively attempt to utilize a growing source of 'professional power' in order to establish their own conditions of existence and reproduction as distinct occupational groups'.[50] Bennell suggests however that ultimately the ability of an occupation to professionalize itself depends on the role of the state, and its willingness to protect monopolistic positions.[51]

The professionalization model helps us to explain how a profession establishes control over a protected market from its skills and the education and training of new recruits. As Larson puts it, 'the double nature of the professional project intertwines market and status orientations, and both tend toward monopoly – monopoly of opportunities for income in a market of services on the one hand, and monopoly of status in an emerging occupational hierarchy on the other. The institutional locus in which both monopolizing tendencies converge is the educational system'.[52] For Larson successful transition of an occupational group to professional status is mediated through the monopoly of a substantial body of knowledge that is shared by all members of the group. The key to professionalization is education.

Professionalization models have their limitations. In a recent study of professions in four Canadian provinces, Tracey Adams tested several key sociological theories of professionalization. She found that no single approach fitted the complex stories of professional creation, and suggested a hybrid approach, although even then with qualifications.[53] Such approaches may add to the debate, but they offer little in the way of practical tools to compare processes of professionalization across diverse geographical regions and extended timescales.

Professionalizing Pharmacy

Because professionalization is itself an historical process it offers a convenient model by which to track developments in pharmacy across several centuries and in different parts of the world. However, whether pharmacy ever fully completed the process, emerging as a fully-fledged profession in its own right, has been debated for many years. In 1968 Denzin and Mettlin described pharmacy as an example of 'incomplete professionalization'.[54] According to these authors, pharmacy did not possess the characteristic traits of a profession as defined by Good.[55] It had failed to recruit altruistic people, exercise adequate control over the sale and manufacture of drugs, develop a unique body of scientific knowledge, or maintain occupation unity. But this analysis has been challenged by other sociologists; Dingwall and Wilson, for example, argue that when these features are examined in relation to both medicine and law, doctors and lawyers are no more 'professional' than pharmacists.[56]

Others such as Birembaum have argued that, in the second half of the twentieth century, pharmacy went through a process of reprofessionalization as a result of changes in the economic and technological environment in which pharmacists worked.[57] He argued that an elite group of pharmacists in a position to practise a more clinical form of pharmacy—clinical pharmacy—engaged in an attempt to establish a new identity for pharmacy, the process he described as 'reprofessionalization'. But this critique has been criticized by other sociologists; Holloway, Jewson, and Mason concluded that 'if professionalization is a dynamic, conditional, interactive process born of group struggle, the concept of reprofessionalization is redundant'.[58]

These authors also considered the value of Larkin's concept of 'occupational imperialism' in relation to pharmacy.[59] Larkin used this term to describe 'a chronic tendency for medical occupations to trespass upon areas of responsibility, or to poach the skills, of neighbouring disciplines'. He saw this as a means by which para-medical occupations pursued collective mobility without challenging the dominance of doctors. But for Holloway, and Jewson and Mason, 'occupational imperialism' failed to capture the nature of the social processes which Larkin described, which were better thought of as the social construction of occupational boundaries or relationships; it implied a specific strategy rather than an overall process. In the case of pharmacy, several strategies were used in combinations which were sometimes self-reinforcing and occasionally mutually contradictory.[60]

Pharmaceutical historians exploring the professional development of pharmacy in Britain and elsewhere have invariably used a professionalization perspective. Juanita Burnby took an implicitly professionalization approach in her study of the development of pharmacy during the early nineteenth century. She noted that the 'Council of the Pharmaceutical Society believed that professionalization could be brought about in two ways'. These were the promotion of education, and the encouragement of research.[61] John Crellin took this approach in his study of the growth of professionalism. He noted that the radical changes that pharmacy went through 'were largely due to the curious situation which allowed the traditional pharmaceutical practitioners – the apothecaries – to become general medical practitioners, who took less and less interest in pharmacy'.[62]

American pharmaceutical historians too have largely rejected trait approaches in favour of professionalization perspectives. Glenn Sonnedecker based his review of the state of American pharmacy in 1960 on the 'five forces of professionalization' adapted by Isador Thorner from the work of Carr-Saunders.[63] More recently Greg Higby, in his study of professionalism in American pharmacy,[64] has followed the advice of sociologist Everett C. Hughes, who wrote that 'the way to understand what professions mean in our society is to note the ways in which occupations try to change themselves or their image, or both, in the course of a movement to become 'professionalized'.[65] Higby also quotes Donald M. Scott, who argues that 'for historians it is more useful to approach profession as a changing social and cultural construct, encompassing different meanings of the idea of profession and different institutional arrangements at different times'.[66]

Professionalizing Pharmacy in the British Empire

If the professionalization of pharmacy is to be compared between the colonies of the British Empire, it is necessary to identify core features that might be expected to be found in all colonies and will be consistent over time. Previous studies of pharmacy and professionalization indicate that most features of the process are shared with other occupational groups seeking professional status. Geoffrey Harding and Kevin Taylor extract four core features of a profession; specialized knowledge and lengthy training; a monopoly of practice; self-regulation, and service-orientation.[67] Similar features have been used in earlier studies of pharmacy in British colonies, most notably that of Bennell in Ghana. These emphasize the need to secure the support of the state in the passage of relevant legislation. From these studies five features which effectively embrace the 'professionalization' of British pharmacy during the nineteenth and early twentieth centuries can be defined.

Markers or 'indicators' of professionalization can now be proposed which can be used to track the development of pharmacy in very different colonies at different times. These can be summarized as follows:

- **Collective action** Members of an occupation work together as a collectivity to pursue the common goal of attaining autonomy with the authority to manage their own affairs. For pharmacy this was usually marked by the formation of professional associations or societies.
- **Role of the state** The ability of an occupation to professionalize depends on the support of the state and its willingness to create monopolistic positions with regard to income and status. This necessitates identifying individuals supportive of the aims and aspirations of the group. For pharmacy this was marked by establishing relationships with key members of local administrations and parliaments.
- **Passage of legislation** The creation of monopolies in opportunities for income and in status in an emerging occupational hierarchy requires that legal authority be established for identifying those

judged to be qualified and the registration of practitioners. For pharmacy this was usually marked by the passage of a Pharmacy Act.

- **Autonomy from other professions** A profession can be distinguished from other occupations when it has attained autonomy from other professions and is not under the control of another. For pharmacy this involved complete separation from medical control, and was marked by only pharmacists being involved in judging the competence and registration of other pharmacists.
- **Extending education** Higher status and remuneration for professionals are rationalized by the higher levels of specialized skills and knowledge that they require. For pharmacy this was marked by the raising of education levels and the establishment of schools of pharmacy.

Pharmacists in British colonies across the world faced similar challenges and used common strategies to deal with them. Many established pharmaceutical institutions, education programmes, journals, and publications, largely based on what had been done in Britain. But there were also large differences, in the response to pharmacy by the medical profession, in the attitudes of colonial pharmacists, and in the actions of the colonial authorities. These five 'markers' of professionalization provide a means of examining these issues in a consistent way, and in exploring how boundaries were established in the British colonies, not only between pharmacy and medicine but also between pharmacists and less qualified practitioners.

The Imperialization of Pharmacy

In Britain and the colonies, control of the sale of poisons was usually of much greater concern than the regulation of pharmacy. How the two became linked, providing pharmacists with a monopoly in supply of certain commodities, was a matter of legislative convenience (Chapter 2). Yet legislation linking the regulation of pharmacy with the control of poisons soon became the basis of legislation elsewhere in the Empire. Colonial authorities sought the advice and guidance of the PSGB, which promoted the approach taken in Britain. In reviewing the draft Sale of Drugs and Poisons Ordinance for Jamaica in 1881 the PSGB was anxious that the sale of certain items be restricted to those holding one of its qualifications.[68] The Ordinance was passed later that year. It was a pattern repeated many times; in 1892 a Drugs and Poisons Ordinance was passed in the Gold Coast (Ghana); it—like the Ordinances passed in many other colonies—was based on the British Pharmacy and Poisons Act of 1868.

Having strengthened its legal foundation through this Act, which also established restricted titles and created a dual profession of chemists and druggists and pharmaceutical chemists, the PSGB extended its influence throughout the Empire. The pharmaceutical historian Leslie Matthews noted in 1962;

> for thirty years or so after passage of the Pharmacy Act in 1868, the Society was engaged in consolidating its position as the body responsible for raising the educational standards of the pharmacists, extending its influence in Great Britain, in the Dominions and the Colonies, and in supporting every endeavour to advance pharmacy in the interest of the public.[69]

It was the issues of education and the recognition of qualifications that were to cause the greatest rifts in pharmacy both in Britain and across the Empire. Practitioners with a wide variety of backgrounds opened for business as chemists and druggists. Some had completed apprenticeships as apothecaries or chemists and druggists; others had few if any qualifications. Those deemed qualified in one colony might not be recognized as such by another. This became a problem when individuals wished to cross colonial boundaries. It was first raised in 1871 when the Council of the Ontario College of Pharmacy 'resolved that certificates of proficiency or diplomas of the Pharmaceutical Society of Great Britain, the Pharmaceutical Association of Quebec, and the Philadelphia College of Pharmacy, should be considered "sufficient" by its Board of Examiners under the terms of the Ontario Pharmacy Act'.[70]

The reciprocal recognition of pharmacy qualifications between the colonies and Britain soon became an issue. In 1881 the Pharmacy Board of Victoria submitted a request to the PSGB for them to recognize the certificates awarded by each other. The request was declined. Over the next 30 years it was the cause of considerable intercolonial conflict and rivalry as well as cross-colonial collaboration. In 1896 the Pharmaceutical Society of Queensland appealed for collective action by pharmacists in colonies across the Empire. The matter was taken up by the Queensland Premier during a visit to London. Collective action by the colonies eventually stirred the PSGB into action. The legislation needed to enable this in Britain was included in the 1908 Poisons and Pharmacy Act of

1908, and the PGB finally published its Terms of Reciprocity Agreement in 1913 (Chapter 12). An imperial pharmaceutical qualification was proposed in 1929, although nothing came of it.[71]

The PSGB was however as much concerned with the imperialization of medicines as with imperialization pharmacy. Work on 'imperializing' the BP had begun soon after publication of the third edition in 1885. The 1898 edition contained many items originating from the colonies, but it did not include many that were in regular use in the colonies themselves. In 1900 an Indian and Colonial Addendum to the BP was published to fill the gap, although further amendments were needed for India, and a Government of India edition appeared in 1901. But by 1899 John Attfield, the BP's editor, was able to report to the Pharmacopoeia Committee of the General Medical Council (GMC) that 'fifty-three of the seventy British Administrations of India and the Colonies, without exception, concur in supporting the broad principle of imperialization of the great national book of medicines'.[72] However, the PSGB considered that many items useful in pharmacy were omitted from the BP, and its Council resolved to prepare a volume which would 'make available in one volume accurate information respecting all drugs and medicines in common use throughout the Empire'. In 1907 it published the *British Pharmaceutical Codex,* describing it as 'an Imperial Dispensatory for the use of medical practitioners and pharmacists'.[73] Empire also became a popular theme in advertisements for medicines, reflecting the pervasiveness of imperial imagery (Fig. 1.2).

In Britain, the development of pharmacy was to be heavily shaped by the testing of the wording of laws in the courts. The 1880 case leading to the rise of company chemists was the first of many, most of which were initiated by the PSGB. In most cases the outcomes were not what the founders of the PSGB either expected or anticipated. Yet it was a pattern that was to be repeated in many parts of the Empire.

Pharmacists in the colonies followed events in Britain with great interest. But being the first to seek legislation in certain areas meant that inevitably mistakes were made. Yet too often lessons were not learnt from mistakes made in Britain, often with serious consequences in the colonies.

Fig. 1.2 Advertisement for Empirin Tablets, 1939 (courtesy of Wellcome Collection)

Pharmacy and Networks

Networks of many kinds played an important part in the development of pharmacy in the colonies, whether religious, scientific, social, or trading. Quaker connections were central to the supply of medicines to America and the West Indian colonies.[74] Trading networks were established between apothecaries in London and Quakers apothecaries and physicians in the Caribbean (Chapter 3).[75] Religious dissenters were the first settlers in some colonies; others such as Christchurch in New Zealand were founded by British church leaders. Pharmacists were often recruited because of their membership of a particular network, and were offered assisted passages for themselves and their families, as with James Butler Swann (Chapter 11). Others, such as those in South Africa, benefited from membership of organizations such as the Freemasons (Chapter 7).

Both formal and informal networks were involved in the transnational movement of pharmaceutical knowledge. Vehicles for the diffusion of knowledge included the movement of people themselves, along with books and journals, and the replication of legislation from metropole to colony. In his study of the early history of the Pharmaceutical Society of Australia between 1857 and 1918, Gordon Boyce noted that its success was largely due to its ability to successfully develop communication channels, and to create strong institutional constructs, such as journals and conferences. He found that setting itself a wide range of objectives and having a broad agenda worked to its advantage, successfully shaping public conceptions of the profession in order to influence an emerging regulatory regime.[76]

If education caused divisions between pharmacists, science helped bring them together. As Richard Drayton points out, science and medicine played important parts in British expansion from the age of Raleigh to that of Curzon and Nehru.[77] Pereira, the PSGB's professor of materia medica,[78] suggested an area where he thought it could take a leading role.[79] He proposed the formation of a scientific committee for the promotion of pharmaceutical knowledge, 'for the elucidation of the natural history and origin of substances used in medicine. No country in the world' he declared.

> possesses so many facilities for carrying on inquiries such as those to which I here allude, as Great Britain. Her numerous and important colonies in all parts of the world, and her extensive commercial relations, particularly fit

> her for taking the lead in investigations of this kind…From her extensive possessions in different parts of the world, we draw a very large portion of the substances now used in medicine.[80]

Pereira's main interest was in the products of Empire, but he was emphasizing its scientific foundations rather than its trading associations, and the importance of scientific networks in the development of pharmacy. A British Pharmaceutical Conference, bringing together those involved in scientific studies in pharmacy, was founded in 1863, and it soon embraced overseas members. At their twenty-first meeting in Hastings in 1883 they agreed to appoint Colonial Secretaries for each of the colonies. It was suggested that India should have one in each of the three presidencies, although it was considered that one would be sufficient for the whole of Canada.[81] Appointments were made in India, Canada, South Africa, Australia, New Zealand, and the West Indies.[82] The Colonial and Indian Exhibition at South Kensington three years later included displays of medicines and materials from across the Empire.[83]

Empire and Britishness

If there was no 'grand plan' for the transplantation of British pharmacy across the colonies, neither was there a 'grand plan' for the development of a British Empire; it grew out of unconnected acts of privateering, exploration, and war. Indeed before 1500 there was no entity called Britain; the countries of the British Isles were separate states. An Act of Union with Wales was passed in 1536, and England and Scotland were separate countries until 1707. A 'Kingdom of Great Britain' existed between 1707 and 1801, when the separate Kingdoms of Ireland and England became the 'United Kingdom of Great Britain and Ireland'. This entity continued until the establishment of the Irish Free State in 1922, and is the name applying to the metropole between 1801 and 1922. With the secession of Ireland, it became the 'United Kingdom of Great Britain and Northern Ireland' in 1927. The distinctions between these descriptions were to have consequences for pharmacy in the Empire.

Andrew Porter suggests that 'in one sense the 'Britishness' of the Empire was inclusive', in that the colonies were conspicuously occupied and ruled by the British rather than just the English.[84] Britain's union with Ireland opened up opportunities for the Irish to engage in the business of Empire. As Thomas Bartlett puts it, 'whether as settlers in Australia and New Zealand, missionaries in Africa and India, or soldiers, administrators, engineers, merchants, and doctors throughout the Empire, the Irish took full advantage of the Imperial opportunities opened up by the Union'.[85] Irish chemists and druggists were no less enthusiastic about taking up the opportunities presented in the British Empire. Scotland too contributed large numbers of its population to the development of the Empire.[86]

But use of the term 'British pharmacy' requires some qualification. There were often differences in the pharmacy practised by English, Scottish, Welsh, and Irish pharmacists; separate national pharmacopoeias were developed in London, Edinburgh, and Dublin.[87] In some colonies the pharmacy of Edinburgh was more prominent than the pharmacy of London. British pharmacy was also different from that elsewhere in Europe. When Britain took over control of colonies from other European powers, they often found that the pharmacy practised there—and the division of activities between pharmacists and doctors—was quite different to that in Britain. France had a lasting impact on pharmacy practice in Canada. Dutch and German influences were apparent in Cape Colony, where pharmacy was already well-developed. In Malta, pharmacy was already an autonomous profession with equal standing to medicine when the British arrived, a legacy of the Edict of Palermo of 1240.

Other factors contributed to the growing sense of Britishness amongst the people of the Empire. Porter notes that by the end of the nineteenth century 'Britishness' was enhanced as a result of better communication links, and the decline of America as an emigrant destination. At the same time there was growing insistence on difference, on increasing diversity, and an acceptance that there were limits to the 'Britishness' that could be stamped on the Empire.[88] Within broadly similar constitutional frameworks, settler communities gradually diverged from each other, and local government administrations responded to differing social and political interests.[89] In some ways, therefore, the attempts by the PSGB and the GMC to standardize pharmacy and medicine use across the Empire in the early years of the twentieth century flew in the face of the overall direction of travel.

Pharmacy, Professionalization, and Empire

What follows then is a narrative history of the professionalization of pharmacy in the disparate territories of the British Empire. Conceptually, the book takes both a cross-national approach, comparing pharmacy development in different colonies within a single country, and a transnational approach, comparing these developments across nations, or across imperial divisions. The chapters are based on the 'divisions of Empire' as they were in 1914. Whilst following a similar pattern, and having professionalization as a central theme, most also feature a secondary theme—an aspect of colonialism with which pharmacy has been associated in some way. Most are not exclusive to one region, but each often played a more prominent role in one region than in others. The slave trade, for example, was an important factor in many places, but its significance was greater for pharmacy in the West Indies than elsewhere. Australia was not the only place in the Empire to be used as a penal colony, but it serves to highlight differences in pharmacy arising from their origins as penal or settler communities. In West Africa, the activities of missionaries played an important part in shaping attitudes to western medicine and pharmacy, and in South Africa previous occupation by other European powers was significant. In Malta pharmacy was already established as an autonomous profession by the time British occupation began, and the Mediterranean islands illustrate the impact on pharmacy of the origin of colonies as naval and military bases.

Although rather narrow in its approach, the book extends over a wide geographical area across nearly two centuries. Although the time frame considered in the book is described as 1780–1970, space limitations mean that it has not been possible to explore each division across the whole period; some accounts start or finish earlier or later than others, with rather less said about the twentieth century and the period after 1945. The time frame has precluded mention of the United States, as it ceased to be part of the British Empire following the 1776 Declaration of Independence. Some British colonies—such as those in East Africa—were not deemed of sufficient significance by the GMC to be referred to in the *British Pharmacopoeia* of 1914 (Table 1.1), and are not covered in this book. Space limitations also prevent detailed accounts of pharmacy in every colony in every division of Empire.

This then is largely a 'top down' history of colonial pharmacy from a metropolitan perspective.[90] I have referenced some important texts from the imperial history literature, but I have not attempted a sustained engagement with current concerns in colonial studies. The voices of indigenous people are rarely heard, whether as customers, support staff, or as the pharmacists of the future. Neither have I made any attempt to tackle issues of gender, exploitation, consumerism, and imperial hegemony. Women too make few appearances, although that is largely because their voices are rarely to be heard in the sources I have used. But the story of pharmacy is first and foremost one about people, and I have included the briefest of details about some of the participants. Detailed biographies of pioneer pharmacists and unusual remedies are to be found elsewhere. Nor does the scope of the book extend to pharmaceutical wholesaling, manufacturing, or marketing, or to practice in hospitals and other locations. Its focus is on policy, legislation and education, and the process by which the 'drug trade' turned itself into the 'pharmacy profession'.

Many of these themes are subjects for future study and analysis. Yet it is perhaps first necessary to have a broad overview of a subject before it is possible to tell the story 'from below'. For colonial pharmacy, such an overview has so far been lacking, and it is hoped that this book offers a framework on which others can build—a baseline for subsequent research and interpretation. It is hoped that it will be of interest to a wide audience, including not only pharmaceutical, medical, and imperial historians, but also past, present, and future practitioners of pharmacy. Despite its limitations, I hope that it provides a useful resource in the history of pharmacy and empire, and offers a firm foundation for further studies.

Notes

1. Bell, Jacob and Redwood, Theophilus. *Historical Sketch of the Progress of Pharmacy in Great Britain* (London, 1880), p. 171.
2. The National Archives (TNA), London, CO-137/502/43. Jamaica Office to Pharmaceutical Society, 11 June (1881), folios 295–301.
3. *British Pharmacopoeia*, fifth edition, 1914 (London, 1914), p. ix.
4. Ibid., p. xxii.
5. *Pharmaceutical Journal and Pharmacist*, 91 (1913), p. 324.
6. Lloyd, T.O. *The British Empire 1558–1995* (Oxford, 1996); James, Lawrence, *The Rise and Fall of the British Empire* (London, 2008);

Brendon, Piers. *The Decline and Fall of the British Empire, 1781–1997* (London, 2008). Darwin, John. *Unfinished Empire: The Global Expansion of Britain* (London, 2012).

7. Marshall, P.J. 'The First British Empire', in Winks, Robin W. (ed.) *Historiography, The Oxford History of the British Empire* (Oxford, 1999), pp. 43–4.
8. Pakenham, Thomas. *The Scramble for Africa* (London, 1991).
9. Horn, James. 'British Diaspora: Emigration from Britain, 1680–1815', in Marshall, P.J. (ed.) *The Eighteenth Century, The Oxford History of the British Empire* (Oxford, 1998), p. 30.
10. Ibid., p. 32.
11. Ibid., p. 48.
12. Fitzpatrick, David. 'Ireland and the Empire', in Porter, Andrew (ed.) *The Nineteenth Century, The Oxford History of the British Empire* (Oxford, 1999), pp. 512–3.
13. Bayly, Susan. 'Colonial Cultures: Asia', in Porter, *Nineteenth Century*, p. 449.
14. Monnais, Laurence. *The Colonial Life of Pharmaceuticals: Medicines and Modernity in Vietnam* (Cambridge, 2019).
15. Arnold, David. *Imperial Medicine and Indigenous Responses* (Manchester, 1988).
16. Marks, S. 'What Is Colonial About Colonial Medicine? And What Has Happened to Imperialism and Health?', *Social History of Medicine*, 10 (1997), pp. 205–19.
17. Ibid., p. 206.
18. Ernst, Waltraud. 'Beyond East and West: From the History of Colonial Medicine to a Social History of Medicine(s) in South Asia', *Social History of Medicine*, 20 (2007), pp. 505–24.
19. Ibid., p. 505.
20. MacLeod, Roy and Lewis, Milton. *Disease, Medicine and Empire: Perspectives on Western Medicine and the Experience of European Expansion* (London, 1988).
21. Bala, Poonam. *Imperialism and Medicine in Bengal: A Socio-Historical Perspective* (New Delhi, 1991).
22. Kumar, Anil. *Medicine and the Raj: British Medical Policy in India, 1835–1911* (Walnut Creek, 1998).
23. Winterbottom, A. and Tesfaye, F. (eds). *Histories of Medicine and Healing in the Indian Ocean World*, Volume 2 (London, 2016).

Payton, P. and Varnava, A. (eds), *Australia, Migration and Empire* (London, 2019).
24. Mahone, S. and Vaughan, M. (eds). *Psychiatry and Empire* (London, 2007).
25. Digby, A., Ernst, W. and Mukharji, P.B. (eds). *Crossing Colonial Historiographies: Histories of Colonial and indigenous medicines in Transnational Perspectives* (Newcastle upon Tyne, 2010).
26. Harrison, Mark. *Medicine in an Age of Commerce and Empire: Britain and Its Tropical Colonies 1660–1830* (Oxford, 2010).
27. Chakrabarti, Pratik. *Medicine and Empire, 1600–1960* (London, 2014).
28. Collin, Johanne. 'French and British Influence in the Birth of a Profession: Pharmacy in Quebec', *Pharmacy in History*, 52 (2010), pp. 100–11. Bothwell, Robert, *Canada and Quebec: One Country, Two Histories*, second edition (Vancouver, 1998).
29. Malleck, Daniel J. 'Professionalism and the Boundaries of Control: Pharmacists, Physicians and Dangerous Substances in Canada, 1840–1908', *Medical History*, 48 (2004), p. 198.
30. Bynum, W.F. and Roy Porter. *Medical Fringe and Medical Orthodoxy 1750–1850* (London, 1987), p. 1.
31. Ibid., p. 2.
32. Loudon, Irvine. 'The Vile Race of Quacks with Which This Country Is Infested', in Bynum and Porter, *Medical Fringe*, p. 123.
33. Holloway, S.W.F. 'The Orthodox Fringe: The Origins of the Pharmaceutical Society of Great Britain', in Bynum and Porter, *Medical Fringe*, p. 154.
34. Holloway, S.W.F. *Royal Pharmaceutical Society of Great Britain 1841 to 1991: A Political and Social History* (London, 1991), p. 51.
35. Ibid.
36. Ibid.
37. Ibid., p. 182.
38. Ibid.
39. Matthews, Leslie G. *History of Pharmacy in Britain* (London, 1962), p. 123.
40. Ibid., p. 132.
41. Anderson, Stuart. 'The Quest for Reciprocal Recognition of Colonial Pharmaceutical Qualifications 1896 to 1914', *Pharmaceutical Historian*, 40 (2010), pp. 13–5.

42. Matthews, *History of Pharmacy*, p. 138.
43. *Concise Dictionary*, fifth edition (Glasgow, 2001).
44. Trease, G.E. 'Introduction', in Poynter, F.N.L. (ed.) *The Evolution of Pharmacy in Britain* (London, 1965), p. 12.
45. Evetts, Julia. 'The Concept of Professionalism: Professional Work, Professional Practice and Learning', in Billets, S., Hartels, C. and Gruber, H. (eds). *International Handbook of Research in Professional and Practice-Based Learning* (Dordrecht, 2014), pp. 29–56.
46. Rushton, Alan. 'Professionalization and Professionalism: Diphtheria and Medical Practice in Minnesota 1850–1910', *Social History of Medicine* (2019), hkz121, https://doi.org/10.1093/shm/hkz121.
47. Bennell, Paul. 'Professionalization: The Case of Pharmacy in Ghana', *Social Science and Medicine*, 16 (1982), p. 601.
48. Hughes, E.C. *Men and their Work* (Glencoe, 1953).
49. Freidson, Eliot. 'The Theory of Professions: State of the Art', in Dingwall, R. and Lewis, P. (eds), *The Sociology of the Professions* (New York, 1983), pp. 32–3.
50. Bennell, 'Professionalization', p. 602.
51. Ibid.
52. Larson, Magali. *The Rise of the Profession: A Sociological Analysis* (Oakland, 1977).
53. Adams, Tracey L. *Regulating Professions: The Emergence of Professional Self-Regulation in Four Canadian Provinces* (Toronto, 2018).
54. Denzin, N.K. and Mettlin, C.J. 'Incomplete Professionalization: The Case of Pharmacy', *Social Forces*, 46 (1968), pp. 375–81.
55. Goode, W.J. 'Encroachment, Charlatanism and the Emerging Profession: Psychiatry, Sociology and Medicine', *American Sociological Review*, 25 (1960), pp. 902–14.
56. Dingwall, R. and Wilson, E. 'Is Pharmacy Really an Incomplete Profession?', *Perspectives on Social Problems*, 7 (1995), pp. 111–28.
57. Birenbaum, A. 'Reprofessionalization in Pharmacy,' *Social Science and Medicine*, 16 (1982), pp. 871–78.
58. Holloway, S.W.F., Jewson, N.D. and Mason, D.J. 'Reprofessionalization or Occupational Imperialism? Some Reflections on Pharmacy in Britain,' *Social Science and Medicine*, 23 (1986), pp. 323–32.

59. Larkin, G. *Occupational Monopoly and Modern Medicine* (London, 1983).
60. Holloway, Jewson and Mason, 'Reprofessionalization', p. 330.
61. Burnby, J.G.L. 'The Professionalization of British Pharmacy', *Pharmaceutical Historian*, 18 (1988), pp. 3–5.
62. Crellin, J.K. 'Pharmaceutical History and Its Sources in the Wellcome Collections. Part 1: The Growth of Professionalism in Nineteenth-Century British Pharmacy', *Medical History*, 11 (1967), pp. 215–27.
63. Sonnedecker, Glenn. 'To Be or Not to Be Professional?', *American Journal of Pharmacy*, 133 (1961), pp. 245–6. See also Carr-Saunders, A.M. and Wilson, P.A. *The Professions* (Oxford, 1933).
64. Higby, Gregory J. 'Professionalism and the Nineteenth-Century American Pharmacist', *Pharmacy in History*, 28 (1986), pp. 115–24. Higby, Gregory J. *In Service to American Pharmacy: The Professional Life of William Procter, Jr* (Tuscaloosa, 1992), p. 3.
65. Hughes, Everett C. *The Sociological Eye* (Chicago, 1971), p. 377.
66. Scott, Donald M. 'The Professional That Vanished; Public Lecturing in Mid-Nineteenth-Century America', in Geison, Gerald L. (ed.) *Professions and Professional Ideologies in America* (Chapel Hill, 1983), p. 14.
67. Harding, Geoffrey and Taylor, Kevin. 'Occupational Status of Pharmacy', in Harding, G. and Taylor, K. (eds), *Pharmacy Practice*, second edition (London, 2016), p. 202.
68. TNA, London, CO-137/502/43. Jamaica Office to Pharmaceutical Society, 11 June (1881), folios 295–301. TNA, London, CO-137/500/3. Pharmaceutical Society to Jamaica Office (1881).
69. Matthews, *History of Pharmacy*, p. 136.
70. *Pharmaceutical Journal and Transactions*, 31 (1871), p. 772.
71. Anderson, Stuart. 'The Proposal for an Imperial Pharmaceutical Qualification 1929,' *Pharmaceutical Historian*, 41 (2011), pp. 27–30.
72. GMC Archives. *British Pharmacopoeia 1898, Indian and Colonial Addendum: A Report of Progress, 1899. Appendix XX,* Pharmacopoeia Committee (1899), p. 754.
73. *British Pharmaceutical Codex, 1907* (London, 1907), Preface.

74. Zahedieh, N. 'Economy', in Armitage, D. and Braddick, M.J. (eds), *The British Atlantic World, 1500–1800* (Basingstoke, 2002), p. 66.
75. Magee, Gary B. and Thompson, Andrew S. *Empire and Globalisation: Networks of People, Goods and Capital in the British World, c.1850–1914* (Cambridge, 2010), p. 135.
76. Boyce, Gordon. 'A Professional Association as Network and Communicating Node: The Pharmaceutical Society of Australia, 1857–1918', *Australian Economic History Review*, 39 (1999), p. 258.
77. Drayton, Richard. 'Science, Medicine and the British Empire', in Winks, *Historiography*, p. 264. Bennett, B. and Hodge, J. (eds) *Science and Empire* (London, 2011).
78. The separation of materia medica into pharmacology (actions and uses of drugs) and pharmaceutics (making suitable dosage forms) was crucial to the professionalization of the pharmacist. Crellin, 'Pharmaceutical History and Its Sources', pp. 215–27.
79. Hudson, Briony and Boyland, Maureen. *The School of Pharmacy, University of London: Medicines, Science and Society, 1842–2012* (London, 2013), p. 16.
80. Bell and Redwood, *Historical Sketch*, p. 171.
81. *Pharmaceutical Journal and Transactions*, 43 (1883), p. 375.
82. *Chemist and Druggist*, 26 (1884), p. 375.
83. *Pharmaceutical Journal and Transactions*, 46 (1886), pp. 245–6, 305–8.
84. Porter, Andrew. 'Introduction', in Porter, A. (ed.) *The Nineteenth Century, The Oxford History of the British Empire* (Oxford, 1999), p. 20.
85. Bartlett, Thomas. 'Ireland and the British Empire', in Marshall, P.J. (ed.) *The Eighteenth Century, The Oxford History of the British Empire* (Oxford, 1998), p. 273.
86. Devine, T.M. *Scotland's Empire, 1600–1815* (London, 2003).
87. Anderson, Stuart. 'National Identities, Medical Politics, and Local Traditions: The Origins of the London, Edinburgh, and Dublin Pharmacopoeias, 1618 to 1807,' in Crawford, Matthew James and Gabriel, Joseph M. (eds) *Drugs on the Page: Pharmacopoeias and Healing Knowledge in the Early Modern Atlantic World* (Pittsburgh, 2019), pp. 199–221.
88. Porter, 'Introduction', p. 20.

89. Ibid., p. 21.
90. Jackson, Ashley. *The British Empire: A Very Short Introduction* (Oxford, 2013), p. 105.

CHAPTER 2

Great Britain: Professionalizing Pharmacy in the Metropole

The history of pharmacy in Britain remains an under-researched field of study. Although it has received some attention from medical as well as pharmaceutical historians, discussion of significant events in the history that pharmacy shares with medicine—which include the 1617 foundation of the Society of Apothecaries, the 1704 Rose Case, and the 1815 Apothecaries Act—usually place greater emphasis on their implications for medicine than for pharmacy, which are often neglected or ignored. There are frequent misconceptions about the development of pharmacy in Britain from the mid-nineteenth century; it is often asserted, for example, that the foundation of the PSGB marked the emergence of organized pharmacy as a separate profession in Britain—the point at which it was no longer a branch of medicine, but a distinct profession with its own discipline and organization.[1] But this too is a gross over-simplification of a protracted story.

This chapter examines the development of pharmacy in Britain with the aim of establishing the extent to which it might be described as a 'profession' by the time the British Empire reached its peak in the early twentieth century. It focusses on the core features of professionalization identified in the last chapter—how pharmacy practitioners came together as a united collective to fulfil common aims by forming associations; how pharmacists engaged with politicians in order to secure state support for

S. Anderson, *Pharmacy and Professionalization in the British Empire, 1780–1970*, Medicine and Biomedical Sciences in Modern History,
https://doi.org/10.1007/978-3-030-78980-0_2

monopolistic positions; how they were able to obtain the passage of legislation to regulate who was entitled to practise; how pharmacy came to be separate and distinct from medicine as a self-regulating profession; and how boundaries with other groups were established by raising educational standards.

The study of pharmacy in the British Empire requires an understanding of the metropolitan experience for several reasons; the professionalization of pharmacy in Britain was still underway at the time attempts were being made to do the same in some colonies; British pharmacists were finding their way and made many errors as they went, some of which went on to be replicated or avoided in the colonies. This chapter explores how this 'British model' of pharmacy evolved in the second half of the nineteenth century, and identifies five 'markers' which describe important ways in which it developed differently in Britain to that elsewhere in Europe. It begins however with a brief account of significant events in the early history of pharmacy in Britain and Europe (Table 2.1).

Pepperers, Spicers and the Edict of Palermo, 1240

Pharmacy in Britain has a long history.[2] Some of the earliest practitioners were spice merchants who dealt in drugs, prepared medicines, and sold perfumes.[3] They later specialized to become pepperers, who were mainly London-based wholesalers, or spicers, who were found in towns across the country. By the thirteenth century 'spicer-apothecaries' involved in the retail trade of drugs were to be found in all large cities,[4] and by the early fourteenth century the pepperers were known as grocers ('grossers'). In London both groups joined the Grocers' Company when it was founded in 1345.[5] By 1497 this was a substantial organization, and a significant proportion of its members were apothecaries. By the early sixteenth century the apothecary in England was an established dealer in drugs and medicines; he had little interest in medicine or surgery, although this slowly changed.[6]

Events taking place elsewhere in Europe were to later have a profound impact on the development of pharmacy in countries that were to become part of the British Empire. It was in Sicily and southern Italy that pharmacy first became legally separated from medicine, in a series of measures enacted between 1231 and 1240 CE. The Edict of Palermo established basic principles for their separation.[7] Doctors were to prescribe but not dispense; apothecaries were to dispense but not prescribe. Business

Table 2.1 Pharmacy in Great Britain: Timeline of significant events, 1240–1970

Year	*Medical or pharmaceutical institution founded*	*State action or court case*	*Pharmacy and poisons legislation*	*Significance*
1240		Edict of Palermo		Separation of pharmacy and medicine
1518	College of Physicians of London			
1617	Society of Apothecaries			Separation of apothecaries from grocers
1704		Rose Case		Apothecaries allowed to diagnose and treat patients
1794	General Pharmaceutical Association of Great Britain			
1802	Association of Druggists			First collective action by chemists and druggists
1812	Association of Apothecaries			
1815	Apothecaries Act			Apothecaries become general practitioners
1819	Association of Chemists and Druggists in Great Britain			
1832	Provincial Medical and Surgical Association (BMA from 1856)			Chemists and druggists not included
1841	Pharmaceutical Society of Great Britain			
1843		PSGB Royal Charter		Legal recognition

(continued)

Table 2.1 (continued)

Year	*Medical or pharmaceutical institution founded*	*State action or court case*	*Pharmacy and poisons legislation*	*Significance*
1851			Arsenic Act	First control of poison
1852	British Association of Chemists and Druggists		Pharmacy Act	Register of pharmaceutical chemists
1858	General Medical Council		Medical Act	Pharmacists not included in medical register
1860	United Society of Chemists and Druggists			
1868			Pharmacy Act	Register of chemists and druggists added
1869			Pharmacy Act	
1876	Chemists and Druggists' Trade Association			
1880		PSGB vs. London and Provincial Supplies		Start of company chemists in Great Britain
1898			Pharmacy Acts Amendments Act	
1908			Poisons and Pharmacy Act	Byelaws provide for reciprocal recognition
1920			Dangerous Drugs Act	
1920	Retail Pharmacists Union	Jenkin Case		PSGB not a trade union
1925			Therapeutic Substances Act	
1929			Pharmacy Act	
1933			Pharmacy and Poisons Act	Membership of PSGB no longer voluntary

(continued)

Table 2.1 (continued)

Year	*Medical or pharmaceutical institution founded*	*State action or court case*	*Pharmacy and poisons legislation*	*Significance*
1941			Pharmacy and Medicines Act	
1965		Dickson Case		
1968			Medicines Act	Medicines no longer regulated as poisons

relationships between physicians and apothecaries were prohibited. Official supervision of the practice of pharmacy was mandated, recognizing the importance of pharmacy as an essential service for the protection of the public's health. Apothecaries were obliged by oath to prepare drugs in accordance with 'skilled art' and to supply drugs of consistent and uniform quality. The number of pharmacies was limited, and the government fixed the prices of remedies.[8]

The Edict applied only to the Kingdom of Sicily, but from 1091 this included the Maltese Islands, later to become part of the British Empire (Chapter 6). Its legacy was to have considerable impact on the professionalization of pharmacy there and in other colonies previously subject to control by European powers. The Edict formed the basis for similar legislation elsewhere in Europe, but not Britain. In most European countries, pharmacy developed as a separate occupation, but in Britain it remained part of medicine until the nineteenth century. Because visiting physicians were very expensive and there were few of them, ordinary people increasingly relied on the apothecary for medical advice. The physicians became concerned about the competition posed by the apothecaries, although they also expected them to make up their prescriptions and to use their shops as consulting rooms.

The principle of no business relationships between physicians and apothecaries was not adopted in Britain. Indeed, some university-educated physicians encouraged the apprenticeship-trained apothecaries to become more medically trained and to see patients on their behalf.[9] An Act in 1553 gave the physicians powers to search apothecaries' shops, to destroy any drugs considered inferior, and to fine anyone who sold 'evil and faulty stuff'. As their numbers increased apothecaries became a significant group within the Grocers' Company, which held the monopoly on the importation of drugs.[10] In 1617 James I granted them a Charter establishing a corporate body under the title 'the Worshipful Society of the Art and Mystery of the Apothecaries'. It specified an apprenticeship of at least seven years, followed by a test of knowledge of preparing medicines.[11] But its rambling prose meant that there was nothing preventing apothecaries from prescribing remedies and attending patients.[12] It was to lay the foundation for a dispute which continued for almost two centuries.[13]

The Medicalization of the Apothecary

Although now separated from the grocers, the apothecaries were not the only group trading in medicines. The trade of some merchants included supplying drugs to apothecaries, with some specializing in the retail sale of mainly plant drugs, becoming known as druggists.[14] They acquired considerable skill in assessing the quality of medicines—something that apothecaries sometimes lacked.[15] The title 'apothecary' increasingly came to mean a general medical practitioner rather than a pharmacist, and with the growing use of chemicals in medicine those engaged in their wholesale trade became known as chemists. Retail trade in plant and chemical medicines slowly converged, with a new group of 'chemists and druggists' emerging to fill the void left by the apothecaries. A knowledge of practical pharmacy was now shared by physicians, apothecaries, and chemists and druggists.

In 1704 the issue of who had the legal right to prescribe and dispense medicines came to a head. In 1701 the Royal College of Physicians (RCP) had taken out a summons against an apothecary, William Rose, for supplying medicines which had not been prescribed by a physician.[16] Although the physicians initially won, the case went to appeal, and the House of Lords found in favour of Rose. It was, they judged, in the public interest to allow apothecaries to advise patients, as well as to supply them with medicines.[17] Rose's actions were normal practice.[18] The Rose Case confirmed the right of apothecaries to practise medicine; they had the legal right to both visit the sick and prescribe remedies. But they could not charge for their services—only for the medicines they prescribed.

The Rose Case has received considerable attention from medical historians. Roy and Dorothy Porter noted that it 'secured the apothecaries' right to prescribe' and 'enjoy the legal right to give medical advice, so long as they charged only for their medicines'.[19] However, Hal Cook has argued that by then it was already the custom 'that physicians themselves were not to charge fees for advice, although they could accept "honoraria"'. In 1704, they rejected charging for advice as a way of making a distinction between practising physic and not doing so. The Rose Case made little difference to what apothecaries did. The RCP had tried to change the status quo through the exercise of the law; the Rose Case illustrated the decline in the College's authority to regulate medical practice in London, and set the apothecaries firmly on the path of medical rather than pharmaceutical practice.[20]

Yet the impact of the Rose Case on the development of pharmacy in Britain can hardly be overstated; had the physicians won their case, the apothecaries would have had to restrict themselves to pharmacy practice, and they would have evolved into pharmacists, as they did in Europe.[21] Indeed, Thompson in his history of the apothecary notes that 'owing to the curious evolution of the apothecary into the medical practitioner, and the chemist combining with the druggist and absorbing the trade side of the apothecary's business, the position of pharmacy in England remained so much behind that of other countries until after the middle of the nineteenth century'.[22] As a consequence of the Rose Case the professionalization of pharmacy in Britain was delayed by over 150 years.

Apothecaries and Chemists and Druggists

By mid-eighteenth century the chemists and druggists were already a serious threat to the apothecaries. Of the 700 apothecaries' and chemists' shops in London, only half were run by members of the Society of Apothecaries. In 1746 that Society drafted a Parliamentary Bill which would oblige all apothecaries, chemists, and druggists selling medicines in the capital to be examined, and then to be admitted as members.[23] If successful, the Bill would have given the Society of Apothecaries a monopoly over the drug trade, increase its members, and improve its finances. The RCP objected and introduced its own Bill, but Parliament was prorogued, and the matter lapsed.

Two years later, in 1748, the Society of Apothecaries succeeded in obtaining authority under an Act of Parliament for its members to sell medicines within a 7-mile radius of the City of London—in addition to the right to dispense the prescriptions of physicians.[24] This new authority was intended to not only prevent chemists and druggists from dispensing, but also to prohibit physicians and surgeons from preparing or selling medicines. Had it been possible to implement, it would have given the apothecaries a monopoly on the supply of all medicines, but it proved impossible to apply in practice.

A group of apothecaries more concerned with pharmaceutical than medical practice came together in 1794 to consider how best to protect their interests. They formed a new body, the General Pharmaceutical Association of Great Britain, which would engage only with those druggists who would relinquish the dispensing of prescriptions, take on no assistants or apprentices, and limit themselves to wholesale business.[25] The Association petitioned Parliament to limit druggists to wholesale business, but without success. They continued with their campaign for some time, but eventually disbanded.[26]

After the Rose Case many apothecaries left the dispensing of prescriptions to unqualified assistants. By the late eighteenth century, the number of chemists and druggists had risen greatly. In 1802, in response to the threat posed by the apothecaries, they formed themselves into an Association of Chemists and Druggists.[27] This had some success in protecting their interests in relation to the Medicine Stamp Duty Acts and in influencing subsequent Stamp Duty legislation.[28] Some apothecaries moved into pharmaceutical wholesaling and manufacturing,[29] encouraged by an increase in population, particularly in towns, and by rising prosperity. But the apothecaries were not willing to let the lucrative drug trade fall into the hands of chemists and druggists. The matter was to again end up in the Courts.

Apothecaries Act and General Medical Practitioners, 1815

By the early nineteenth century, a variety of pharmaceutical practitioners offered services, with titles that were fluid and overlapped from one generation to the next. The term 'apothecary' was never a restricted title in law. For the Society of Apothecaries, the title referred to someone licensed by them, but besides these 'properly educated' apothecaries there were also many 'uneducated apothecaries' in practice. Whilst apothecaries increasingly encroached on the physicians' domain, chemists and druggists increasingly took over the dispensing activities of the apothecary. The chemists were accused of using impure drugs, the apothecaries of making monstrous profits and using dishonest practices, and the physicians of prescribing vast amounts of medicines for the benefit of the apothecaries, who in turn recommended those physicians who made a habit of 'multiplying their nauseous superfluities'.[30]

The underlining cause of this situation was the disorganized state of medical education. Calls for an Act to regulate medical practice, particularly that of apothecaries in England and Wales, began in 1793. The Society of Apothecaries were eventually successful in getting a Bill for regulating the practice of apothecaries passed by Parliament, having obtained the support of both the physicians and the chemists and druggists; the result was the Apothecaries Act 1815. The RCP readily accepted it, as the powers it contained were vested in the Society, whose Charter stressed its trading activities and placed it firmly under the thumb of the RCP. A clause made it an offence for an apothecary to refuse to dispense a physician's prescription. Their place was 'to be the physician's cooke'.[31]

The Association of Chemists and Druggists withdrew their opposition to the Act when the apothecaries agreed to insert a clause exempting chemists and druggists from its operation. It sealed the apothecaries' fate. Under the Act that was passed the chemists and druggists secured 'a continuance of all their former privileges', and a clause protected them from prosecution by the Society for over twenty-five years.[32] The Apothecaries Act failed to achieve its proposers' aims. Instead of elevating the status of apothecaries as general medical practitioners, they found themselves members of a London company still very active in the wholesale drug trade and under the indirect supervision of the RCP. The Society could prosecute those judged to be 'practising medicine' but not everyone who simply called themselves an 'apothecary'. Its prosecutions made little difference; the Act failed to prevent 'wholly ignorant and incompetent persons from endangering the health and lives of the community'.[33]

The Apothecaries Act clarified the roles of the different branches of medicine, but it was left to the courts to spell it out. In 1820 Mr. Justice Williams defined an apothecary as a pharmaceutical practitioner 'who mixed and prepared medicines prescribed by a physician, or by any other person, or by himself'. According to Mr. Justice Park in May 1828, 'the object of the Act was to keep the business of apothecary distinct from the other branches of the profession'. There were, he said 'four degrees in the medical profession, physicians, surgeons, apothecaries, and chemists and druggists'. The RCP were keen to extend their powers to cover the whole of medical practice throughout England and Wales. The physicians would be at the top, with the surgeons, apothecaries, and chemists and druggists in descending order beneath them. 'Each is protected in his own branch', the judge noted, 'and neither must interfere with the province of the other'.[34]

The Act did not define the role of apothecaries other than that their duty was to dispense physicians' prescriptions, but it neither limited their activities to this nor prevented chemists and druggists from doing so. By 1834 Mr. Justice Cresswell defined the apothecary as 'one who professes to judge of internal disease by its symptoms and applies himself to cure that disease by medicine'.[35] The transition of the apothecary from dispenser of prescriptions to general medical practitioner was largely complete.

Pharmacy and Medical Reform

Attempts to reform the medical profession continued unabated throughout the early nineteenth century. The prospect of its imminent reorganization prompted a sense of urgency amongst chemists and druggists, which led to the foundation of the Pharmaceutical Society of Great Britain. It was widely assumed that medical reform would include the regulation of chemists and druggists. On 5 February 1841 a Bill was introduced in the House of Commons by Benjamin Hawes MP to 'amend the laws relating to the medical profession in Great Britain and Ireland'. It gave definitions for 'practising medicine', 'medical practitioner', and 'chemist and druggist', which was defined as 'a person who shall sell, deal in, mix, or dispense for sale, any drug or medicine for the cure or relief of any bodily disorder, ailment or illness' and who had 'obtained a certificate to practise medicine'.[36] The chemists and druggists saw this as an attempt to place them firmly under medical control.

The proposed Act provided for the making of byelaws to regulate education, examination, and qualification not only to practise medicine but also to carry on the trade of chemist and druggist. As Jacob Bell, the founder of the PSGB, pointed out: 'chemists and druggists would be under the jurisdiction of a body in which they were not represented, and in the election of which the apothecaries had the largest number of votes'[37] (Fig. 2.1). Five days after the Bill was introduced a small group of chemists and druggists met to consider its implications. They presented

Fig. 2.1 Jacob Bell, founder of the Pharmaceutical Society of Great Britain, 1852 (courtesy of Royal Pharmaceutical Society)

their objections to Hawes, and on 19 February he withdrew the Bill, telling them that he would bring in a new Bill excluding all reference to chemists and druggists. He did so on 26 February, but the chemists and druggists found it unacceptable because it still gave the apothecaries power over them, and the new Bill was dropped.

The chemists and druggists concluded that a permanent association was needed to fight future threats. A meeting at the Crown and Anchor

Tavern in the Strand on 15 April 1841 approved a resolution 'that, for the purpose of protecting the permanent interests, and increasing the respectability of chemists and druggists, an association be now formed under the title Pharmaceutical Society of Great Britain'.[38] It had three main aims, but from the beginning the emphasis was on science and education. 'It would be absurd to lay great stress on the importance of science and skill in writing the prescription' wrote Jacob Bell 'and at the same time to leave the preparation of it to chance'.[39]

Bell insisted that the regulation of chemists and druggists should be included in any medical reform legislation.[40] He declared that

> we have always maintained that our body is and must be considered a branch of the medical profession, and that whatever regulations respecting education, registration or protection may be considered necessary for medical practitioners, the same or similar enactments are no less requisite in our department.[41]

But his hopes were quickly dashed; when the Medical Reform Bill came before Parliament in 1844, the Home Secretary decided that the task of developing suitable arrangements for regulating the physicians, surgeons, and general practitioners was 'difficult enough without including the chemists and druggists'.[42]

An important step in the professionalization of an occupation was to exclude the 'unqualified'. To chemists and druggists, the 'unqualified' were grocers and others who sold medicines; but for the general medical practitioners (formerly 'apothecaries') the 'unqualified' were the chemists and druggists.[43] Medical reform centred around education, examination, and registration. The PSGB had been formed 'to unite the chemists and druggists of Great Britain into one ostensible recognised and independent body, for the protection of their general interests and the advancement of the art and science of pharmacy'.[44] Education, examination, and registration would provide the basis for such unity, but such aspirations were not shared by the majority of chemists and druggists. They were to remain disunited until the end of the nineteenth century.

Select Committees and Pharmacy Acts, 1852

Bell had always intended to seek an Act of Parliament as early as possible to give the PSGB the necessary legal status. The first step was to obtain a Royal Charter, as this would put it in a better position to secure an Act; this was granted in 1843. By 1846 it was clear that the government had no intention of including pharmacy in any medical reform Bill, and the PSGB concluded that it would have to prepare its own Pharmacy Bill. The first draft required all dispensing chemists to pass a qualifying examination, and for the PSGB to enforce it. But it spoke for only a small minority of chemists and druggists.[45] Parliament would not allow an unrepresentative body to have a regulatory role, or for the same body to be both an educational and examining authority.[46] As the Charter already defined the Society as an educational body, a second draft of the Bill proposed the creation of a 'College of Pharmacy', independent of the PSGB, to take on the job of examination and registration.

The proposal was approved by its members in 1846, although some had misgivings about the proposed 'College of Pharmacy', fearing that its creation would 'reduce the Pharmaceutical Society to insignificance'. But the RCP objected to the very idea of a 'College of Pharmacy', and it was hastily dropped.[47] Had the Bill passed into law, the British model of pharmacy would have included a representative pharmaceutical society and an independent 'Pharmacy Board'. Instead, the representative and regulatory functions were to be combined in a single organization, the PSGB. The issue of separating functions between a pharmaceutical society and a Pharmacy Board was to trouble pharmacists in many colonies over many years.[48]

In 1847 and 1848 medical reform was considered at length by two parliamentary select committees, and in November 1847 a deputation from the PSGB met the Home Secretary to press their case for a Pharmacy Act. However, his view was that pharmaceutical legislation should follow, not precede, medical reform.[49] The PSGB then focussed on securing its own Act. It drafted a new Bill providing for the registration of chemists and druggists, and for the examination of all future entrants to the profession. It would give them a monopoly in the drug trade by establishing a clear boundary between themselves and unregistered persons, who would be prohibited from preparing, compounding, or dispensing medicines for sale or gain. Despite this, support from its members was muted.

The Society found itself unable to recruit a sponsor to take the Bill through parliament. Bell's response was to seek election as a Member of Parliament (MP) himself, and he became the MP for St Albans between December 1850 and July 1852.[50] On 12 June 1851 he introduced his Bill 'for regulating the qualifications of pharmaceutical chemists, and for other purposes in connection with the practice of pharmacy'. The PSGB would compile a register, which would include the names of its members plus those producing evidence of already being in practice as chemists and druggists. Thereafter it would be necessary to pass the Society's examinations in order to be admitted. Medical practitioners could continue dispensing provided they did not describe themselves as 'pharmaceutical chemists, chemists and druggists or dispensing chemists'.[51] Bell moved the second reading on 2 July 1851.

The Bill came under close scrutiny in parliament. Some MPs thought that the responsibilities proposed should be given to the Society of Apothecaries. Others complained that the Bill made no reference to the need to control the availability of patent and proprietary medicines.[52] Various amendments were made, and Bell introduced an amended Bill on 12 February 1852.[53] The second reading took place on 17 March, when it came under further scrutiny. The Home Secretary considered the powers the Bill gave the PSGB to make byelaws and regulations were excessive, and that penalties for those falsely assuming the business of chemist and druggist were too severe. The Bill was referred back to a select committee, who gathered evidence from physicians, surgeons, pharmacists, and others.

The purpose of the Act had been to define the boundary between the chemists and druggists and other retailers, rather than between them and medical practitioners. But the Bill would allow untrained persons to join the register simply because they were in business before it was passed. Bell had originally hoped to secure a monopoly in the right to sell medicines for chemists and druggists, but he opted not to link the regulation of pharmacy with the control of poisons in his Bill after the Home Secretary made it clear that medical reform would need to be based on free trade principles.[54] The main opposition came from the Society of Apothecaries, who thought chemists and druggists might gain a competitive advantage over their members, and might become a new class of medical practitioner. The Bill decreed that no member of the medical profession could be registered, and that if a pharmaceutical chemist obtained a medical qualification their name would be removed from the Register.

With amendments, the Bill received its Royal Assent and became the Pharmacy Act 1852. Under the Act, the PSGB itself kept the register of 'pharmaceutical chemists'. It now combined the roles of education provider, examining body and registration body. This situation remained unchanged until 1925 when its school of pharmacy became part of the

University of London.[55] What was enacted bore little relation to what had originally been intended. The titles 'pharmaceutical chemist' and 'pharmaceutist' were reserved for PSGB members and those who had passed the Major examination, but the title 'chemist and druggist' was not restricted, and anyone could still set themselves up as such without any training. Whilst the Act enabled chemists and druggists already in business to register, only 800 of the over 5,000 in business at the time chose to do so.[56] An 'institutional rift' had begun between the registered pharmaceutical chemists and the unregistered chemists and druggists.[57]

Regulating Pharmacy and Controlling Poisons

Whilst the 1852 Pharmacy Act did not establish a link between professional regulation and the sale of poisons, neither did it unite the profession of pharmacy. Chemists and druggists who chose not to join the PSGB could not use the title 'pharmaceutical chemist' and were excluded from other privileges of membership. There were dissenting voices even within the PSGB Council. William Bastick and William Dickinson supported a new organization, the British Association of Chemists and Druggists (BACD), which was launched by two London chemists and druggists, Charles Linder and John Rossiter.[58] Members would receive a certificate declaring that they were duly qualified chemists and druggists, and an independent College of Pharmacy would maintain a register.[59] But despite its apparent attractions the new association did not survive much beyond its inaugural meeting.

The conflict between the chemists and druggists and the pharmaceutical chemists continued for over 40 years. A new trade journal, the *Chemist and Druggist*, was founded in 1859. In August 1860 it carried the prospectus for yet another society, the United Society of Chemists and Druggists (USCD). It was rather more successful than the BACD, surviving until 1867. Both associations emerged from deep dissatisfaction amongst chemists and druggists with the PSGB; they believed it to be far too concerned with science, education, and the dispensing of prescriptions at the expense of trade and business. The *Chemist and Druggist* thought it needed to be 'more tolerant of tradesman-like views'.[60] It was seen as elitist by many chemists and druggists. The 'institution which was established to protect the general interests of the chemist and druggist is exclusively devoted to those whose lot is to dispense prescriptions and retail perfumery'.[61] Most never saw a prescription, and the sale of medicines, nostrums, and ingredients was often only a small part of their business. Whilst the BACD had aimed to improve the social position and professional status of dispensing chemists, the USCD described itself as 'a general trade union' representing the interests of chemists and druggists 'as a trading community'.[62] It considered its aims complementary rather than competitive to those of the PSGB.

One of the benefits of inclusion in the Register of Pharmaceutical Chemists was exemption from jury service; by then this had come to be seen as a hallmark of a profession. In 1862 a Bill had been introduced in Parliament on the mode of summoning juries. The PSGB petitioned to have its members—then numbering around 2000—included in the list of those exempt. The response of the USCD was to seek exemption for all chemists and druggists, of which it claimed there were then over 40,000. Parliament balked at this, and when the Jury Act was finally passed, exemption was given to the pharmaceutical chemists but not the chemists and druggists. Pharmaceutical chemists thus received privileges and legal recognition of professional status not available to chemists and druggists.[63] Exemption from jury service was later to emerge as an indicator of professional status in the colonies.

It became increasingly clear to the chemists and druggists that education, examination, and registration were indeed the route to public recognition and privileges. The USCD became the poor man's PSGB.[64] But the medical profession had not entirely relinquished its claim to control pharmacy. In 1863 a General Medical Council (GMC) committee proposed that 'pharmaceutical education, examination and practice' should be brought under its control.[65] It drafted a Bill defining pharmacy as a branch of medicine, a view still fully supported by the PSGB's Council. Daniel Hanbury thought that 'medicine, surgery and pharmacy should be regulated by one comprehensive Act of Parliament'. This would, he suggested, 'transform pharmacy into a responsible and important profession'.[66]

The USCD saw the proposal as an attempt to prevent chemists and druggists from dispensing medicines. The *Chemist and Druggist* reported that they had agreed 'that neither the Medical Council nor the Pharmaceutical Council would reign over them'.[67] But the GMC committee's proposal received a cool response from members of its own Council. There was little support for it taking on responsibility for regulating pharmacy. Professor Christison 'disapproved of any attempt to introduce pharmacy into the Medical Act'. Dr. Storrar concluded that 'the Council should drop pharmacy and encourage the Pharmaceutical Society to go on with an independent measure'.[68] In early 1864 the GMC decided that pharmacy legislation should be left to pharmacists. It heralded the ending of the medical profession's bid to control pharmacy in Britain, although the Society of Apothecaries had not given up entirely.

In March 1864 the PSGB considered introducing a new Pharmacy Bill which would protect the interests of those already in business but exclude the chemists and druggists; the USCD responded by producing its own Bill, which included them. Both Bills were submitted to Parliament. The PSGB Bill would entitle anyone in business before a chosen date to be registered, but in future they would have to pass the Society's Minor examination. 'Chemist', 'druggist', and 'chemist and druggist' would be added to the list of restricted titles. The USCD Bill was more extensive, as the society lacked resources and wanted to ensure the demise of the PSGB. But it had one great strength; it proposed restrictions on the sale of poisons.[69] Widespread public concern about deaths from poisoning had resulted in an Arsenic Act in 1851, which restricted its sale to chemists and druggists.[70] There was a case for controlling other poisons in the same way, and this could be linked to the regulation of pharmacy.[71]

Publication of the two Bills prompted extensive debate; both had second readings on 29 March 1865 and were referred to a Select Committee. The evidence presented favoured legal restrictions on the sale of some poisons, and recognized the important role played by pharmacists. The GMC supported the PSGB Bill, but the Select Committee abandoned both and recommended that the government bring in its own. The experience exposed divisions in both societies.[72] The PSGB incorporated points from the USCD Bill that had been favourably received by the Select Committee, and in early 1866 it submitted its revised proposals to the Home Office. After further negotiations 'a Bill to regulate thesale of poisons and alter and amend the Pharmacy Act 1852', to be known as the 'Pharmacy Act 1868',was passed on 11 May 1868. It was an Act that linked the regulation of pharmacy to the sale of poisons, and was to form the basis for much of the legislation subsequently enacted across the British Empire.

Prescribing and Dispensing Medicines

The 1868 Pharmacy Act maintained the distinction between pharmaceutical chemists and chemists and druggists. The restricted titles 'chemist', druggist', 'chemist and druggist', 'pharmacist', and 'dispensing chemist or druggist' were added to those listed in the 1852 Act. The register indicated whether those listed held the Major (pharmaceutical chemist) or Minor (chemist and druggist) certificate. Most could only describe themselves as 'chemists and druggists' who the Act inadvertently defined as 'persons keeping open shop for the compounding of medical prescriptions'.[73] Those listed on the Register were qualified only 'to sell or keep open shop for retailing, dispensing or compounding poisons'; chemists and druggists were licensed to sell poisons, not to dispense prescriptions. The PSGB had failed to secure a monopoly in dispensing prescriptions for its members (the pharmaceutical chemists). Unqualified drug sellers took up titles such as 'herbalist' or 'medical botanist', and their shops became 'drug stores' or 'medical halls'. But use of the word 'pharmacy' to describe a shop where medicines were sold was not restricted, and it was first used by unregistered practitioners.[74]

The battle over the right to dispense prescriptions continued between the PSGB, the RCP, and the Society of Apothecaries. The latter sought to prevent its licentiates from dispensing prescriptions, in order to move away from association with trade. But the physicians were anxious to retain the subservient role of the apothecaries as dispensers of their prescriptions, and in 1861 the RCP initiated a reform whereby licentiates of the Society of Apothecaries would be permitted to dispense as well as prescribe medicines. The Society of Apothecaries had threatened to prosecute licentiates found to be both prescribing and dispensing medicines but were unable to do so, and a new breed of doctor emerged who were not only qualified to attend patients but also to prescribe and dispense for them.[75]

The PSGB was still anxious to maintain the distinction between pharmaceutical chemists and chemists and druggists. Full membership was available only to pharmacy owners, and there was a growing dissonance between educational attainment and employment status; young pharmacists who had passed the Society's Major examination could only be associates. The PSGB finally abolished the property qualification in 1872, when all pharmaceutical chemists (but not chemists and druggists) could become full members. It was at this point that 'a society of proprietor chemists became an association of professional pharmacists'.[76] But the PSGB's continued reluctance to get involved in the protection of trade interests led to further calls to establish an organization that would. In July 1876 a group of pharmaceutical wholesalers and retailers came together to form a Chemists and Druggists' Trade Association (CDTA), claiming that its aim was to supplement the PSGB's activities rather than replace them.[77]

By the mid-1890s the PSGB's finances were such that there was an urgent need to recruit new members. A Pharmacy Acts Amendments Act was passed in 1898 to give all chemists and druggists—as well as all pharmaceutical chemists—the right to become full members. Only then could all pharmacy practitioners in Britain be described by the single title 'pharmacist'. But even then some differences remained; only pharmaceutical chemists were still exempt from jury service.

The Emergence of Company Chemists, 1880

The PSGB considered that professional practice required pharmacists to own and control their own businesses. Following the 1868 Act it believed that corporate bodies would be unable to operate chemists' businesses since they could not sit examinations or be registered as pharmaceutical chemists. But in the 1870s limited companies began to sell medicines, using the term 'chemist' to describe the part of the shop where this took place. The issue of whether companies could own pharmacies was tested in the courts, in the case of the Pharmaceutical Society v. The London and Provincial Supply Association, which ran a department store in London. Its owner, William Mackness, was not registered, but the business included a pharmacy department, managed by a registered chemist and druggist supported by two qualified assistants. In January 1878 the PSGB prosecuted Mackness for the illegal sale of a poison. He paid the fine and converted the business into a limited liability company.[78]

The 1868 Act referred only to 'persons'; Section XV declared that 'any Person who shall sell or keep an open shop for the retailing, dispensing or compounding of Poisons…without being duly qualified, shall be liable to pay a penalty'.[79] The Act did not explicitly prevent companies from doing so, and the legal argument was about whether 'person' could include a company. If it could not, companies would not be able to own pharmacies. The case was initially heard in a County Court, where the judge gave judgement in favour of Mackness. The Society appealed to the High Court, which gave judgement in its favour—that companies could not legally own pharmacies.[80] But Mackness appealed to the Court of Appeal, who again ruled in his favour. The Society appealed again, and in July

1880, the Law Lords finally judged that the carrying on of a pharmacy business by a limited company was indeed legal.

Titles restricted to chemists and druggists could now legally be used by companies, provided that a qualified person undertook the sale of poisons. Businesses were free to open as many branches as they wished. The first was operated by an unqualified druggist, Jesse Boot, in Nottingham. By 1883 he had ten shops, and by 1900 he already had by far the largest retail chemist chain, with more than 250 branches.[81] But by 1890 three other firms had opened ten or more branches; Taylor's of Leeds, Warhurst of Liverpool, and Timothy White of Portsmouth.[82] By the mid-1890s more than 200 companies were registered for retail trade in drugs and dispensing around the country. A Chemists' Cooperative Society was established with 15 pharmacies in London and 25 in provincial cities.

The wording of the Act and its testing in the courts had irreversible consequences for pharmacy in Britain. The PSGB made at least one attempt to slow the growth of company chemists. The judgement failed to state explicitly that individual members of a company who were not themselves registered chemists and druggists were not liable to penalties. In June 1887, the PSGB opened proceedings against seven partners in a limited company trading as chemists and druggists in Edinburgh. In the Summary Court judgement was made against the shareholders, making them liable to penalties. But the defendants appealed to the High Court, where once again the judgement of the lower court was overturned.[83] No further appeal was possible, and the PSGB had to concede that the battle against company chemists was lost.

The loss of the 1880 case taught pharmacy leaders that the only way to correct a weakness or omission in legislation was by passage of either new or amending legislation. This was easier said than done, although pharmacists in Victoria, Australia, were subsequently able to secure amending legislation concerning company chemists (Chapter 10). It was not until 1901 that the PSGB circulated a draft Bill to amend the wording and to make other changes, and they discovered that proposed changes to existing legislation tend to come under particularly close Parliamentary scrutiny, as the Bill 'suffered much change when the government decided to take a hand in it by superimposing its own views'.[84] It was a fate that was to be shared by New Zealand pharmacists when they faced the same issue (Chapter 11). It was much better to avoid the need for amending legislation in the first place.

Separating Prescribing and Dispensing

The 1868 Act was 'more a patchwork quilt of amendments than a seamless web of legislative thought' and caused confusion for half a century.[85] It linked the development of pharmacy with the sale of poisons by placing their sale and supply in the hands of chemists and druggists.[86] Precautions to be taken in the sale of arsenic were extended to fifteen other poisons, many of which were used as medicines; and it mandated that medicines listed in the *British Pharmacopoeia* must be compounded only in forms stated in its formularies. These stipulations were to have profound implications in many parts of the Empire as the 1868 Act became the basis of pharmacy legislation.

The shortcomings of the Act were subjected to extensive debate in the profession and Parliament. The sale of poisons, the actions of company chemists, and the registration of pharmacists qualified in the colonies, all came under close scrutiny. The PSGB and the government both submitted draft Bills to Parliament, but neither was approved. The position was transformed with a change of government in 1906 with many newly elected MPs supporting the PSGB's Bill. As with the 1852 Act, a pharmacist MP, Richard Winfrey, played a leading role in taking it through Parliament.[87] Following several amendments a new 'Act to regulate the sale of certain Poisonous Substances and to amend the Pharmacy Acts' was passed in 1908.[88] Its provisions included regulation of the use of titles by corporate bodies, who could use certain descriptions; company chemists were now formally written into the legislation. It also brought the title 'pharmacist' into prominence.

With plans to introduce a National Health Insurance (NHI) scheme in the early twentieth century, the government re-examined the roles of doctors and pharmacists with regard to prescribing and dispensing prescriptions. This was one of the PSGB's concerns from its foundation; in 1885 Thomas Greenish, a former president, stated its position succinctly as 'let the doctor prescribe and the druggist dispense'. But it was left to Lloyd George, the Chancellor of the Exchequer, to 'separate the drugs from the doctors'.[89] He considered that paying doctors to supply medicines encouraged excessive prescribing. In future 'all the drugs should be dispensed by somebody else rather than by the doctor. It should be for the chemist to dispense…there should be a compulsory separation of the two'.[90] It was only with the 1911 National Insurance

Act that a legal distinction was finally made between the prescribing and dispensing of medicines in Britain.[91]

Exceptions were allowed in rural areas where no chemist was available, although some doctors' leaders were in favour of stopping doctor-dispensing provided that the pharmacists gave up counter-prescribing.[92] In 1914 Alfred Cox, the British Medical Association's medical secretary, conceded that 'there is a good deal to be said for all the dispensing being taken out of the hands of the doctors, and of all the prescriptions being taken out of the hands of the chemists'.[93] But the suggestion was declined by the pharmacists; by then doctors and pharmacists had a common enemy in patent medicine manufacturers.[94] Proprietary preparations usually cost less than what dispensing chemists could make them for. Even so, many pharmacists regarded dispensing doctors more harmful to their interests than these manufacturers.[95]

Alone amongst the medical bodies the Society of Apothecaries retained an interest in regulating pharmacy. In 1913 it considered introducing a Bill to extend its powers under the 1815 Apothecaries Act in respect of the apothecaries' assistant qualifying examination.[96] They approached Charles Bathurst MP for assistance in taking it through Parliament. He advised them to come to an agreement with the PSGB, but the Society of Apothecaries prepared a draft Bill anyway that would give it powers to create a register of assistants, to award the restricted title 'certified dispenser of the Society of Apothecaries', and to remove names from it. Such dispensers would be authorized to dispense prescriptions, but not to keep open shop, or to sell medicines and poisons. But before the draft Bill could be introduced into Parliament war broke out.[97]

From Trade to Profession

By the time the First World War was over, Britain and pharmacy looked very different. The government set about reforming work practices and industrial relations in many occupations including pharmacy. The PSGB proposed to set up a Joint Industrial Council for pharmacy, but following opposition decided to test whether it had the legal authority to do so; it found it did not. In the 1920 'Jenkin Case' the judge ruled that it was 'not within the powers or purposes of the Pharmaceutical Society to take part, or expend any of its funds, in the formation, establishment, maintenance or work of an Industrial Council'.[98] At the height of Empire, the PSGB was left wondering about its role. It chose not to appeal, and within two months a new organization, the Retail Pharmacists' Union (RPU), had been formed as a 'union of retail employer chemists for the protection of trade interests'.[99]

The Society of Apothecaries resurrected its Bill in 1920, when the Association of Certified Dispensers urged it to amend and extend its powers.[100] But the Privy Council advised it that such a course of action was not advisable. Nevertheless, the Association's secretary and chairman met the Apothecaries' Private Court to press for a new Act, after which the Society's representatives met with Sir John Anderson, permanent under-secretary at the Home Office, to discuss it. The Association pressed the case for an increased period of training, a modified examination, and the creation of a Statutory Register of assistants, so that their qualification would correspond to the PSGB Minor examination.[101] Such a proposal was never likely to be acceptable to the government, and the matter was not pursued. It was the last attempt by a medical body in Britain to seek a role in the regulation of pharmacy.

Following the Jenkin Case, the PSGB set about redefining its purpose and having this legally recognized. A 1933 Pharmacy and Poisons Act clarified the relationship between the PSGB Council, the Privy Council, and PSGB members.[102] The distinction between registration as a pharmacist and PSGB membership ended; a single fee covered both. Holloway states that, with the 1933 Act, 'professional regulation triumphed over protection and trade unionism'.[103] It ended any hope of it amalgamating with the RPU and the Chemists' Defence Association to form a 'British Medical Association for Pharmacy'. Holloway later suggested that the Retail Pharmacists' Union (now National Pharmacy Association) was the true successor to the PSGB in terms of its original aims.[104]

The PSGB's objectives were formally changed through a Supplemental Charter in 1953. The words 'the protection of those who carry on the business of chemists and druggists' were replaced by 'to maintain the honour and safeguard and promote the interests of the members in the exercise of the profession of pharmacy'. For the PSGB at least, the transition from trade to profession was complete.

Pharmacy Education in Britain

In his first address as PSGB president in 1841, William Allen announced its aim of establishing a 'uniform system of education, which will promote the advancement of science and the elevation of the profession of pharmacy'.[105] It would begin with an entrance examination, consisting of a test in the classics, and would be followed by a Minor examination to be taken at the end of a four or five-year apprenticeship. Successful candidates could be associates of the PSGB, and act as assistants to chemists and druggists. To become full members and to open their own pharmacy they would need to take a more advanced Major examination. Over the following years this two-tier approach to pharmacy qualifications was to play a significant role in many parts of the British Empire.[106]

The model took some time to become a reality in Britain. There were no schools of pharmacy, although they had long existed in France and Germany, and America already had five.[107] In 1841 pharmacy subjects were taught only in medical schools, at Apothecaries' Hall in London, and at the Edinburgh Royal Dispensary.[108] The PSGB founders recognized the need for a school and undertook to 'furnish the means of proper instruction'. It opened its school in 1842, and in 1843 lectures were arranged at Manchester's Royal Medical Institution.[109] In the same year the PSGB awarded grants to establish 'Branch schools' in Manchester, Norwich, Bath, and Bristol, but these proved unsustainable. In 1845 the London school was designated a 'national school' intended to meet the needs of the whole country, although it was only available to those living in London. With the failure of the provincial schools, the initial opportunity to develop a uniform system of pharmaceutical education in Britain was lost.[110]

The preliminary examination could be conducted by approved examiners in the provinces. The Minor examination was taken as soon as apprentices felt ready. It was entirely oral and was held at the PSGB headquarters at monthly intervals (Fig. 2.2). Candidates were examined in materia medica, botany, chemistry, and Latin.[111] The Major examination consisted of the same subjects along with toxicology; the standard was higher and tested with oral and written questions. This examination could be taken soon after the Minor, although an interval of not less than three months was recommended. Initially, taking examinations was voluntary. Plans to make them compulsory were considered in 1851 but were only enacted in the 1868 Pharmacy Act. Following pressure from the USCD, the standard of the qualifying examination was lowered, and passing the Minor examination qualified candidates for registration.[112] This prompted action in the provinces; by 1870 twenty-two local pharmacy associations had schemes to provide education, usually in local

Fig. 2.2 Headquarters of the Pharmaceutical Society of Great Britain, 17 Bloomsbury Square, London, 1883/4 (courtesy of Royal Pharmaceutical Society)

colleges, although not all provided classes in all subjects. But by 1872 most schemes had fizzled out through lack of support.

Although the examination was now compulsory, there was no agreed curriculum. This led to a proliferation of 'crammer' schools which prepared candidates for the examination. In 1872 reports suggested that some students had passed the examination after learning 500 answers to 500 questions.[113] In 1870 there were two such schools, but by 1880 there were five. The Westminster College of Chemistry and Pharmacy was founded by George Wills in 1874.[114] He claimed in 1899 that 4000 chemists and druggists had trained at his college.[115] Many came from overseas to attend his courses, take the PSGB's examinations, and obtain its certificate. By 1890 there were seven such schools and by 1900 twenty-two, with ten in London and the rest in the provinces. Overall, the standards exhibited by the candidates were poor; in 1900 the London Board of Examiners reported that of 1371 candidates presenting, 993 (72.4 percent) were failed.[116]

The lack of a curriculum was addressed in 1881. A PSGB committee considered the relationship between examination and education and recommended a systematic course of instruction for the Minor examination, but the Council was unable to support it. The matter was discussed again when the Pharmacy Act Amendment Bill was being considered in 1891, but it was again subject to widespread criticism. George Wills fumed that such action would deny access to the profession by those who could not afford to attend a recognized college. 'What can it matter to the examiners or to the public' he argued, 'where or how a candidate obtained his knowledge? Is he qualified? that is the important question'.[117] So the move to enforce a period of study failed again. The consequence was substantial variation in the content and teaching of pharmacy courses and in the facilities available.

Only after the First World War was a compulsory course of study adopted for pharmacy in Britain. Only then did a process of rationalization begin in pharmacy schools, with increasing resources being required to ensure that education kept pace with pharmaceutical knowledge. The PSGB's school of pharmacy began offering a degree course in 1925.[118] The last of the private schools closed in 1949; and pharmacy training by apprenticeship followed by completion of the Pharmaceutical Chemist Diploma Course ended in 1967, when pharmacy in Britain became a degree-entry only profession.

The British Model of Pharmacy

This chapter has shown that professionalizing pharmacy in Britain was a lengthy and difficult process. Although the foundation of the PSGB in 1841 was a major turning point, the origins of professionalization date from much earlier. Actions taken to secure the support of the state and the passage of legislation were just first steps in this journey. Separation from medicine was not the founder's intention; Bell had hoped to develop pharmacy as a discrete section of medicine, but passage of the 1852 Pharmacy Act made the exclusion of pharmacists from the medical profession inevitable. Before it, chemists and druggists could still be regarded as an integral part of the medical profession, along with the physicians, surgeons, and apothecaries. Without it the 1859 Medical Register might have been published in four sections rather than three. As Holloway notes, 'there is nothing inevitable about the present division between the practice of medicine and that of pharmacy'.[119] It was socially and politically constructed by three Acts of Parliament.

Getting pharmacy practitioners to work collectively towards common goals took rather longer. Deep divisions between pharmaceutical chemists and chemists and druggists continued until the early twentieth century. The gulf between trade and professional interests seemed unbreachable, with angry voices on both sides. In 1890 a PSGB member declared: 'the idea that the profession and the trade of pharmacy are distinct and incompatible is incorrect, for they are closely interwoven, and the most successful man from a professional point of view will probably draw the largest returns from his business'.[120] But another shouted 'it is pharmacy, pharmacy, pharmacy, not trade, trade, trade! They are incompatibles; they will never mix. An explosion will result by and by if you try to mix them'.[121] But the PSGB Council eventually concluded that professionalism alone could not withstand the forces of market capitalism.[122] In the end it was in the interests of both sides to work together.

It is now possible to identify core features which constitute the 'Britishness' of British pharmacy for which 'indicators' can be proposed. These markers of the British model can then be used to track the Britishness of pharmacy in the colonies. They can be summarized as follows:

- **Self-regulation established through Pharmacy Act** The separation of pharmacy from medicine and control by doctors through a Medical Board was ended by means of a Pharmacy Act which recognized the legal foundation of a pharmaceutical society.
- **Pharmaceutical Society determines register of qualified practitioners** Responsibility for education and examination, qualification and registration was vested in a professional organization; no separate and independent Pharmacy Board was considered necessary.
- **Two-tier register of qualified practitioner** Between 1868 and 1954 pharmaceutical chemists and chemists and druggists in Britain were identified as such in the register. This normalized an internal hierarchy suggesting that some were more qualified than others.
- **Regulation of pharmacy linked to control of poisons** After the initial 1852 Pharmacy Act it became necessary subsequently to link the regulation of pharmacy with the control of poisons in Pharmacy and Poisons Acts in order for legislation to be enacted.

- **Establishment of chains by company chemists** The ownership of multiple pharmacy outlets by corporate bodies became possible as a consequence of the wording of the 1868 Pharmacy and Poisons Act, and pharmacy chains became a key feature of British pharmacy.

Pharmacy in the colonies was strongly influenced by legislation and practice in Britain. The chapters that follow explore the professionalization of pharmacy and the extent to which the 'British model' was followed in the territories that made up the British Empire in the early twentieth century.

Notes

1. See for example Stieb, Ernst W. *Drug Adulteration: Detection and Control in Nineteenth-Century Britain* (Madison, 1966), p. 144. Hunting, Penelope. *A History of the Society of Apothecaries* (London, 1998), p. 217.
2. Anderson, Stuart (ed.) *Making Medicines: A Brief History of Pharmacy and Pharmaceuticals* (London, 2005).
3. Trease, G.E. *Pharmacy in History* (London, 1964), p. 43.
4. Ibid., p. 62.
5. Ibid., p. 53.
6. Trease, *Pharmacy in History*, p. 88.
7. Sonnedecker, Glenn. *Kremers and Urdang's History of Pharmacy*, fourth edition (Madison, 1976), pp. 34–5 and Note 32, p. 502.
8. Ibid., p. 35.
9. Ibid., p. 89.
10. Ibid., p. 109.
11. Copeman, W.S.C. *The Worshipful Society of Apothecaries of London: A History 1617–1967* (London, 1967).
12. Hunting, *Society of Apothecaries*, p. 34.
13. Cripps, Ernest C. *Plough Court: The Story of a Notable Pharmacy* (London, 1927), p. xv.
14. Trease, *Pharmacy in History*, p. 105.
15. Barrett, C.R.B. *The History of the Society of Apothecaries of London* (London, 1905), p. 55.
16. For the dispute between the apothecaries and the physicians see Hunting, pp. 47–55.
17. Trease, *Pharmacy in History*, p. 151. See also Hunt, John A. 'Echoing Down the Years: The Tercentenary of the Rose Case', *The Pharmaceutical Journal*, 266 (2001), pp. 191–195.

18. Barry, Jonathan. 'John Houghton and Medical Practice in London c.1700', *Bulletin of the History of Medicine*, 92 (2018), pp. 575–603.
19. Porter, Roy and Porter, Dorothy. 'The Rise of the English Drugs Industry: The Role of Thomas Corbyn', *Medical History*, 33 (1989), p. 280.
20. Cook, Harold J. 'The Rose Case Reconsidered: Physicians, Apothecaries, and the Law in Augustan England', *Journal of the History of Medicine and Allied Sciences*, 45 (1990), pp. 527–55.
21. Hunt, 'Echoing Down the Years', pp. 191–195.
22. Thompson, C.J.S. *The Mystery and Art of the Apothecary* (London, 1927), p. 280.
23. Hunting, *Society of Apothecaries*, p. 171.
24. Bell, Jacob and Redwood, Theophilus. *Historical Sketch of the Progress of Pharmacy in Great Britain* (London, 1880), p. 26.
25. Hunt, John A. and Jones, Ian F. 'The First Pharmaceutical Society?', *Pharmaceutical Journal*, 259 (1997), pp. 997–9.
26. For the General Pharmaceutic Association of Great Britain see Good, John Mason. *The History of Medicine, So Far as It Relates to the Profession of the Apothecary, from the Earliest Accounts to the Present Period* (London, 1795).
27. Trease, *Pharmacy in History*, p. 181.
28. Stebbings, Chantel. *Tax, Medicine and the Law: From Quackery to Pharmacy* (Cambridge, 2018), p. 41.
29. Burnby, Juanita G.L. *A Study of the English Apothecary from 1660 to 1760* (London, 1983), p. 49.
30. Holloway, S.W.F. 'Significance of the Apothecaries' Act, 1815'. *Pharmaceutical Historian*, 1 (1970), p. 7.
31. Ibid., p. 8.
32. Holloway, S.W.F. *Royal Pharmaceutical Society of Great Britain 1841 to 1991: A Political and Social History* (London, 1991), p. 85.
33. Holloway, 'Significance', p. 9.
34. Holloway, S.W.F. 'The Apothecaries' Act, 1815: A Reinterpretation', *Medical History*, 10 (1966), pp. 124–9.
35. Ibid.
36. Bell and Redwood, *Historical Sketch*, p. 87.
37. Ibid., p. 88.
38. Holloway, *Royal Pharmaceutical Society*, p. 92.

39. *Pharmaceutical Journal and Transactions*, 2 (1842–43), p. 678.
40. Holloway, *Royal Pharmaceutical Society*, p. 147.
41. *Pharmaceutical Journal and Transactions*, 4 (1844–45), p. 101.
42. Holloway, *Royal Pharmaceutical Society*, p. 147.
43. Ibid.
44. Cripps, *Plough Court*, p. 62.
45. Holloway, *Royal Pharmaceutical Society*, p. 148.
46. When University College London was founded its aim was to be both educational institution and degree awarding body. Its request for a Charter was declined on the basis that it could only have one function. It opted for educational institution; the University of London was created to award degrees.
47. Holloway, *Royal Pharmaceutical Society*, p. 150.
48. In Britain the government ultimately resolved that no corporate body could be both representative body and regulatory authority. In 2010 an independent Pharmaceutical Council was created, and the PSGB reverted to being a voluntary body 169 years after its foundation. It was designated 'Royal' in 1988.
49. Holloway, *Royal Pharmaceutical Society*, p. 151.
50. For an account of Bell's career as an MP see Holloway, *Royal Pharmaceutical Society*, pp. 151–63.
51. Ibid., p. 163.
52. Homan, P.G., Hudson, B. and Rowe, R.C. *Popular Medicines: An Illustrated History* (London, 2007).
53. Holloway, *Royal Pharmaceutical Society*, p. 166.
54. Ibid., p. 185.
55. Hudson, B. and Boylan M. *The School of Pharmacy, University of London: Medicines, Science and Society, 1842–2012* (London, 2013), p. 130.
56. Holloway, *Royal Pharmaceutical Society*, p. 186.
57. Ibid., p. 181.
58. Ibid., p. 188.
59. *Annals of Pharmacy and Practical Chemistry*, 3 (1854), pp. 17–9, 83–4.
60. *Chemist and Druggist*, 2 (1861), pp. 193–5.
61. *Chemist and Druggist*, 3 (1862), pp. 144–5.
62. Holloway, *Royal Pharmaceutical Society*, p. 199.
63. Ibid., p. 203.

64. *Chemist and Druggist*, 5 (1964), pp. 134–6; 3 (1862), pp. 181–2; 4 (1863), pp. 151–9.
65. Holloway, *Royal Pharmaceutical Society*, p. 205.
66. *Pharmaceutical Journal*, 5 (1863–64), pp. 484–5.
67. *Chemist and Druggist*, 5 (1864), pp. 134–6.
68. Holloway, *Royal Pharmaceutical Society*, p. 208.
69. Ibid., p. 211.
70. Ibid., p. 180.
71. Anderson, Stuart. 'From 'Bespoke' to 'Off-the-Peg': Community Pharmacists and the Retailing of Medicines in Great Britain 1900 to 1970', *Pharmacy in History*, 50 (2008), pp. 43–69.
72. *Chemist and Druggist*, 7 (1866), p. 91.
73. Holloway, *Royal Pharmaceutical Society*, pp. 240–1.
74. Ibid., p. 241.
75. Hunting, *Society of Apothecaries*, p. 207.
76. Holloway, *Royal Pharmaceutical Society*, p. 272.
77. Ibid., p. 256.
78. Holloway, *Royal Pharmaceutical Society*, pp. 274–80.
79. Ibid., p. 274.
80. Ibid., p. 275.
81. Chapman, Stanley, *Jesse Boot of Boots the Chemists* (London, 1974).
82. Holloway, *Royal Pharmaceutical Society*, p. 278.
83. Ibid.
84. Matthews, Leslie G. *History of Pharmacy in Britain* (London, 1962), p. 371.
85. Holloway, *Royal Pharmaceutical Society*, p. 239.
86. Glyn-Jones, W.S. *The Law Relating to Poisons and Pharmacy* (London, 1909).
87. Holloway, *Royal Pharmaceutical Society*, p. 302.
88. Linstead, Hugh N. *Forensic Pharmacy* (London, 1929), pp. 99–103.
89. Holloway, *Royal Pharmaceutical Society*, p. 333.
90. Hunt, John A. and Jones, Ian F. 'David Lloyd George: His Influence on Pharmacy in Britain', *Pharmaceutical Journal*, 253 (1994), pp. 912–5.
91. Anderson, Stuart. 'Community Pharmacy and the Rise of Welfare in Great Britain 1900 to 1986', *Pharmaceutical Historian*, 36 (2006), S10–7.

92. *Pharmaceutical Journal* 86 (1911), p. 806; 88 (1912), p. 717.
93. *Report from Select Committee on Patent Medicines.* Command 630 (1914). No. 11845-11851, p. 25.
94. Ueyama, Takahiro. *Health in the Marketplace: Professionalism, Therapeutic Desires, and Medical Commodification in Late-Victorian Britain* (Palo Alto, 2010), p. 48.
95. Ibid., p. 55.
96. Society of Apothecaries Archive, Court Minute Book 1913–26, (1913), p. 8. Cited in Adams, Derek. *The Rise and Fall of the Apothecaries' Assistants 1815–1923*, PhD Thesis, University of Hertfordshire (2011), p. 296.
97. Society of Apothecaries Archive, Box 11, E/4/4/2/8. Bill to amend Apothecaries Act (1923).
98. Holloway, *Royal Pharmaceutical Society*, p. 363.
99. Ibid., p. 367.
100. The Association of Certified Dispensers was established in response to perceived threats to their livelihoods under NHI arrangements.
101. Society of Apothecaries Archive, Box 10, E/4/4/1/2 (1923).
102. Linstead, *Poisons Law*, pp. 246–80.
103. Holloway, *Royal Pharmaceutical Society*, p. 390.
104. Holloway, S.W.F. 'Towards a History of the National Pharmaceutical Association', *Supplement to the Pharmaceutical Journal*, 265 (1996), p. N3.
105. *Pharmaceutical Journal*, 1 (1841), p. 80.
106. Earles, M.P. 'The Pharmacy Schools of the Nineteenth Century', in Poynter, F.N.L. *The Evolution of Pharmacy in Britain* (London, 1965), pp. 79–95.
107. Earles, 'Pharmacy Schools', p. 80.
108. Matthews, *History of Pharmacy*, p. 163.
109. Robinson, Brian. *The History of Pharmaceutical Education in Manchester* (Manchester, 1986). pp. 32–43.
110. Earles, 'Pharmacy schools', p. 82.
111. Trease, *Pharmacy in History*, p. 191.
112. Earles, 'Pharmacy schools', pp. 85–7.
113. Ibid., pp. 87–91.

114. Kurzer, Frederick. 'George S.V. Wills and the Westminster College of Chemistry and Pharmacy: A Chapter in Pharmaceutical Education in Great Britain', *Medical History*, 51 (2007), pp. 477–506.
115. Wills, George S.V. *A Jubilee Souvenir: The Work of G.S.V. Wills and the Westminster College of Chemistry and Pharmacy* (London, 1899), pp. 201–2.
116. Earles, 'Pharmacy Schools', p. 91.
117. *Pharmaceutical Journal*, 10 (1900), p. 366.
118. Hudson, *School of Pharmacy*, p. 89.
119. Holloway, *Royal Pharmaceutical Society*, pp. 181–2.
120. *Pharmaceutical Journal and Transactions*, 50 (1890), pp. 1027–8.
121. Ibid., p. 51 (1891), p. 967.
122. Ueyama, *Health in the Marketplace*, p. 40.

CHAPTER 3

Canada and Newfoundland: French and American Connections

In the North American Division pharmacy had its origins with early settlers, and it was subject to many influences beyond metropolitan control, particularly those resulting from the French presence and the proximity of the United States. This chapter considers significant stages in the professionalization of pharmacy in Canada and Newfoundland: forming professional associations; establishing relationships with members of local administrations and parliaments; securing the passage of Pharmacy Acts; achieving autonomy from the doctors; and raising education levels with the founding of schools of pharmacy. Early developments in Quebec and Ontario strongly influenced those in other provinces, where British authority was often modified as a result of factors including geography and size of population.

Pharmacy in the Canadian provinces developed in phases. In her study of pharmacy in Quebec, Johanne Collin suggests that the relative impact of French and British influence on pharmacy there can be considered in three stages: a period following the first Ordinance between 1788 and 1815, in which pharmacy in Quebec was characterized by the superimposition of the British model on the pre-existing French one; a stage of evolution between 1815 and 1837, marked by the emergence of pharmacy as a distinct discipline; and a third period of expansion in commercial and industrial pharmaceutical activity, largely resulting from

S. Anderson, *Pharmacy and Professionalization in the British Empire, 1780–1970*, Medicine and Biomedical Sciences in Modern History,
https://doi.org/10.1007/978-3-030-78980-0_3

the increasing numbers of pharmacists between 1837 and 1870.[1] These stages provide a useful guide to developments elsewhere in Canada.

Many significant political events took place during this period. In 1841 the French region of Lower Canada (Quebec) was joined with the British region of Upper Canada (Ontario) to form a single British colony, the Province of Canada, where many early steps in the professionalization of pharmacy were carried out.[2] This was followed by federation with other Canadian provinces in 1867, after which pharmaceutical associations were established and the profession was legally recognized. But pharmacy's origins in Canada predate 1788, and details about many of the early pioneers survive; they give an insight into their backgrounds, lives, and practice.

Before 1788

Britain and France fought for control in Northeast America over several centuries.[3] When John Cabot made landfall there in 1497, he claimed the territory for the English king, although no attempt was made to create a permanent settlement.[4] In 1534 Jacques Cartier sailed up the St. Lawrence River and claimed the land for the French king. Various attempts at settlement were made from 1541, but these all failed. French fishing and fur trading continued, and in 1608 Samuel de Champlain founded Quebec City as a permanent settlement and the capital of 'New France'. But after frequent battles and treaties most of the region fell into British hands. Britain acquired Nova Scotia under the 1713 Treaty of Utrecht and the rest during the Seven Years War, following the 1760 fall of Montreal ('the Conquest'). Under the 1763 Treaty of Paris, New France formally surrendered to Britain. By then Canada had about 70,000 French-speaking inhabitants, along with numerous indigenous groups. The 1774 Quebec Act, passed by the British Parliament, extended the frontiers of the colony westwards into what became Ontario, gave French Canadians freedom of worship, and confirmed the continuance of French practice in all civilian matters.[5] Migration extended westwards from the main population areas around Toronto and Montreal (Fig. 3.1).

The first apothecary to reach the region was not British but French. Louis Hébert settled in Port Royal, Nova Scotia, before moving to Quebec in 1617.[6] When France first settled in New France, there were no laws governing the practice of medicine,[7] although by the early eighteenth century some control over medicine, pharmacy, and the sale of

Fig. 3.1 Map of the North American division of the British Empire, 1914

medicines was in place. After 1763 French regulations were gradually displaced and British ones imposed; military doctors and practitioners from Britain began to assume responsibility for the health of the inhabitants.[8] They brought with them their own dispensatories, and the drugs used became those listed in the London and Edinburgh Pharmacopoeias rather than that of Paris. The 1774 Quebec Act was passed before the separation of Upper and Lower Canada, and thus applied to both French and English practitioners. As such it was one of the first which sought to regulate the practice of pharmacy in what became Ontario.

New Brunswick was founded in 1784 when Nova Scotia was partitioned to become the Provinces of Nova Scotia and New Brunswick; Nova Scotia had originally been founded as a Scottish colony in 1629. The earliest apothecary shop was opened in Halifax around 1759 by an English physician, Dr. Phillips. Others soon followed, with two more in Halifax in 1760 and one in Picton County in 1767.[9] It has been claimed that the history of pharmacy in British Columbia on the Pacific coast can be traced to James Cook's visit in 1778,[10] and to the medicine chests which accompanied the fur company explorers on their initial voyages, particularly those of MacKenzie in 1795 and Simon Fraser in 1808, along with those of naval surgeons. However, it was many years before the first druggist settled there. One of the earliest chemist's shops in Ontario was the Apothecary's store run by F.W.A. Hirst in Niagara. On 23 November 1796 he advertised in the *Upper Canada Gazette* that he had just received a delivery from overseas and was offering for sale a wide variety of remedies.[11]

Early British Legislation

The first British legislation which defined the practice of medicine and pharmacy in Canada was a Royal Order signed by King George III in 1788; it applied only to Quebec.[12] It forbade the practice of medicine, surgery, or midwifery in Quebec City or Montreal without a licence.[13] But the preamble also declared that no person

> under any circumstances, shall retail or distribute medicine, prescribe medicine to the sick, or derive any profit from such activity...without having first obtained permission from His Excellency the Governor, or the Commander in Chief at the time.[14]

Requirements for the licence were 'an adequate knowledge of physic, or skill in surgery, or pharmacy or midwifery'. Military surgeons and those holding a degree were exempt from obtaining a licence.

In 1791 the new Province of Upper Canada (Ontario) was created, and Quebec was renamed Lower Canada. The Legislative Council replaced the Quebec Act with a new Upper Canada Act in 1795. This established a Board of Surgeons to examine the qualifications of applicants, and to grant licences to practice physic, surgery, and midwifery.[15] The Act extended to the supervision of pharmacy and pharmacists in Ontario, although an exception was made for some medicines: 'nothing in this Act shall extend...to any person vending, selling, or distributing drugs or compounded medicines for which a patent hath been obtained'.[16] Such drugs could be sold by retailers without restriction. Several apothecaries practised as physicians with limited powers; others were licensed to practise as barber-surgeons and doctors.

In Quebec, the boundaries between pharmacy and medicine underwent significant change between the late eighteenth century and 1831.[17] Initially, apothecaries sought to use the title 'physician', but candidates specifically seeking to practise 'pharmacy' first appeared in 1815 after the Apothecaries Act had passed in Britain. Enforcement was always difficult, and some immigrants were able to open drug stores without any qualifications. One such was William Lyon Mackenzie; in 1820, at the age of 25, he moved to Canada from Scotland. He first lived in York, near Toronto, before settling in Dundas, where he opened a drug and bookstore, apparently without any qualifications.[18]

Elsewhere few restrictions applied to the sale of medicines. In Manitoba, the first settlements were under the control of a company formed by the 1821 merger of the Hudson's Bay and North-West Fur Trading Companies. The company required every trader who sold drugs to take out a permit, for which no fee was charged. This was intended as a check on indiscriminate selling, the first settlers having arrived in 1812. Its trading posts carried a few drugs, but most compounding of medicines was done by doctors in their surgeries.[19] In the 1860s Dr. John C. Schultz established a drugs business near Fort Garry. His stock was later bought by a chemist and druggist, James Stewart, who opened the first pharmacy in the Province. The first drug store to be operated by someone other than a doctor in Halifax, Nova Scotia, was opened in 1828 by J.D.B. Fraser, a chemist and druggist. Two of his brothers were also chemists and druggists in Halifax, where doctors continued to practise pharmacy. In 1843 a drug store was opened in Amherst by Dr. Charles Tupper, an Edinburgh medical graduate. Tupper went on to become Prime Minister of Canada in 1896, and his business was taken over by his brother, Dr. Nathan Tupper.[20]

The Province of Canada, 1841

In 1841 the French region of Lower Canada and the British region of Upper Canada were brought together into a single British colony, the Province of Canada. The number of people obtaining the titles 'apothecary' or 'chemist and druggist' in Quebec grew steadily. There were three during the 1830s, 19 in the 1840s, and 51 between 1850 and 1870.[21] Most of the new arrivals had English names, although Collin highlights the difficulty of establishing whether English-speaking practitioners in Quebec were British, American, or Canadian. British pharmacists included William Rexford, a chemist and druggist who obtained a licence to practise from the Medical Bureau of Montreal in 1845. In 1851, H.J. Lamplough was accepted without examination having presented the certificate awarded by the PSGB. In 1857, so too was Hyacinth Cuniffe, who had qualified at Apothecaries' Hall in Dublin. Licentiates of the Society of Apothecaries in London generally opted for the title 'physician', and most were successful in obtaining it.[22] This followed the pattern in Britain where after 1815 most apothecaries opted to practise as medical practitioners rather than pharmacists.

The result was a shortage of pharmaceutical practitioners, despite a rise in the number of local residents obtaining qualifications. Most chemists and druggists in Quebec were trained through apprenticeship, but in Ontario the requirements of the 1795 Act regarding the licensing of pharmacists proved unworkable because of the shortage of suitably qualified candidates.[23] In 1845 Robert Gourlay noted that

> there was quite enough difficulty in getting qualified men to settle in the country districts without inquiring too closely into their qualifications. Besides which, the country people were supposed to be able to look after themselves. How absurd to think of preventing remotely scattered people from choosing whom they liked to draw their teeth, bleed, or blister them, or that a woman should not have the assistance of a handy neighbour, without fear of a fine.[24]

In practice the medical needs of rural communities were met by an assortment of individuals who might include midwives, clergy, and unqualified practitioners.

In the mid-nineteenth century medicines were freely available in Ontario. They might be obtained from a duly registered doctor who dispensed drugs in a shop; from an apothecary who had completed an apprenticeship; from a practitioner qualified in another country; or from unqualified individuals who simply put up signs saying 'open for business'.[25] But the numbers were small; in 1840 only seven apothecaries and vendors of drugs were registered with the College of Physicians and Surgeons for the whole province.[26] William Lyon MacKenzie, an apothecary, opened a Drug and Book Store in Dundas in 1823, where he sold a wide range of merchandise including dry goods, hardware, stationery, fancy goods, and crockery, along with medicines and drugs.[27] This was typical of pharmacy businesses at the time.

Dominion of Canada, 1867

On 1 July 1867, the Province of Canada was united with the other British colonies into a federation—the Dominion of Canada—consisting of Ontario, Quebec, Nova Scotia, and New Brunswick. As a Dominion, it was initially a semi-independent country under the British Crown, but in 1907 its status changed to that of a fully self-governing colony. With Canada's new status as the first Dominion in the British Empire emigration from Britain was encouraged, with assisted passage schemes offered.[28] Immigrants usually arrived at east coast ports, and many moved on along rivers such as the St. Lawrence to new regions. Settlement extended ever westwards, with occasional surges such as the Klondike Gold Rush when an estimated 100,000 prospectors migrated to the Yukon region between 1896 and 1899 (Fig. 3.1). As the political, economic, and social landscape of the country evolved Canada's vast territory was gradually organized into territories and provinces, largely based around the towns and cities that grew up around the early colonists. The first independent druggist to settle in British Columbia was Arthur Langley. In 1858—a year before the colony was formally established and 11 years before British Columbia became part of the Federation—he opened for business as Langley & Co. Druggist (Wholesale and Retail) at Fort Victoria.[29]

Provinces established their own legislatures according to local circumstances, and differences emerged between them. Pharmacy in Canada proceeded largely on a local basis, with most pharmacists working in isolation from each other; they were kept informed of developments in Britain and elsewhere through the weekly journals. Many became increasingly dissatisfied with the numbers of unqualified practitioners with whom they had to compete, with their lack or recognition and status, and above all with their subservience to the doctors. Most physicians considered pharmacy to be an integral part of their professional domain. Regulations concerning the practice of pharmacy were embedded in medical legislation, a fact that gave the doctors considerable power over the pharmacists. The largest numbers of pharmacists were to be found in the cities of Toronto and Montreal, and it was there that the first steps in collective action were taken.

Escaping Control by Doctors

It was the threat of being engulfed by the medical profession that persuaded Canadian pharmacists of the need to press for recognition as an autonomous profession. Dan Malleck has shown how pharmacists were able to curtail the 'professional rapacity of physicians' in Canada by collective effort. They slowly acquired a level of social and cultural authority that enabled them to claim a central role in the regulation and control of medicines.[30] There was recognition that, if change were to be effected and the power of the doctors to control pharmacists curtailed, it would need to be through legislation. Committed individuals were needed to drive the process forward; there needed to be widespread support from practitioners across the colony; and, crucially, the support of people in government in a position to see legislation through to enactment would be needed. In Canada, these processes began in earnest, province by province, following confederation.

In many ways the circumstances were similar to those that had led to the foundation of both the PSGB in 1841 and the American Pharmaceutical Association in 1821. The threats were tackled at the provincial level, and great energy went into the foundation of pharmaceutical associations in each province. But although there were many parallels between the British and Canadian experience, there were also differences. In Canada changes in pharmacy organizations had to adapt to changes in the boundaries and names of Provinces. Whilst some pharmaceutical associations arose as voluntary membership organizations, others were embedded in legislation or Orders in Council. At the same time as plans were being made for confederation, pharmacists in some provinces were seeking autonomy and self-regulation for members of their profession.[31]

Whilst in the early part of the century doctors had actively sought to control all aspects of pharmacy, in the 1840s physicians in both Lower Canada and Upper Canada stopped their attempts to do so. But these attempts recommenced in the late 1850s, with doctors seeking to regulate the practice of pharmacists in Quebec and Ontario. They culminated in the Medical Act of 1865 in Upper Canada, in which qualifications would be subject to approval by a Medical Board, and chemists and druggists would need to be registered with it. Passage of the Act encouraged physicians elsewhere to seek legislation which would give them the right to control the practice of pharmacy. Some were undoubtedly motivated by a desire to raise standards and improve public benefit. But it was this direct threat to their independence that persuaded pharmacists in Quebec and Ontario to form associations, and to seek the right to self-government of the profession.[32]

Pharmaceutical Associations, 1867

Chemists and druggists in Canada had begun to explore forming associations soon after the foundation of the PSGB in 1841, although little was achieved before the threat from the doctors reappeared. As in Britain, their main objectives were to secure legislation to provide pharmaceutical associations with legal recognition, to provide for the registration of pharmacists by pharmacy bodies rather than Medical Boards, and to improve the educational qualifications of pharmacists. The foundation of associations, the pursuit of legislation through provincial parliaments, and the quest for improved education were closely linked and often progressed simultaneously.

Early attempts to persuade pharmacists to work together were unsuccessful. In Quebec, an attempt was made in 1848 to mobilize pharmacists to obtain legal recognition and autonomy from the doctors, and a Pharmaceutical Society of Canada East was formed.[33] This proved to be short-lived and ineffective, as the threat from the doctors at that time was not perceived to be great. But in response to the renewed threat a Montreal Chemists' Association was formed in 1867.[34] Its members were mainly assistant pharmacists (those not owning their own shops) and apprentices; it had members only in Montreal.[35] But the Montreal pharmacists acted quickly; by 1869 they had submitted a Pharmacy Bill, and in 1870 the Quebec Pharmacy Act was passed.[36]

A similar situation existed in Ontario. In 1867, just before confederation, a Toronto Chemists' and Druggists' Association was formed as a call to arms against 'the consolidation of medical interests' (i.e. the Ontario Medical Council). The doctors were seeking to regulate pharmacy by compelling all those engaged in it to pass examinations, to go before a Medical Board, and to pay fees.[37] The new association quickly embraced members from neighbouring towns and changed its name to the Ontario Pharmaceutical Association. By the end of 1867 fourteen Ottawa druggists joined and it changed its name again, to the Canadian Pharmaceutical Society.[38] By the end of its second year it had about 350 members, and a first issue of the *Canadian Pharmaceutical Journal* appeared in May 1868. In 1869 J. Baker Edwards reported to British readers that 'second annual reports have been presented by the two pharmaceutical societies in Canada [those in Quebec and Ontario]'.[39] But, like the Pharmaceutical Society of Canada East, the Canadian Pharmaceutical Society as a national organization was short-lived; it became the Ontario Pharmaceutical Association once more.

A New Brunswick Pharmaceutical Society was founded in 1873 when a group of pharmacists recognized that a statutory register of chemists and druggists was necessary to ensure that only qualified people were permitted to make and sell medicines.[40] In 1884 a Bill was passed by the province's Legislative, and the new society was legally established. A Nova Scotia Pharmaceutical Society was formed two years later, in 1875. But Nova Scotia pharmacists were more willing than colleagues elsewhere in Canada to engage with doctors still heavily involved with pharmacy. In 1879 they agreed to allow physicians to become members of their society 'if they were already in business and intended to remain associated with pharmacy'. Even if they were not, physicians often continued to favour certain drug stores; professionalism, like the legal framework, had grey zones. By 1882 membership of the Nova Scotia Pharmaceutical Society had increased to 62.[41]

A Manitoba Pharmaceutical Association was founded in 1878, and by 1909 it had its own journal, the *Western Canadian Pharmacist*.[42] The last provinces to establish pharmaceutical associations were Alberta and Saskatchewan, although a Territories one had been founded earlier. In 1911 this was divided into separate Alberta and Saskatchewan Pharmaceutical Associations. Saskatchewan became a province in 1905, having been carved out of the North-West Territories. The Order in Council which established the Alberta association declared that 'all persons registered as members of the North West Territories Pharmaceutical Association shall constitute the Alberta Pharmaceutical Association', although those living in Saskatchewan joined their own.[43] Each was established along similar lines, supported by legislation passed by the provinces' parliaments. In both, a Pharmacy Act gave pharmacists protection from medical interference, authority to examine and register qualified candidates, and a monopoly over the supply of named medicines and poisons.[44]

Securing State Support

The goal of securing legal recognition demanded strong support within parliaments to see the legislation through. This necessitated establishing good working relationships with sympathetic parliamentarians, or in the absence of such to have pharmacist members of the association stand for parliament themselves (as Jacob Bell had done in Britain, Chapter 2). The next step for the Ontario pharmacists was to secure a Pharmacy Act through Parliament, and the Society's secretary therefore submitted a draft for a Druggist's Qualification Act to Francis Hurdon, a supportive member of the Canadian Parliament. Hurdon, together with the Hon. Mr. (later Sir) Leonard Tilley, arranged to see the Prime Minister, Sir John Macdonald, who assured them that he proposed to introduce a measure based on the recent Pharmacy Bill in Britain, which became the Pharmacy Act 1868.[45]

Hurdon promised to watch out for the interests of the chemists and druggists and endeavoured to have all the Society's suggestions embodied in the Bill. Sir John Macdonald himself introduced the first draft to 'regulate the sale of poisons' on 1 April 1868, but following considerable criticism it was withdrawn on 13 May, after which it was subjected to extensive behind the scenes negotiation. Following extensive amendment, the Ontario Pharmacy Act was finally passed in February 1871. It had several significant differences from the British and Quebec Acts; unlike their colleagues in Quebec (below) the doctors of Ontario had no objection to the use of the title 'College of Pharmacy'. The new Act provided for the establishment of an Ontario College of Pharmacy, which was both a professional association and a regulatory body with authority to set examinations. It was also given powers to establish a teaching function, and this was eventually founded in 1882.[46]

Pharmacy and Poisons

Access to poisons and dangerous drugs still lay largely in the hands of the medical profession.[47] In 1859 the Upper Canada legislature passed a Poisons Act to regulate the sale of 'deadly poisons'. This predated the British 1868 Pharmacy and Poisons Act and followed the pattern of the 1851 Arsenic Act. It provided that no chemist or druggist could sell poisons unless presented with appropriate certification issued by a physician, magistrate, or minister, indicating that the customer could safely be in possession of the item concerned. The 1859 Poisons Act represented a first step in regulating the dispensing of medicines containing poisons in Ontario, although doctors and apothecaries retained the right to dispense them along with chemists and druggists.[48]

The link between pharmacy and poisons was repeated in Quebec. In 1868 the Montreal pharmacists prepared a draft Bill asking the Legislature to establish a 'Quebec College of Pharmacy' as a legal entity 'for the purposes of registration, examination and licensing graduates in pharmacy, and prohibiting the sale to the public of certain poisons of a dangerous character, except by persons duly qualified and registered under the Act'.[49] But, as with the physicians in Britain, the medical profession objected to the name, and an alternative had to be found. The 1870 Quebec Pharmacy Act created a 'Quebec Pharmaceutical Association' which would accept full and associate members. Full members had to have passed written and oral examinations involving 'reading written prescriptions, translating them into English, detecting unusual doses, compounding medicines, and demonstrating knowledge of the Pharmacopoeia and antidotes' having completed an apprenticeship of at least four years at a drug store.[50]

As in Quebec, the British 1868 Pharmacy Act was the model for the revised Ontario Pharmacy Act passed in 1871.[51] As in Britain and Quebec, it was both a pharmacy and a poisons act, giving pharmacists a monopoly in the sale and supply of certain substances. It laid down regulations for registration and educational requirements, and controlled poisons through lists and schedules. But it lacked adaptation to conditions in Canada. J. B. Edwards, commenting on the Ontario Pharmacy Bill, noted that 'it is so closely a copy of the [British] Pharmacy Act that some of its provisions may not prove best adapted to the circumstances of thinly populated country districts'.[52] Nevertheless, very similar legislation was enacted by the other Canadian Provinces and Territories over the following years. Pharmacy Acts had been enacted in seven provinces before 1900.[53] In Nova Scotia a Pharmacy Bill was turned down on first presentation to the Provincial Legislature in 1875, but finally became law in 1876.[54] Manitoba passed its Pharmacy Act in 1878, New Brunswick in 1884, British Columbia in 1891, and North-West Territories in 1892 (Table 3.1).

A second more comprehensive Pharmacy Act was passed in Quebec in 1875. This constituted a significant turning point in Canadian pharmacy, marking not only a shift in the power the doctors had over pharmacy practice but also the emergence of pharmacy as an independent profession. Responsibility for the qualifying examinations passed from a Medical Board to a Pharmacy Board consisting solely of pharmacists. But physicians operating pharmacies were exempted from registration under the

Table 3.1 Professionalizing pharmacy in North American colonies

Province	*British colony*	*Joined federation*	*Pharmaceutical society or association*	*Early pharmacy legislation*	*Early poisons legislation*	*Pharmacy education*
Alberta	1870	1910	Pharmaceutical Association 1911	NWT Act 1891	NWT Act 1891	Pharmacy school 1915
British Columbia	1859	1869	Pharmaceutical Association 1891	Pharmacy Act 1891	Pharmacy Act 1891	Apprenticeship
Manitoba	1763	1870	Pharmaceutical Association 1878	Pharmacy Act 1878	Pharmacy Act 1878	Manitoba College of Pharmacy 1899
New Brunswick	1784	1867	Pharmaceutical Society 1873	Pharmacy Act 1884	Pharmacy Act 1884	Pharmacy school 1917; joint with Nova Scotia
North West Territories (NWT)	1713	1870	Pharmaceutical Association 1892	NWT Act 1891	NWT Act 1891	Apprenticeship
Nova Scotia	1629	1867	Pharmaceutical Society 1875	Pharmacy Act 1876	Pharmacy Act 1876	Classes 1911; joined with New Brunswick 1917
Ontario	1763	1867	Toronto Chemists' and Druggists' Association 1867	Upper Canada Act 1859	Upper Canada Act 1859	1871; Ontario College of Pharmacy
Prince Edward Island	1763	1873	Charlottetown Association 1895; PEI Society 1905	Pharmacy Act 1905	Pharmacy Act 1905	Apprenticeship
Quebec	1763	1867	Montreal Chemists' Association 1867	Quebec Pharmacy Act 1870	Quebec Pharmacy Act 1870	Pharmacy school Montreal 1868; in French at Laval 1905

(continued)

Table 3.1 (continued)

Province	*British colony*	*Joined federation*	*Pharmaceutical society or association*	*Early pharmacy legislation*	*Early poisons legislation*	*Pharmacy education*
Saskatchewan (from NWT)	1818	1905	Pharmaceutical Association 1911	NWT Act 1891	NWT Act 1891	Apprenticeship
Yukon	1898	1898	Pharmaceutical Association 1902	NWT Act 1891	NWT Act 1891	Apprenticeship
Newfoundland	1583	1949	Pharmaceutical Society 1910	Pharmaceutical Act 1910	Pharmaceutical Act 1910	Apprenticeship

Act. Further clarity and definitions were given in an 1885 Quebec Pharmacy Act, and another revision in 1890 required that physicians operating or planning to open a pharmacy must register with the Quebec Pharmaceutical Association and pay a fee. In 1903 an amendment to the Act stipulated that physicians planning to open pharmacies in Quebec would not be allowed to practise medicine at the same time.[55] Over 33 years the position of the doctors in Quebec shifted from being in control of pharmacy to having to choose between practising pharmacy or medicine.

The first Medical Act in Manitoba in 1871 established a Medical Board which had powers to regulate the study of medicine, surgery, midwifery, and pharmacy, and it restricted dealing in medicines to doctors and those holding a licence obtained from the Lieutenant-General. The Manitoba Pharmaceutical Association subsequently persuaded the provincial government to bring in a Bill placing regulation of the practice of pharmacy in the hands of pharmacists, and the Manitoba Pharmacy Act was passed on 2 February 1878.[56]

In Prince Edward Island the Pharmacy Act of 1905, which incorporated the Prince Edward Island Pharmaceutical Association, established a category of products that could only be stocked, compounded, or sold by a qualified person. It allowed registered physicians, surgeons, or vets to dispense or compound drugs or medicines in their own practices, but were otherwise subject to the provisions of the Act.[57] Further pharmacy and medicines legislation was enacted across Canada throughout the twentieth century. As in Britain, extravagant claims made for medicinal products by manufacturers in the early decades of the century led to demands for remedial legislation requiring appropriate labelling and the limitation of certain ingredients.

Pharmacy Education

In the late nineteenth century pharmacy was still taught in medical schools, as many doctors needed to compound and dispense medicines for their patients. The Halifax Medical College in Nova Scotia was giving lectures in pharmacy to medical students in the 1870s. In Ontario pharmacy had been taught in its medical school for some years before confederation; a professor of materia medica, pharmacy, and botany was appointed in 1843.[58] For pharmacy students British education was the benchmark; Canadians were told that

> of all the schemes that were discussed by the Board of Examiners of the Ontario College of Pharmacy, none appeared to be so good as that of adopting, verbatim, the regulations relating to the Minor examination held under the Pharmacy Act of Great Britain. Considering the relation of this Province to the mother country, and possible relations that might exist between this Society and that at home, and also the advantage of having uniform standards in both places, this plan has been adopted.[59]

The Ontario pharmacists took great pride in the fact that the education they provided was up to British standards. An editorial in the *Canadian Pharmaceutical Journal* noted that 'we may congratulate ourselves that our students are well up to the mark...we have every reason to conclude that our educated pharmacists will at least bear comparison with those of Great Britain'.[60] The Ontario College's school was not the first in Canada; In Quebec, the first organized pharmacy courses were started at the Montreal College of Pharmacy in 1868; it organized lectures in chemistry, botany, and toxicology. At first all lectures were presented in English, but by 1888 French-speaking students predominated, and the College started giving lectures in chemistry and materia medica in both English and French. Separation of English and French courses occurred in 1905, when Laval University opened a School of Pharmacy for French-speaking students. The Montreal College of Pharmacy became the College of Pharmacy of McGill University in 1918.[61]

In Manitoba, the 1878 Pharmacy Act required attendance on courses of instruction as a prerequisite to registration. An arrangement was made in 1889 with the Manitoba Medical College for pharmacy students to attend their pharmacy lectures, but in 1894 responsibility for pharmacy education was assumed by the Manitoba Pharmaceutical Association and a course of lectures was arranged. A Manitoba College of Pharmacy was established in 1899, becoming affiliated with the University of Manitoba in 1902. In 1914 its Department of Pharmacy introduced its own course, and the Association's college closed.[62] The Nova Scotia Pharmaceutical Society began offering classes in pharmacy in 1911. Formal education in pharmacy was not available in New Brunswick until 1917, when the Pharmaceutical Society became co-sponsors of the Nova Scotia College of Pharmacy.[63] The two societies later agreed that its name should be changed to Maritime College of Pharmacy. In 1917 it merged with the New Brunswick College to form the Maritime School of Pharmacy.

Inter-Provincial Collaboration, 1872

With federation pharmacists in Canada initially showed little interest in intercolonial cooperation or unity. The plan by the Ontario pharmacists in the 1860s to create a Canadian Pharmaceutical Society fizzled out through lack of support.[64] But in 1872 they raised the issue of intercolonial recognition of pharmacy qualifications. The Ontario College of Pharmacy 'resolved that certificates of proficiency or diplomas of the Pharmaceutical Society of Great Britain, the Pharmaceutical Association of Quebec, and the Philadelphia College of Pharmacy, should be recognised by the College', provided that holders had been in business for at least four years.[65] Such qualification would be considered 'sufficient' by the Board of Examiners under the terms of the Ontario Pharmacy Act. But these were not reciprocal arrangements; the Quebec, British, and American bodies did not recognize qualifications awarded by the Ontario College. The issue of reciprocal recognition of qualifications dragged on for several decades, although some progress was made; by 1911 an agreement existed between pharmacy associations in Alberta and Ontario.[66]

In the late 1890s further attempts were made at intercolonial collaboration. In 1893 the Quebec Association approached the Nova Scotia Pharmaceutical Society with two proposals which would establish a national identity for pharmacy in Canada. The first was a proposal to form a national association of pharmacy; the second was a request for support in seeking federal legislation regarding the publication of a national pharmacopoeia. The Nova Scotia Society gave its fullest support to both suggestions. But yet again neither was pursued because of lack of agreement from other provincial pharmacy organizations. It was not until 1907 that delegates from all provinces finally met and agreed to form a Canadian Pharmaceutical Association.[67] It had taken 40 years since confederation for a truly national Canadian Pharmaceutical Association to be formed. It prompted greater collaboration between pharmacists in different provinces; in 1917 a first joint meeting was held between the Nova Scotia and New Brunswick Pharmaceutical Societies. The Association took on the role of professional body for pharmacists across Canada, and in 1923 the *Canadian Pharmaceutical Journal* became its official organ.[68]

Greater collaboration led to the identification of differences in the pharmacy and poisons legislations that had been passed in the various provinces. Four provinces still did not have Pharmacy Acts, although the Northwest Territories Pharmacy Act of 1891 was still in effect in the recently created provinces of Saskatchewan and Alberta.[69]

Pharmacy in Newfoundland

St. John's in Newfoundland was claimed by Humphrey Gilbert for the English Queen in 1583 as the first north American English colony, but it was many years before the population grew to the point of sustaining pharmaceutical practitioners. Pharmacy in Newfoundland emerged as a distinct occupation in the 1820s, by which time St. John's had a sufficient population to support specialist drugstores.[70] One of the early drugstore owners was a British naval surgeon, James Dobie; although essentially a general medical practitioner he used his shop as both an 'open shop' retail store and a 'closed shop' consulting room where he saw his patients.[71] By the 1830s it had been taken over by Thomas McMurdo, a chemist and druggist who arrived in St. John's having recently completed an apprenticeship in Scotland.[72] It was originally called 'Apothecaries Hall'. John J. Dearin, a druggist and dentist, opened a drugstore in Water Street in the 1840s.[73] By the 1850s—in Newfoundland as in Britain—a clear distinction had been made between the 'open' and 'closed' shop activities of the apothecary. Chemists and druggists (often described just as 'druggists' reflecting the influence of United States practice) kept 'open shop' for the sale of medicines; general practitioners kept 'closed shop' where they prescribed and usually dispensed medicines.[74] The first pharmacist to open a retail pharmacy in Bay Roberts on Conception Bay was D.G. Fraser; his father had been a doctor who operated his own dispensary in the community.

In 1873 a Newfoundland-born chemist and druggist, John T. O'Mara, opened another pharmacy on Water Street, St. John's (Fig. 3.2). It was destroyed in the great fire of 1892, which left the city in ruins and destroyed many records relating to the history of pharmacy.[75] But one week after the fire O'Mara opened a temporary pharmacy in a shed on King's Road, and a few months later he opened a permanent pharmacy in Rawlins Cross.[76] In the 1890s pharmacy in Newfoundland continued to be controlled by doctors; pharmacists were not self-regulating and held no monopoly. There was little control over the sale of poisons, little training available for aspiring pharmacists beyond apprenticeship—although this was often very effective—and little progress had been made

Fig. 3.2 O'Mara's Drugstore, St. John's, Newfoundland, c.1887 (Courtesy of City of St. John's Archives)

in forming pharmaceutical associations to enable pharmacists to pursue common goals.

When proposals were put forward for a federation of Canadian provinces in the late 1860s, Newfoundland rejected being part of it, despite being the first British territory to be colonized. It remained a separate British colony until it acquired separate Dominion status in 1907. Change came slowly, and Newfoundland pharmacists were amongst the last to form a society. Their links to colleagues in the neighbouring states of Nova Scotia and New Brunswick and those on the west coast tended to be closer than those to colleagues in Quebec or Ontario. They also had close ties to colleagues in Boston, as this was an important source of drug imports.

Societies and Legislation

In Newfoundland, the only laws relating to pharmacy were part of the statute law of the colony. These were considered hopelessly out of date, and in early 1910 a group of St. John's pharmacists met with the Prime Minister, Sir Edward Morris, to seek his advice about the best way to achieve autonomy from the doctors, to raise the status and recognition of their profession, and to develop pharmacy education. Morris, having sought the advice of the PSGB, advised them to form a society as a first step towards obtaining a Pharmacy Act.[77] The St. Johns group prepared a draft Bill based on the Nova Scotia Pharmaceutical Society Act. It specified the constitution of the new Society and provided for the regulation of pharmacists by pharmacists. In June 1910, the Newfoundland Pharmaceutical Society Act became law.

The Act differed from the British model in one important way; it provided not only for the creation of a pharmaceutical society, but also for the establishment of an independent Pharmacy Board, whose main functions would be to register pharmacists, set examinations for apprentices, and to ensure that pharmacy was practised in an ethical manner. This was the model introduced some years earlier in Victoria (Chapter 10). But the numbers of both druggists and drugstores were so small that demarcation between the Board and the Society was indistinct and remained so for many years. The Act did however follow the British model in another way, by linking pharmacy practice with the sale of poisons. It contained schedules listing drugs and chemicals which could only be purchased through retail and wholesale pharmacies.[78]

Newfoundland remained outside the Canadian federation for the first half of the twentieth century. Labrador became part of Newfoundland in 1927, and the province of Newfoundland and Labrador became the tenth province of Canada in 1949. Pharmacy then became part of a much larger community; a Pharmacy Act was passed in 1954 to bring it into line with other provinces, and in 1955 the Newfoundland Pharmaceutical Association was invited to join the Canadian Pharmaceutical Association. Newfoundland pharmacy finally became integrated into Canadian pharmacy after its delegates joined the Canadian Pharmaceutical Association's Council in 1958.

Education in Newfoundland

For settlers in Newfoundland, pharmacy training involved apprenticeship. Those wishing to become chemists and druggists received training based on the British system. Indentures laid down the conditions under which apprentices would work. Apprenticeship normally lasted five years, after which apprentices received a 'wooden ticket' enabling them to set up on their own.[79] Few opportunities existed for pharmacists beyond retail practice. Apprentices were most often taken on by doctors, most of whom had been trained in England, and the focus was on the practical aspects of pharmacy. Pharmacists made their own tinctures, decoctions, pills, and powders; compounding skills were needed in more than half the prescriptions filled.

Although in Britain and in Canada apprenticeship was usually supplemented by college lectures, in Newfoundland college lectures were not available. In Canada in 1867, pharmacy students attended a local college two or three evenings each week to hear lectures on chemistry, materia medica, and pharmacy.[80] The length of the course became standardized at all the schools, with 40–44 weeks of attendance at lectures being necessary for graduation. Instruction was linked to an apprenticeship usually lasting four years. Pharmacy education in Newfoundland was therefore different; in some places, most chemists and druggists initially received their training under a small number of doctors. In St. John's, Newfoundland, almost all chemists and druggists in the late nineteenth century were trained under the supervision of Thomas McMurdo or his son-in-law John McNeil.[81]

Although advances had been made in the larger Canadian provinces by the end of the nineteenth century, in Newfoundland there was still no formal education available for pharmacists outside of apprenticeship.[82] In

1910 the lack of examination and regulation convinced legislators and others in Newfoundland that, unlike elsewhere, the public had no real way of assessing the professional state of pharmacy, and this played an important part in enabling the Newfoundland Pharmaceutical Society Act to become law.[83]

By the early twentieth century scientific progress relating to drugs emphasized the need for pharmacists to acquire a high level of specialist knowledge of pharmaceutical chemistry, materia medica, and other subjects, and classes were introduced in St. John's; salesmanship was often included in a druggist's formal education.[84] In Newfoundland essential texts included the *British Pharmacopoeia* and *The Art of Dispensing*, along with textbooks on chemistry and botany. These were often supplemented by a correspondence course provided by the Practical Druggist Institute of Boston.[85]

French and United States Influence

British pharmacy in Canada was influenced both by the legacy of its French connection and the proximity of its neighbour, the United States. In 1869 J. Baker Edwards told British readers that 'there is very little pure pharmacy among the French Canadians, the French druggists generally being qualified practitioners of medicine'.[86] The French apothecaries who had evolved into 'pharmaciens' during the eighteenth century had little incentive to make a new start in a British colony in the nineteenth century. But in Quebec the influence of the 'mother country' was at least partially modified by the impact of developments in pharmacy in France.[87] French medical practitioners supplied and dispensed medicines themselves. French practices may also have contributed to a more positive attitude to women pharmacists in Canada than in Britain (below).

Whilst Britain was the greatest influence on the professionalization of pharmacy in Canada, the impact of developments in the United States was strong.[88] Relations between pharmacists in the two countries were cordial and cooperative. In 1877 the American Pharmaceutical Association held its Annual Meeting in Toronto. The Nova Scotia Pharmaceutical Society posted the New York periodical, *Druggist Circular*, to all its members.[89] But the greatest impact of American pharmacy was through education. Sonnedecker suggests that after confederation the evolution of pharmaceutical education in Canada had striking parallels with that in the United States.[90] Pharmacy textbooks were a vital resource for apprentices, as courses were usually self-taught.[91] Articles aimed at students in the *Canadian Pharmaceutical Journal* indicated that, whilst British pharmaceutical textbooks were widely used, American textbooks such as Remington's *Science and Practice of Pharmacy* were also heavily promoted.

The dominance of British texts was not surprising given that Canadian pharmacy education largely mirrored that of Britain. In Canada, the *British Pharmacopoeia* was not accepted as the sole authority on all matters concerning medicines; in 1878 the Nova Scotia Pharmaceutical Society adopted a resolution to accept it as an official reference, except 'where some other more reliable process was warranted'.[92]

Women in Pharmacy

Women practitioners of pharmacy played an important part in the initial European settlement of Canada. Nursing Sisters of St. Augustine worked in the dispensary of the Hotel Dieu in Quebec. They were 'cultured and excellently trained by the best doctors in France'.[93] As late as 1734 there was no other dispensary in Montreal, and the Sisters compounded, kept, and dispensed all medical supplies for the sick of the district. With transition from French to British control in 1763 practices and personnel changed, although positive attitudes to the role of women in pharmacy survived. By the late nineteenth century different attitudes to women in pharmacy were apparent in Britain and Canada. In 1869 Fanny Elizabeth Potter became the first woman to register with the PSGB as a chemist and druggist, and in 1875 Isabella Skinner Clarke was the first woman to pass its Major examination and register as a pharmaceutical chemist in Britain.[94] But whilst the right of women to practise pharmacy was by then enshrined in law, many members of the PSGB wanted it to remain a male preserve, and her application to register was at first refused.

The *Canadian Pharmaceutical Journal* noted the contrast between the British and Canadian positions:

> However great may be our respect for the Council of the Pharmaceutical Society of Great Britain, we certainly must confess that our estimate of the gallantry of that body is by no means high… [The] gallant Ontario College of Pharmacy… not only admits, but is proud to rank among its members, those of the gentler sex. Let the lady-pharmacists of unappreciative Great Britain make a note of this, nor longer 'waste their sweetness on the desert air'.[95]

Elsewhere in Canada the role of women pharmacists took longer to be recognized; the first woman to complete the qualifying examination in Nova Scotia was Bertha Archibald in 1917.[96] Mary E. Johns became the first woman to register in Newfoundland in 1927.[97]

The voices of successful women pharmacists from overseas were welcomed by the pharmaceutical community. The special guest speaker at the Canadian Pharmaceutical Association's Convention in 1921 was Margaret Buchanan.[98] She had been the driving force behind the foundation of the Association of Women Pharmacists in Britain in 1905, and in 1909 had founded the Gordon Hall School of Pharmacy for Women. She was described by the *Chemist and Druggist* as holding 'the front rank among the women pharmacists of the British Empire'.[99] In 1918 she had been elected as the first woman member of the PSGB's Council and travelled extensively in Canada.[100] She blazed a trail for Canadian women pharmacists, including Rosa Bell who served on the Council of the Prince Edward Island Pharmaceutical Association during the 1950s.[101]

Professionalizing Canadian Pharmacy

This chapter has explored common features in the professionalization of pharmacy in Canada and Newfoundland. Despite the profession's ambiguous position, and the contest for authority with the orthodox medical profession, pharmacists in nineteenth-century Canada were largely successful in securing their professional autonomy. They secured the support of the state for the passage of legislation giving pharmaceutical associations legal recognition and the authority to specify criteria for the education, examination, and registration of pharmacists, who would be regulated by other pharmacists rather than a Medical Board. By defining the skills and qualifications needed for registration, pharmacists established boundaries between themselves, the doctors, and the unqualified.

Efforts by physicians to secure the legal right to regulate the practice of pharmacy were ultimately defeated by the collective efforts of pharmacists.[102] Malleck suggests that in the 1870s the failure of the doctors to secure control over the education and licensing of pharmacists caused doctors to change their view of their own role in relation to pharmacy and medicines, and that by the turn of the century this was limited to a role in scrutinizing and restricting the patent medicines trade.[103] Yet occupational autonomy had its limits in a crowded medical marketplace;

professional aspirations had many barriers to overcome. The ambivalent nature of pharmacists' work linked it to both professions and trades.[104]

British pharmacy had a powerful impact on Canadian pharmacy during the years following confederation. The early local and provincial pharmaceutical associations were generally modelled on the PSGB; a single body would serve pharmacy as the professional organization, as the licensing and regulatory body, and as an educational institution, although Pharmacy Boards were established in Quebec and Newfoundland. Agency between British and Canadian pharmacy operated through several channels; provincial parliaments sought advice on legislation and other matters from the British authorities; the PSGB provided advice and support through official channels as well as through individuals; and a small number of dedicated individuals made significant contributions to developing pharmacy in the British image.

But pharmacists in Canada also learnt from the mistakes made by the PSGB in Britain. Whilst the British link between pharmacy and poisons legislation was replicated in Canada, Canadian pharmacies remained firmly in the hands of individual proprietors rather than corporations, at least until the 1950s, a situation which more closely followed the French model of pharmacy than the British.[105] And although British pharmacy education formed the basis of training in Canada, the two-tier register of chemists and druggists and pharmaceutical chemists was not transplanted.

The close relationship between pharmacy in Canada and Britain continued well into the twentieth century. The BP and the Canadian Formulary were key sources for drug standards in Canada, but by the 1950s the official formularies of other countries, especially the United States and France, also enjoyed full legal status under the Canadian Food and Drugs Act.[106] Medicine standards became international; in 1956 copies of the *International Pharmacopoeia* were sent to all current and former British territories.[107] The British influence on pharmacy in Canada was just one of many.

Notes

1. Collin, Johanne. 'French and British Influence in the Birth of a Profession: Pharmacy in Quebec', *Pharmacy in History*, 2 (2010), p. 101.
2. Bothwell, Robert. *Canada and Quebec: One Country, Two Histories* (Vancouver, 1998).

3. Morton, Desmond. *A Short History of Canada* (London, 2017). Martin, Ged. *Britain and the Origins of Canadian Confederation, 1837–67* (Vancouver, 1995). Woodcock, George. *A Social History of Canada* (London, 1989).
4. The Kingdom of Great Britain was formed following passage of the Act of Settlement in 1701.
5. The Canadian Encyclopedia. https://www.thecanadianencyclopedia.ca/en/article/conquest.
6. Martin, Audrey M. *Pharmacy in Canada: Highlights of Its History from Early to Recent Times* (Vancouver, 1955), p. 3.
7. Paterson, G.H. 'Canadian Pharmacy in Pre-Confederation Medical Legislation', *Journal of the American Medical Association*, 200 (1967), pp. 849–52.
8. Martin, *Pharmacy in Canada*, p. 4.
9. Raison, Arnold V. *A Brief History of Pharmacy in Canada* (Ottawa, 1967), p. 65.
10. Martin, *Pharmacy in Canada*, p. 18.
11. Raison, *Brief History*, p. 71.
12. Collin, 'French and British Influence', p. 102.
13. 28 George II, Chapter 8, 1788. The Act was also published in the *Gazette de Quebec*, 1 (1788), p. 187.
14. Raison, *Brief History*, p. 71.
15. Kennedy, D.R. 'One Hundred Years of Pharmacy Legislation', in Stieb, E.W. (ed.) *One Hundred Years of Pharmacy in Canada 1867–1967* (Toronto, 1969), pp. 25–37.
16. Act of 1795. Kennedy, p. 26.
17. Collin, 'French and British Influence', p. 106.
18. Raison, *Brief History*, p. 71.
19. McDougall, D. (ed.). *The History of Pharmacy in Manitoba 1878–1953* (Winnipeg, 1954), p. 3.
20. Raison, *Brief History*, p. 66.
21. Collin, 'French and British Influence', p. 106.
22. Ibid., p. 107.
23. Jackson, S.W. 'The First Pharmacy Act of Ontario', in *Centennial Publication of the Ontario College of Pharmacy* (Toronto, 1967), pp. 2–3.
24. Gourlay, Robert. 'Chemistry, Materia Medica and Pharmacy', *British American Journal of Physic and Medical Sciences*, 4 (1845), pp. 105–6.

25. Wilkinson, Ron. 'Pharmacy 1867–1967', *Drug Merchant*, 48 (1967), pp. 3–15.
26. Raison, *Brief History*, p. 71.
27. Beales, J.D. and Austin, Z. 'The Pursuit of Legitimacy and Professionalism: The Evolution of Pharmacy in Ontario', *Pharmaceutical Historian*, 36 (2006), p. 23.
28. Harper, Marjory. 'British Migration and the Peopling of the Empire', in Porter, Andrew (ed.) *The Nineteenth Century, the Oxford History of the British Empire* (Oxford, 1999), p. 80.
29. Fullerton, Frederick H. (ed.). *Pharmacy through 60 Years in British Columbia 1891–1951* (Vancouver, 1951).
30. Malleck, Daniel J. 'Professionalism and the Boundaries of Control: Pharmacists, Physicians and Dangerous Substances in Canada 1840–1908', *Medical History*, 48 (2004), pp. 175–98.
31. Steib, E.W. 'One Hundred Years of Organized Pharmacy', in Stieb, E.W. (ed.) *One Hundred Years of Pharmacy in Canada 1867–1967* (Toronto, 1969), p. 11.
32. Kennedy, 'One Hundred Years,' p. 28.
33. *British American Journal of Physic and Medical Sciences*, 4 (1848–1849), pp. 220–1.
34. Raison, *Brief History*, p. 80.
35. Collin, 'French and British Influence', p. 108.
36. Raison, *Brief History*, p. 80.
37. Ibid., p. 71.
38. Beales and Austin, 'Pursuit of Legitimacy', pp. 22–7.
39. *Pharmaceutical Journal and Transactions*, 11 (1869–1870), pp. 354–5.
40. Raison, *Brief History*, p. 56.
41. Ibid., p. 66.
42. McDougall, *History of Pharmacy in Manitoba*, p. 41.
43. Cameron, Donald M. 'Early Legislation Governing Pharmacy', in Cameron, Donald M. (ed.) *The History of Pharmacy in Alberta: The First One Hundred Years* (Edmonton, 1993), p. 7.
44. Ibid.
45. Raison, *Brief History*, p. 71. Tilley himself operated a pharmacy in Saint John, New Brunswick, from 1853 to about 1860.
46. For progress of Ontario Pharmacy Bill through Parliament see Raison, *Brief History*, p. 72.

47. Malleck, Daniel J. *Refining Poison, Defining Power: Medical Authority and the Creation of Canadian Drug Prohibition Laws 1800 to 1908* (Kingston, 1998).
48. Beales and Austin, 'Pursuit of Legitimacy', p. 24.
49. Raison, *Brief History*, p. 80.
50. Ibid.
51. Saunders, W. 'Pharmacy in Canada before 1871', *Canadian Pharmaceutical Journal*, 75 (1941), pp. 10–11. Originally published in *Proceedings of the American Pharmaceutical Association* (1871).
52. *Pharmaceutical Journal and Transactions*, 11 (1869–1870), pp. 354–6.
53. Matthews, Leslie G. *History of Pharmacy in Britain* (London, 1962), p. 145.
54. Raison, *Brief History*, p. 66.
55. Ibid., p. 81.
56. McDougall, *History of Pharmacy in Manitoba*, p. 3.
57. Raison, *Brief History*, pp. 77–8.
58. Raison, *Brief History*, p. 68–71.
59. *Canadian Pharmaceutical Journal*, 9 (1875–1876), p. 292.
60. Ibid., p. 140.
61. Raison, *Brief History*, p. 82.
62. McDougall, *History of Pharmacy in Manitoba*, pp. 17–22.
63. Raison, *Brief History*, p. 57.
64. *Canadian Pharmaceutical Journal*, 20 (1886–1887), p. 98.
65. *Pharmaceutical Journal and Transactions*, 3 (1871), p. 772.
66. Cameron, 'Early Legislation', p. 9.
67. Raison, *Brief History*, pp. 66–7.
68. Paterson, Garnet R. 'Canadian Pharmacy 1906–1956', *Canadian Pharmaceutical Journal*, 89 (1956), pp. 71–4.
69. Ibid., p. 2.
70. Crellin, John K. *Newfoundland Drugstores: A History* (St. John's, 2013), p. 1.
71. Ibid., p. 2.
72. Raison, *Brief History*, p. 60.
73. Crellin, *Drugstores*, p. 3.
74. Ibid.
75. The Great Fire of St. John's in July 1892 followed earlier great fires in 1819 and 1846.

76. Raison, *Brief History*, p. 60.
77. Ibid., p. 61.
78. Ibid.
79. O'Mara, J.J. *History of Pharmacy in Newfoundland: A Paper Dealing with the History of Pharmacy in the Province of Newfoundland* (St. John's, 1969).
80. Sonnedecker, Glenn. 'Pharmaceutical Education, 1867 and 1967', in Stieb, E.W. (ed.) *One Hundred Years of Pharmacy in Canada 1867–1967* (Toronto, 1969), p. 5.
81. Raison, *Brief History*, p. 60.
82. Crellin, John K. *A Social History of Medicines in the Twentieth Century: To Be Taken Three Times a Day* (Binghamton, 2004), p. 283.
83. Ibid., p. 121.
84. Crellin, *Drugstores*, p. 116.
85. Raison, *Brief History*, p. 61.
86. *Pharmaceutical Journal and Transactions*, 11 (1869), pp. 354–5.
87. Exhibition Notes. *British and Canadian Pharmacy 100 Years Ago* (London, 1967), p. 4.
88. Steib, 'One Hundred Years', p. 11.
89. Raison, *Brief History*, p. 66.
90. Sonnedecker, 'Pharmaceutical Education', p. 7.
91. Exhibition Notes, p. 1.
92. Raison, *Brief History*, pp. 61–6.
93. Martin, *Pharmacy in Canada*, p. 3.
94. Holloway, S.W.F. *Royal Pharmaceutical Society of Great Britain 1841 to 1991: A Political and Social History* (London, 1991), p. 262.
95. *Canadian Pharmaceutical Journal*, 9 (1875–1876), pp. 299–300. The quote is from Thomas Gray's *Elegy* and is a metaphor for ordinary people who do extraordinary things that are never recorded in history.
96. Raison, *Brief History*, p. 69.
97. Ibid., p. 62.
98. Hudson, Briony. 'Buchanan, Margaret Elizabeth (1865–1940)', *Oxford Dictionary of National Biography* (Oxford, 2020).
99. *Chemist and Druggist*, 74 (1909), p. 377.
100. Holloway, *Royal Pharmaceutical Society*, p. 268.
101. Raison, *Brief History*, p. 79.

102. Paterson, 'Medical Legislation', p. 852.
103. Malleck, 'Professionalism', p. 198.
104. Clark, R.J. 'Professional Aspirations and the Limits of Occupational Autonomy: The Case of Pharmacy in Nineteenth-Century Ontario', *Canadian Bulletin of Medical History*, 8 (1991), pp. 43–64.
105. Exhibition Notes, p. 1.
106. Paterson, 'Canadian Pharmacy', p. 74.
107. Cartwright, Anthony C. *The British Pharmacopoeia 1864 to 2014: Medicines, International Standards and the State* (Farnham, 2015), p. 51.

CHAPTER 4

West Indies: The Impact of Slavery

The early history of pharmacy in the British West Indies is closely associated with slavery and can be traced to the development of plantations in the early seventeenth century. During the slave trade, between about 1650 and 1807, those involved in the drug trade supplied slave ships, slave forts in Africa, and plantations in America and the West Indies.[1] By the middle of the eighteenth-century arrangements for the supply of medicines to the West Indies were well established. Quaker networks played an important part in the trade with plantation societies, and for apothecaries and druggists in London it continued well into the nineteenth century. From the mid-nineteenth century a transition took place in both the medical and pharmaceutical services available, with plantation hospitals closing and medicine supply increasingly shifting from apothecaries on plantations to pharmacies and drug stores on high streets and street corners run by chemists and druggists, dispensers, and general traders.

In 1914 the West Indian Division of the British Empire consisted of 16 territories, all of which were in British hands during the age of slavery. They were scattered across vast distances, stretching over more than 1500 miles both east to west and north to south (Fig. 4.1). They included British Guiana (Guyana) on the South American coast, British Honduras (Belize) on the coast of Central America, and the Bermuda

S. Anderson, *Pharmacy and Professionalization in the British Empire, 1780–1970*, Medicine and Biomedical Sciences in Modern History,
https://doi.org/10.1007/978-3-030-78980-0_4

Fig. 4.1 Map of the West Indian Division of the British Empire, 1914

Islands in the North Atlantic. Thirteen were Caribbean islands, including Jamaica, Trinidad, and Barbados, along with five Leeward Islands and three Windward Islands.

Slavery had a lasting impact on the professionalization of pharmacy in the region beyond its economic one. Few aspects of professionalization were in place during the period of British rule; chemists and druggists remained under the control of the doctors until after independence. In the British West Indies, Pharmacy Acts providing pharmaceutical societies with legal recognition, allowing pharmacists to regulate themselves, and giving pharmacists monopolies in the supply of certain services, came only in the second half of the twentieth century. The PSGB was founded only in 1841, four years after slave apprenticeship ended in the British Empire. What pharmacy there was in the West Indies could be described as 'British', but it was a pharmacy that privileged the European over the non-European population.

The British in the West Indies

The origin of English involvement in the Atlantic slave trade is normally traced to the middle of the sixteenth century.[2] By 1660 the Royal African Company was supplying the English sugar colonies with 3000 slaves a year at an average price of £17.[3] Almost 11,000 ships were fitted out in England between 1698 and 1807, transporting around three million Africans into slavery before the trade was abolished.[4] The need for slave labour was driven by the growth in demand for sugar. The Portuguese developed plantations in Brazil and began importing slaves from Africa to work on them. Sugar cane was introduced to Barbados in the mid-1640s by Dutch merchants and planters who had been expelled from Brazil.[5] The island came into British hands in 1663, and by 1670 it had 900 plantations. Sugar cultivation spread rapidly to other islands, and reached Jamaica and St. Domingue in the early eighteenth century. By 1776 there were an estimated 1800 sugar plantations in the British West Indies alone.[6]

At first, the new occupants of the West Indian islands included relatively few Europeans; in the 1640s most of the settlers in the Bahamas came from Virginia; the Caribbean islands were then considered a more attractive prospect that the North American plantations.[7] But the number of Europeans was quickly overtaken by the number of slaves. Before 1670, whites exceeded blacks; by 1780 there were 489,000 slaves and 48,000 whites; and by 1830 the British West Indian population consisted of 684,996 slaves, 102,980 free blacks, and 54,772 white people.[8] The white population of Jamaica fell sharply after 1815, when free port trading with the Spanish colonies ended. Later settlement schemes designed to bring in new Europeans as family farmers proved ineffective.

Around 60 per cent of slaves were based on sugar plantations, where increasing numbers trained as craftsmen. Another 20 per cent worked on coffee, cotton, and livestock estates, with 10 per cent in other rural activities and about 10 per cent in towns.[9] In places like Kingston and Bridgetown they worked as shopkeepers, mariners, street sellers, and domestic servants.[10] Slaves were unequally distributed across the islands; in smaller ones including the Bahamas, Anguilla, and Barbuda slaves were put to work on maritime activities or livestock rearing rather than sugar plantations, and were present in relatively small numbers. But on islands dominated by sugar plantations slave density was huge. Before 1700 Barbados had the largest number of slaves on the British Caribbean islands, but thereafter Jamaica became the most important destination for slave ships.[11] By 1834 there were 500 per square mile in Barbados, 260 in Antigua and St. Kitts, and 245 in Nevis.[12]

Medicine and Pharmacy During Slavery

Medical care on plantations varied considerably island to island. Slave medicine in Barbados was well established by the start of the eighteenth century, but health care in Trinidad and Tobago developed only slowly.[13] Death and disease were a constant presence on the Jamaican slave plantations.[14] Where medical care was available its purpose was to preserve the fitness of the labour force and its economic productivity, and to support its natural increase. Slave masters generally showed little regard for elderly or incapacitated slaves although attitudes varied.[15] On some estates the medical arrangements were substantial; some employed apothecaries who vaccinated the slaves against smallpox, performed surgical procedures and treated a wide range of ailments.[16]

The first British hospitals on the islands were those built for the army and navy, but their standard varied greatly. The hospital at Monks Hill in Antigua bore 'more resemblance to a plantation sick house of an inferior description for [slaves] than a hospital for British soldiers'.[17] But there were exceptions; the military hospital at St. Joseph's on Trinidad was said to have had ample and well-ventilated accommodation for the sick, despite being made of wood.The Navy established hospitals in Jamaica, Antigua, and Barbados. The first was built at New Greenwich in Jamaica in 1745 close to marshes and had to be rebuilt at Port Royal in 1756. The naval hospitals played an important role in maintaining both manpower capacity and morale.[18]

Many plantations later established their own hospitals. The English Quaker chemist and druggist Thomas Harvey, along with the reformer Joseph Sturge, visited several estate hospitals in Antigua following the end of apprenticeship in 1837. Many estates drastically reduced spending on medical care, as a result of which many of the hospitals had become useless. They noted:

> On a Monday morning, during slavery, the doctor would find eight, ten or even twenty in the sick-house. Now, he has comparatively nothing to do. He is paid one-third less per head than before; but his duties have diminished in a much greater ratio.[19]

Later, they were shown round the buildings on the Hyde Hall estate near Falmouth in Jamaica, and judged its hospital to be one of the best they had seen.[20]

The value of plantation hospitals was often negated as absentee proprietors left them in the charge of plantation managers who often treated both slaves and doctors badly in order to maximize labour and profits.[21] Overall, the impact of European medical practice on slave conditions was small. Richard Sheridan concludes that physicians probably did more harm than good, with many patients being exposed to massive doses of opium, mercury, and antimony.[22] But on the plantations slaves also treated themselves, using a variety of plant medicines, along with the spiritual healing and religion practices they brought with them from Africa.[23] They were often very knowledgeable; in 1886 the Trinidad assistant commissioner reported at a meeting of pharmacists in London that 'the Africans who came over to Trinidad were well up in the use of drugs'.[24]

Slavery, Apothecaries, and Networks

The slave trade presented enormous opportunities for apothecaries and druggists engaged in the supplying and making of medicines; it was potentially a very profitable business for those prepared to take the financial risk.[25] As well as supplying the slave ships and the plantations in America and the Caribbean, some apothecaries and druggists settled in the West Indies, and a small number of them owned plantations. By the early eighteenth century there was already a substantial planter class of mainly British expatriates, and the European population grew steadily during the course of the century.[26] Traditional medical practice and commercial pharmacy overlapped, and a large export trade in emetics, purgatives, cathartics, and other established remedies soon built up.[27]

Religious groups such as the Jews and Quakers provided a sound basis for successful trading networks.[28] They established effective channels of communication across the British world that built trust and provided support in times of need. By the mid-eighteenth-century trade links were well established between London Quaker apothecaries and druggists and physicians, apothecaries and plantation owners across America and the Caribbean. One such was the Quaker apothecary, Thomas Corbyn.[29] By the 1750s he was already trading with large numbers of dealers in American states and the Caribbean. These included Robert James in Antigua, Cadwallader Evans in Jamaica, and Dr. Joseph Gamble in Barbados. Large quantities of drugs were made up, packed into chests or casks, insured, and sent to Bristol to await shipment.[30] Corbyn had strong competition; William Jones owned a retail and wholesale pharmacy business near Drury Lane, although it also embraced banking. He traded with the East India Company as well as with agents in Tobago and Jamaica.[31]

The Plough Court pharmacy in London also had extensive links to the slave trade. Within 10 years of its foundation in 1715 its proprietor, Silvanus Bevan, was supplying drugs and other goods to apothecaries, surgeons, and traders overseas.[32] Most were fellow Quakers, and some of his best customers were in Jamaica, Antigua, and Barbados.[33] His grandson, Joseph Gurney Bevan, expanded the overseas trade, particularly that to the West Indies. For Quakers, direct involvement with the slave trade was forbidden. Bevan's letters contain frequent references to his abhorrence of what he called 'the man-trade'.[34] He supplied the best quality drugs but asked high prices for them. Some of his orders were huge, one in 1776 to Dr. John Moodie in Jamaica being for over £1000. But by the late eighteenth century the economics of sugar production was changing, planters' profits were decreasing rapidly, and by 1776 the West Indian colonies were already in decline.[35]

West Indies After Slavery, 1840s

The period following abolition of the slave trade by Britain in 1807 was one of unrest in the West Indies. Slave rebellions broke out in Barbados in 1816, in Demerara in 1823, and in Jamaica in 1831, where up to 20,000 slaves were involved.[36] When slavery was finally abolition in 1834 some 750,000 slaves were freed across the British West Indies.[37] But in order to maintain the structure of plantation society a system of apprenticeship was devised. Unskilled field workers would be tied to their employers for 6 years, whilst skilled slaves would be tied for 4 years. However, Antigua and Bermuda opted to allow full freedom immediately, and Barbados recruited freeholders to grow tobacco, cotton, indigo, and ginger. They farmed small areas with the help of indentured labourers, who worked for 3–10 years in return for a paid passage and the promise of land when their contract expired. Many struggled to survive.[38] With fewer labourers available planters turned to mechanization. In Trinidad and British Guiana nearly all plantations installed steam engines; St. Kitts had 33 steam mills by 1846, and in Tobago a third were steam driven by 1849.

A Sugar Duties Act was passed in Britain in 1846 to equalize the duty paid on imported sugar. This meant that West Indian sugar then had to compete against sugar from Cuba and Brazil, where slavery continued until 1888. For Britain's West Indian colonies, it contributed to the economic decline that was to last well into the next century, from which those trading in medicines were not immune. They included a few apothecaries and chemists and druggists along with unqualified individuals. The title 'apothecary' was by then associated with medical rather than pharmaceutical practice, and those previously employed on plantations usually set themselves up as general medical practitioners. A few opened druggist shops in the growing towns, but they were soon joined by chemists and druggists from Britain, as the drug trade slowly developed.

Trade Depression, 1880s

In Jamaica, the upheaval and unrest that had been widespread during the time of slavery continued in the post-emancipation era. Crop cultivation expanded rapidly, with much of the population dependent on agriculture. There was a marked contrast between the lives of those living in the towns and those in rural areas, where life was very hard for most of the population, which by then included some poor white families—those of indentured labourers who stayed on. After 1884 there were reports of widespread destitution and malnutrition, and an increasing incidence of disease.[39]

For all those involved in the making, dispensing, and supply of medicines in the West Indies the years following the ending of slavery were difficult. The situation was made worse by the large-scale migration of plantation workers to other parts of the region. In March 1884, the *Chemist and Druggist's* correspondent reported that 'there has been marked depression in trade generally, owing in the main to the migration of thousands of the labouring classes to work on the Panama Canal…labour strikes are being heard of nowadays'.[40] France had started work on a canal in 1881, but had to stop because of engineering problems and high mortality.

The situation got progressively worse during the year. In November the correspondent reported that action had been taken to replace the missing labourers.

> The general trade depression has attained rather large dimensions, and as regards sugar and rum-the main staples-it may be said to be at its climax. The labour question is still a vexed one. The natives continue flocking to the Isthmus of Panama, where they succeed in earning a living. [Chinese workers have] been imported at lavish expense to supply the vacuum caused by the exodus of native labourers, but [they do] not see the reasonableness of adapting [themselves] to our loyal mode of living...a feeling of discontent is gaining ground among our people.[41]

Most drugs used in the West Indies continued to be imported from England, including substantial quantities of patent medicines, which were extremely popular. Some chemists and druggists sent laudatory testimonials to British manufacturers. An 1890 advertisement for *Elliman's Royal Embrocation* referenced a letter from H. A. Deschamps of Kingston, praising their use of illustrations: 'being very attractive they increase the sale of your valuable preparation'.[42] Imperial imagery was good for sales. From the mid-nineteenth century trade with other Caribbean islands increased, and new suppliers entered the market. By 1880 the *Chemist and Druggist's* correspondent was lamenting that 'the specialties advertised in the Jamaica journals are nearly all French. We miss the almost ubiquitous advertisements of our well-known British patent medicine proprietors'.[43]

By the mid-1880s English pharmacies could survive only by diversifying both their product range and customer base. The correspondent noted that they depended on 'Jamaicans, especially the peasantry, being a physic-taking people. Druggists deal largely in grocery, and your correspondent has seen malt liquors, common soap, and even matches form a portion of a druggist's stock'. They also faced stiff competition from drug stores. 'The trade in patent medicines is chiefly confined to the quality dispensers, but here and there the more harmless articles, such as *Holloway's Pills and Ointment*, can be obtained at the stores' (Fig. 4.2). Price cutting of medicines was a problem. 'The "cutting system" is carried on to some extent. One form of druggist, unqualified men, offer through their qualified manager all the English articles... at cut prices'.[44] Counter-prescribing, mainly of items for the treatment of minor ailments, was also an important part of the chemist's business.

Although some British chemists and druggists in the West Indies were involved in wholesaling or manufacturing, most were engaged in retail pharmacy. Amongst them was Charles Grant, a chemist and druggist

Fig. 4.2 Advertisement for *Holloway's Pills and Ointment*, 1903 (Courtesy of Wellcome Collection)

from Britain who had been established in Kingston, Jamaica, for many years before his death in 1885.[45] Whilst some chemists and druggists settled permanently on the islands, others returned to Britain after a few years. The death of John G. Leach, a chemist and druggist of Bridgetown, Barbados, was announced by the *Chemist and Druggist* in 1880.[46]

Whilst retail pharmacies mainly obtained their supplies from local wholesalers, medicines for Government hospitals and dispensaries had been supplied under a contract held by the Society of Apothecaries for many years. But in 1897 the high prices charged by the Society were a concern to the colonial authorities. The Governor of Jamaica, Sir Henry Blake, wrote to the head of the Island Medical Department with new instructions: 'in future stores should be procured from Messrs Ferris & Co., in preference to the Society of Apothecaries'.[47] It was accompanied by a Statement of Requirements showing the substantial savings that could be made by making such a switch.[48]

Colonial Authorities and Pharmacy

The West Indian colonies came into British possession over a period of years and had been subject to different forms of colonial government. Older ones had their own legislatures, with nominated Legislative Councils and elected Assemblies, which had significant powers over legislation and taxation.[49] Freed slaves had been elected to the Jamaican House of Assembly since the 1830s, although for the British authorities this was an unwelcome development. For most of the freed slaves, circumstances remained desperate in the decades following emancipation. By the early 1860s there were regular public protests; in Jamaica these culminated in the 1865 Morant Bay Rebellion, a violent protest against injustice and widespread poverty. After the Rebellion, the House of Assembly was abolished and replaced by direct rule from London, with Jamaica becoming a Crown Colony.

Prior to direct rule, the Colonial authorities took little interest in the regulation of pharmacy or the sale of poisons. Even in Britain at the time there was no control over the sale of any substance, however lethal; yet the products available from drug stores varied from harmless flavoured waters to dangerous poisons.[50] In Britain nothing was restricted to prescription by a medical practitioner; there was no need to ask about the intended use of a poison, no need to label it in a particular way or present it in a particular container; no need to keep records of any sale, and no penalties for inappropriate supply. The same applied in the British West Indian colonies, but by the late 1870s concerns began to surface about the consequences of uncontrolled sale of drugs and poisons.

The first rudimentary controls in the West Indies were implemented through Medical Ordinances, although on some islands it took several years for these to be passed. Most mandated a Medical Board to maintain a register not only of qualified medical practitioners but also of chemists and druggists. Pharmacists were firmly under the control of the doctors; no Pharmacy Ordinances were passed before 1880, by which time Britain had passed several Acts concerning the regulation of pharmacy and the control of the sale of poisons.

Pharmacy Legislation

Once the need for stronger legislation was recognized the colonial authorities turned to the PSGB for advice and guidance. The PSGB based its advice on its own experience: only those holding the British qualification should be registered; a link should be made between access to poisons and the regulation of pharmacy; and registered pharmacists should have a monopoly in the sale of certain items. The various Sale of Drugs and Poisons Ordinances passed in the West Indian colonies were all based on the British Pharmacy and Poisons Act of 1868, largely word for word. They not only restricted the sale of poisons to chemists and druggists, but also required that those included in the *British Pharmacopoeia* must only be sold in preparations made up according to formulas given in that volume. Such formulas often included expensive or unobtainable items and required skill in preparation. If strictly enforced, the Ordinances meant that many medicines would not be available from the drug stores, and the local population would be denied access to them.

The first legislation passed was a Pharmacy Ordinance in Barbados in 1880. Its progress through the Barbados Legislature was followed by the *Chemist and Druggist*, as were developments in pharmaceutical legislation throughout the Empire. 'Two Bills of pharmaceutical interest have been discussed' it reported in January 1880, including the second reading of a Bill 'for the registration of medical men on the island'.[51] It was referred to a Select Committee. The Act, which was passed some months later, followed the pattern on other islands in mandating the Medical Board to keep a register of qualified chemists and druggists, who were to be given a monopoly in the supply of certain drugs classified as poisons.[52]

Further Ordinances followed on other islands, all based on the British 1868 Pharmacy and Poisons Act. A draft of the Sale of Drugs and Poisons Ordinance for Jamaica was sent to the PSGB in 1881 for comment.[53] The PSGB was anxious that the sale of certain items be restricted to those chemists and druggists holding its own qualification.[54] It was passed a few months later. The *Chemist and Druggist* reported that 'a new Pharmacy Act has been passed and will come into operation on 1 July 1881'. They were not impressed with the PSGB's efforts. 'It confines trade in drugs and medicines to certain licensed persons, but the exceptions are so numerous that it does not seem to grant any substantial monopoly'. However, they were reassured by the fact that 'any duly certified British chemist and druggist is entitled to a licence without examination'.[55]

The Pharmacy Act repealed a previous Colonial Act passed in 1852 which was mainly concerned with the registration of medical practitioners. The granting of licences was entrusted to the superintending medical officer of the island, and all registered medical practitioners were to be regarded as licensed. A list of all licensed persons was to be published in the *Jamaica Gazette* once a year. The superintending medical officer was authorized to visit and search any drug store, to examine the drugs and medicines, and to take samples for further examination. The Jamaican Poisons List was a copy of the British list plus eight others, including antimony, copper sulphate, and carbolic acid. The Act left no doubt as to who was in control of pharmacy in Jamaica in 1881.[56]

The award of monopoly privileges to chemists and druggists in the sale of certain substances brought with it closer scrutiny of their regulation. This was provided by the doctors, who were granted increased powers over the chemists and druggists and their qualifications, at a time when such powers were diminishing elsewhere in the Empire. The PSGB was asked to comment on a proposed Apothecaries Ordinance for St. Lucia in 1885.[57] It was not the first time it had been involved; the PSGB president reminded his Council that in 1882 the Colonial Secretary had sent them a copy of an Ordinance relating to pharmacy and the sale of poisons in St. Lucia. The PSGB had made a series of recommendations to the Government. The new Ordinance received from the Colonial Office was examined by the Council, after which the president reported that 'their recommendations had been practically carried out'.[58] It was passed later that year, establishing a Medical Board to maintain registers of both medical practitioners and chemists and druggists.

The influence of the PSGB was seen again when a new Pharmacy Ordinance was proposed for Jamaica in 1894. This would strengthen the qualification and registration requirements for pharmacy, and the draft Ordinance was referred to the PSGB for comment. They made several suggested changes to bring practice more into line with that in Britain, insisting that only the British qualification should be accepted for registration.[59] Appropriate changes were made, and the Ordinance was passed. A draft Druggists Ordinance for Barbados, largely based on the Jamaica one and again insisting that only the British qualification should be accepted for registration, was sent to the Society later in 1894. The Society replied that the 'Act appears admirably suited to purpose'.[60]

By the late 1890s at least some colonial authorities in the British West Indies appointed an experienced British pharmacist as their Government pharmacist, to manage pharmacy in the government hospitals and dispensaries, and to oversee the implementation of pharmacy and poisons Ordinances. But most of these were short term appointments, with appointees returning to Britain at the end of their tenure. In 1901 it was reported that Mr. J. R. Johnston, pharmaceutical chemist, had recently returned from Jamaica, where he had been the Government pharmacist.

He was appointed resident dispenser at the Society of Apothecaries in London.[61]

Prosecuting Poison Sellers, 1902

The wording in the Pharmacy and Poisons Act 1868 that had caused such difficulties in Britain in 1880 re-emerged in Jamaica in 1902, although this time the problem was caused by different words. The Jamaican authorities instituted a series of prosecutions against traders in Kingston for infringements of the local pharmacy laws. Messrs J. Cresswell & Co. were prosecuted for selling a poison in a bottle which was labelled with the 'name and address of the firm' instead of the 'name and address of the person who handed the article to the customer'. According to the wording in the British Act on which the Jamaican Ordinance was based, the 'name and address of the seller' meant the 'name and address of the person on whose behalf the sale was made'.

The case was taken up enthusiastically by the local newspaper, the *Jamaican Gleaner*, which considered that the wording must mean 'the firm' and not 'the assistant', as the poison was not the assistant's property to hand over. Yet Section 21 of the Jamaican drugs and poisons law required the assistant's name and address: when sold it must include the word 'poison' and the 'name and address of the assistant' written or printed on the label. Yet at the same time, the paper pointed out, 'an unrestricted and unchecked sale of the deadliest poisons goes on in ironmongery-stores and bookstores'. The Legislative Council was asked to review the wording of the Act.[62]

In a separate case James Taylor, who kept a drugstore in East Queen Street, Kingston, was prosecuted for selling drugs and poisons without a licence, and without there being a qualified person licensed to sell them or to manage the store. The case related to the sale of rhubarb. The wording of the law referred to 'any drug', and rhubarb was used medicinally. In fact, a certified dispenser, James G. Surgeon, was in charge on the day, but the drug was sold by an unqualified shop assistant, without the dispenser's knowledge. The case was dismissed.[63]

These cases illustrate that British laws tended to be enacted in colonies without sufficient attention to how they might be interpreted or implemented in different countries. In Britain cynicism extended to how the law was implemented in Britain itself. In 1888 the *Chemist and Druggist* reported that legislation on the lines of the English Merchandise Marks Act (relating to the use of trade names) had already been passed in eleven colonies and was about to be implemented in eighteen more. 'Let us hope' they declared 'that the colonial authorities will work the Act in a less idiotic manner than is done by the Customs in this country'.[64]

Forming Associations

Chemists and druggists in the West Indies, as elsewhere in the Empire, followed pharmacy developments in Britain through the pages of the *Pharmaceutical Journal* and the *Chemist and Druggist.* Although their numbers were small, from the mid-1880s they began to hold meetings and form pharmaceutical associations. The first to act were pharmacists in Trinidad and Tobago (Fig. 4.3).Chemists and druggists formed Pharmaceutical Associations in 1886 on each island separately, three years before the islands were amalgamated for administrative purposes. It took another nine years for the two bodies to come together. In 1898 members of both voted for the associations to be merged, to form the Pharmaceutical Society of Trinidad and Tobago (Table 4.1).

Fig. 4.3 W. C. Ross & Co., dispensing chemists, Port-of-Spain, Trinidad, c.1895 (Courtesy of Alamy Images)

Table 4.1 Professionalizing pharmacy in the West Indies

West Indian Colony	*Settled by Britain*	*Independence or British Overseas Territory*	*Pharmaceutical society or association*	*Pharmacy Ordinance*	*Poisons/Drugs Ordinance*	*Pharmacy education*
Bermuda Islands	1609 (self-governing colony 1620)	British Overseas Territory	Bermuda Pharmaceutical Association founded 1978	1926 Bermuda Pharmacy Council 1979	1972/1979	Apprenticeship
St. Kitts (Saint Christopher)	1623	Independent 1983	St. Kitts and Nevis Association of Pharmacists	Ordinance	Ordinance	Apprenticeship
Nevis	1628 (joined with St. Kitts 1883)	Independent 1983	St. Kitts and Nevis Association of Pharmacists	Ordinance	Ordinance	Apprenticeship
Barbados	1627	Independent 1966	Barbados Pharmaceutical Society	Pharmacy Ordinance 1880	Druggists Ordinance 1894	Apprenticeship
Antigua	1632 (joined with Barbuda 1860)	Independent 1981	Antigua and Barbuda Pharmaceutical Society	Ordinance	Ordinance	Apprenticeship
Saint Lucia	1635	Independent 1979	Pharmaceutical Association of St. Lucia	Apothecaries Ordinance 1885	Ordinance	Apprenticeship
British Honduras	1638	Belize 1973. Independent 1981	Belize Pharmaceutical Association	Ordinance	Ordinance	Apprenticeship

(continued)

Table 4.1 (continued)

West Indian Colony	*Settled by Britain*	*Independence or British Overseas Territory*	*Pharmaceutical society or association*	*Pharmacy Ordinance*	*Poisons/Drugs Ordinance*	*Pharmacy education*
Bahama Islands	1649	Independent 1973	Bahamas Pharmaceutical Association	Ordinance	Ordinance	Apprenticeship
Montserrat	1667	British Overseas Territory	none	Ordinance	Ordinance	Apprenticeship
Virgin Islands	1672	British Overseas Territory	British Virgin Islands Pharmacists' Association	Ordinance	Ordinance	Apprenticeship
Turks and Caicos Islands	1681	British Overseas Territory	none	Ordinance	Ordinance	Apprenticeship
Tobago	1763 (joined with Trinidad 1889)	Independent 1962	Tobago founded 1886. Pharmaceutical Society of Trinidad and Tobago	Ordinance	Ordinance	Apprenticeship
Grenada	1763	Independent 1974	Grenada Association of Pharmacists	Ordinance	Ordinance	Apprenticeship
Saint Vincent	1763 (joined with Grenadines)	Independent 1979	St. Vincent and Grenadines Pharmaceutical Association founded	Ordinance	Ordinance	Apprenticeship
The Grenadines	1763 (joined with Saint Vincent)	Independent 1979	St. Vincent and Grenadines Pharmaceutical Association founded	Ordinance	Ordinance	Apprenticeship
Jamaica	1802	Independent 1962	Jamaica Pharmacists' Association 1920s. PSJ founded 1944	Pharmacy Ordinance 1894	Sale of Drugs and Poisons Ordinance 1881	Utech School of Pharmacy 1962

West Indian Colony	*Settled by Britain*	*Independence or British Overseas Territory*	*Pharmaceutical society or association*	*Pharmacy Ordinance*	*Poisons/Drugs Ordinance*	*Pharmacy education*
Trinidad	1802 (joined with Tobago 1889)	Independent 1962	Trinidad founded 1886. Pharmaceutical Society of Trinidad and Tobago	Pharmacy Board 1960	Ordinance	Certificate and Diploma B.Sc. 1995
Dominica	1805	Independent 1978	Dominica Pharmaceutical Society	Ordinance	Ordinance	Apprenticeship
British Guiana	1814	Independent 1966	Guyana Pharmacists' Association Founded 1895	Pharmacy Act No. 9 2003	Ordinance No. 36 1956	Apprenticeship. School of Pharmacy 1972

The registration of chemists and druggists on both islands was then in the hands of the doctors, and quarterly meetings were held between representatives of the two groups to discuss matters of mutual concern.[65] Over the following decades, the association made repeated attempts to secure pharmacy as a separate autonomous profession with its own Pharmacy Board, with responsibility for the education, examination, and registration of pharmacists along British lines. But it had no legal status, only being officially registered under the Company's Ordinance in 1925. Its efforts were entirely unsuccessful during the period of British rule. When the Society tried again in 1936 its lack of legal authority and the opposition of the medical profession resulted in the continuing regulation of the profession of pharmacy by the Council of the Medical Board of Trinidad. The Society focused instead on professional matters, establishing its own journal.

A Pharmacy Board of Trinidad and Tobago, giving pharmacy self-regulation, was only established in 1960, on the eve of independence, with passage of a Pharmacy Board Act in 1960. Until then the Society's role was limited to representing pharmacists on matters relating to the profession of pharmacy, ranging from employment issues to regulating the pharmaceutical trade in the colony. As the Pharmacy Board gained in authority, the role of the Pharmaceutical Society of Trinidad and Tobago diminished, and it became defunct. However, it was later revived and again registered as a company with the Registrar of Companies.[66] A Pharmacy Council of Trinidad and Tobago—responsible for the regulation and registration of pharmacists, pharmacy students, pharmacy owners, and authorized sellers of poisons in the two islands—was finally established in 1975.[67]

Other pharmacy associations were founded in the closing years of the nineteenth century, some with official sanction in the form of Ordinances. In 1890 an Ordinance established a Chemists and Druggists' Association in British Guiana, but this placed the registration of chemists and druggists firmly in the hands of a Medical Board. The Association changed its name to the British Guiana Pharmacists' Association in 1895, and set about seeking autonomy from the doctors and the right to maintain its own Register. It had a little more success than most of the other associations in the West Indies. In 1905 a new Ordinance was passed, repealing the earlier one, placing responsibility for the examination and the registration of chemists and druggists with a Pharmacy Board.[68]

In Jamaica, a Pharmacists' Association had originally been founded early in the twentieth century with fifteen members, but in 1944 the name was changed to the Pharmaceutical Society of Jamaica (PSJ).[69] As elsewhere, pharmacy remained under the control of the doctors until after independence. A Pharmacy Act in 1966 finally authorized the establishment of a Pharmacy Council of Jamaica which would have responsibility for the regulation of pharmacists, pharmaceutical students, pharmacy owners, and authorized sellers of poisons, in accordance with the Act. But it was another nine years before the Council was finally established by the Ministry of Health, in August 1975. The Council was tasked with maintaining registers for all the above groups, and set standards for pharmacists' education, scopes of practice and conduct.[70]

Elsewhere, pharmaceutical associations were formed only after independence, or once they had become a British Overseas Territory. Although in Bermuda the first Pharmacy Act had been passed in 1926, further change to the regulation of both the pharmacy profession and of poisons occurred only in the 1970s. A Misuse of Drugs Act was passed in 1972, a Bermuda Pharmaceutical Association was founded in 1978, and a Bermuda Pharmacy Council was established in 1979, following passage of a Pharmacy and Poisons Act in 1979. In Barbados too pharmacy and medicines remained firmly under medical control until after independence. The chief medical officer rather than the chief pharmaceutical chemist was the person empowered under the 1950 Therapeutic Substances Act to act as the licensing authority responsible for 'control of the manufacture for sale or supply of any drug or therapeutic substance' to which the Act applied. With passage of a Pharmacy Act in 1986 a Pharmacy Council was established to control and regulate the practice of pharmacy, to register and control persons admitted to practice, and to register pharmacy premises for the selling of drugs and poisons.[71]

Twin-Track Qualifications

By the 1880s there were three main groups engaged in the supply of medicines in the West Indies, although only two were officially authorized to do so: chemists and druggists, 'dispensers' and unqualified traders. British chemists and druggists, holding certificates from the PSGB, operated 'English' pharmacies in accordance with what was effectively British pharmacy and poisons law. These were few in number and mainly operated retail pharmacies. Newly arrived British chemists and druggists were augmented by apprentices, often family members such as sons and nephews, taken on by those already there. Some were sent back to England to attend one of the crammer schools in London before taking the examinations of the PSGB. By this means the British model of pharmacy education and training was strengthened and perpetuated.

With the establishment of Government Medical Services, a need was recognized to provide at least a basic level of training for those needed for pharmaceutical duties in government hospitals and dispensaries. In 1885 a correspondent, writing to the *Chemist and Druggist* from Jamaica, reported that 'the dispensing school at the public hospital continues flourishing, and pupils from there are to be found in nearly every hospital and Government dispensary throughout the island'.[72] By 1894 pharmaceutical lectures were being given by the Government Analyst, Mr. P. Carmody.[73] Similar arrangements were made elsewhere in the West Indies. In British Guiana, provision was made for the training of 'dispensers' through an Ordinance in 1886 which had formally established a Government Medical Service.[74] Dispensers constituted the second group of pharmaceutical practitioners. This group was locally recruited and trained by the British authorities to order and supply the medicines used in Government hospitals and dispensaries.

The West Indies this operated a dual register of pharmacy qualifications, but there was a crucial difference to that operating in Britain. Whilst in Britain the pharmaceutical chemist certificate was a higher-level qualification than that of chemist and druggist, in the West Indies the dispenser qualification was at a much lower level than that of chemist and druggist; dispensers could not open their own pharmacies. This twin model of pharmacy training and qualification—with British-trained chemists and druggists and locally trained dispensers—remained in place throughout the first half of the twentieth century on most of the islands.

The third group were the 'unqualified'. During slavery, some slaves were engaged in making and supplying medicines, both on the plantations and in drug stores. When it ended, the numbers increased as more drug stores opened. Such druggists were unqualified, having had no formal training, although they were often very knowledgeable about medicines and provided an important service to the Black community. They sold patent medicines and local remedies but were unable to supply—at least legally—medicines containing poisons, or to dispense prescriptions. Unqualified druggists usually outnumbered both chemists and druggists and dispensers.

By the end of the nineteenth century, most of the British West Indian colonies had arrangements in place for the registration of chemists and druggists, but they were in the hands of Medical Boards, who specified the qualifications required for registration. The doctors were still firmly in control of pharmacy. When the PSGB carried out its survey of pharmacy qualifications throughout the Empire, it was the Medical Boards on the West Indian islands who confirmed that only the PSGB certificate was accepted for registration as a pharmacist in their colony. Those holding it could thus work in any of the West Indian colonies as well as in the metropole; those holding any other qualification could not practice as a chemist or druggist. In British Guiana the education and training of chemists and druggists was in line with those of the other British colonies by 1905, in that only the PSGB's qualification was acceptable for registration. It could therefore respond positively to the PSGB's request in that year for information about the training of pharmacists and its readiness for reciprocity with other colonies.[75]

Once arrangements were in place for the reciprocal recognition of qualifications between Britain and her colonies following passage of the 1908 Poisons and Pharmacy Act, the PSGB wrote to the colonies seeking confirmation of qualification procedures. In 1915 *The Pharmaceutical Journal* reported that in Barbados, British Guiana, and British Honduras, amongst other colonies and dependencies, 'the British certificate was accepted for full registration as a chemist and druggist'.[76] Chemists and druggists generally had cordial relations with their local medical community, despite being under their control, and held regular meetings to discuss issues of common concern. Towards the end of the nineteenth century local pharmaceutical associations in the West Indies were consulted by their medical colleagues following a request from London to all colonies for suggestions about what local drugs might be considered suitable for inclusion in an Indian and Colonial Addendum to the BP 1898.[77]

British Retail Pharmacies and Colonial Cultural Capital

By the early twentieth century a well-established pattern had emerged in many of the British West Indian colonies and elsewhere in the Empire of British pharmacies run by British pharmacists. These provided the services and products that were generally available in a typical English pharmacy.

Their customers were mainly the white expatriate community and a small number of middle-class locals. Alongside these were a large number of drug stores run by largely unqualified locals providing essential services to the local population.

In 1922, a correspondent described pharmacy in British Guiana for readers of the *Chemist and Druggist*. Pharmacies there were designed along British lines, he reported, often with fixtures, fittings, and stock all supplied from Britain.

> The business of which the author was the manager was run on up-to-date lines: he had a well-fitted pharmacy that compares more than favourably with most seen in Great Britain, with wholesale, retail and dispensing departments...Though most drug stores handle many lines other than the usual stock (groceries, tobacco, chocolates, etc) a few of the better class places are run as good pharmacies only. The stocks held by the larger drug stores are very varied, and on a much larger scale than one meets in Great Britain.[78]

By this stage local staff were trained to work in pharmacies. In this pharmacy 'three dispensers were kept going at it hard all day, two qualified and one unqualified; these were local men, East Indians or Blacks who had been trained in the colony and had taken the examination of the British Guiana Pharmacy Board'. There were nevertheless differences in the type of work involved. In the British Guiana pharmacy 'the daily average number of prescriptions was in the vicinity of 100, and usually entailed much more work than the ones met with [in Britain]; capsule-filling, for things like quinine, aspirin, and calomel, in a great measure replaced cachet-making'.[79]

There were other differences too, not least that expatriate chemists and druggists often felt more valued in the colonies than they did back in Britain.

> Prices are different, customers of all colours, and language – whilst mostly English – have a brogue one has difficulty in understanding. The shop is called a store, or very often – in the case of our trade – is spoken of as a doctor's shop. The old-time chemist here was also the doctor, and even today the amount of prescribing is greater and more varied than one meets in the home chemist's business. It need hardly be added that the chemist in British Guiana owes much to his success as a prescriber. An assistant is spoken of as a clerk, the manager or proprietor is variously addressed

according to the customer, and, if a successful man, is usually treated with more respect than in the home country.[80]

It is a description that could equally have been applied to many British pharmacies in colonies around the world in the early twentieth century. The presence of modern, Western-style retail pharmacies throughout the Empire presented a powerful image of colonial dominance. They were potent symbols of 'colonial social and cultural capital', a phrase that, in his analysis of colonial British America, Jack Greene describes as referring 'to all of those elements of the larger inheritance that cultures pass along to succeeding generations for later members to perpetuate, modify, discard, or reconstitute in new places'.[81] Retail pharmacies were a small but highly visible part of the 'massive increase in colonial social and human capital which fostered the emergence of a European-style civil society' not only in Britain's American colonies but throughout the Empire.[82]

For the expatriate community English pharmacies offered welcome reassurance that what they could get at home was also available in the colonies. This was especially true in the so-called 'tropical' colonies of Africa, India and the West Indies. Simply by their presence English retail pharmacies presented a powerful statement demonstrating the ascendency and dominance of Western science, technology, and medicine; they provided a clear demarcation between western medicine and indigenous practice, not least by their reliance on imported products; and by selling toiletries, cosmetics and grooming products alongside medicines they presented an image of what it meant to be 'civilized'. Retail pharmacies signalled a clear distinction between 'modernity', 'progress' and 'civilization', as represented by the colonial power, and the 'tradition', 'stagnation', and 'lack of civilization' supposedly represented by indigenous people.

The Coming of Independence

By the 1950s, with the prospect of imminent independence, efforts were made by the colonial authorities to update pharmacy legislation. In British Guiana, a Pharmacy and Poisons Ordinance was passed in 1956, although a Pharmacy Act establishing an independent Pharmacy Board passed only in 2003. Politically, the world had changed. Towards the end of the colonial period, in 1958, efforts were made to encourage the British Caribbean islands to work together as a West Indies Federation. By 1962 the initiative had failed, and most of Britain's Caribbean territories opted for separate independence. Jamaica became the first to be granted independence, from 6 August 1962. Trinidad and Tobago followed 25 days later. Thereafter other islands gained their independence in a process extending over 21 years; the last to gain independence was St. Kitts and Nevis, in 1983.

After the Second World War significant numbers of students from the colonies travelled to Britain for their training. In 1953, with only 32 places available, the University of London's School of Pharmacy had five undergraduate and eleven postgraduate students from overseas, including some from the West Indies and British Guiana.[83] It was only with the imminent prospect of independence that the first steps were taken to open Schools of Pharmacy in the Caribbean. In Jamaica in 1961 the Pharmaceutical Society of Jamaica spearheaded the establishment of a pharmacy school at the College of Arts, Science and Technology in Kingston (now University of Technology, UTech). The training of pharmacists for Jamaica and other Caribbean islands began in 1962. A separate School of Pharmacy was created in 2009, becoming the largest in the English-speaking Caribbean.

Elsewhere, a School of Pharmacy was founded in British Guiana in 1972. Pharmacy education became available in Trinidad after independence, with successful candidates being awarded either a certificate or diploma in pharmacy; a separate School of Pharmacy was established in 2007. In 1995 a B.Sc. Pharmacy degree programme was launched at the University of the West Indies in Trinidad. Nearly forty years after independence the Curacao Accord in 2001 agreed to establish a Caribbean Pharmacy Examination Board to standardize pharmacy education across the West Indies.[84]

During the British period, pharmacists in the various colonies had few opportunities to meet to discuss common concerns. But following the creation of the Caribbean Community and Common Market (Caricom) in 1973, renewed efforts were made to bring them together. A Caribbean Association of Pharmacists (CAP) was eventually founded in 1976.[85] The inaugural meeting was attended by representatives from seven former British colonies. Its objectives reflected just how little progress had been made in the professionalization of pharmacy during the years of British rule. It aimed to encourage the establishment of national pharmaceutical bodies in Caribbean countries where none existed; to foster high standards of control over the quality and distribution of medicines; and to stimulate and maintain high standards of pharmacy education at all levels. It also sought to promote a uniform system of education for pharmacists practising in the region, to guarantee a high level of professional practice, and to enable reciprocal recognition of qualifications.[86] But it was a voluntary body, with no legal standing in any state. It faced familiar challenges, and many of its aims would have to wait until the twenty-first century to be fulfilled.

Professionalization Delayed

When considered against the markers identified in Chapter 1, it is clear that little progress was made in the professionalization of pharmacy in the West Indies during the period of British rule. Rather than securing the support of the state, the colonial administrations imposed a model of control based on medical superiority. Although on some islands chemists and druggists came together to form associations for collective action, they failed to secure legal recognition for them, and they largely remained toothless voluntary organizations. Legislation was passed regulating the profession, but this involved placing pharmacists under the control of doctors, and it was shaped not by local pharmacists in the colonies but centrally according to recommendations by the PSGB.

In the British West Indies chemists and druggists holding the British qualification were a privileged group. Legislation gave them legal rights not available to locally trained dispensers, and gave them a monopoly in the supply of substances not legally available to unqualified traders. Yet they were not an autonomous profession separate from medicine, still being subject to control by the medical authorities. The establishment of a lower tier of education available to dispensers, which did not qualify them for registration as chemists and druggist, reinforced the hierarchical nature of the occupation and established clear boundaries between the chemists and druggists and others.

But in many ways pharmacy in the West Indies can be described as 'British'. Its hierarchical nature mirrored the dual nature of the profession operating in Britain, where a distinction was made between pharmaceutical chemists and chemists and druggists, although in the West Indies the distinction was between chemists and druggists and lesser qualified dispensers. Although Pharmacy Acts were passed, colonial pharmacists played little part in drafting them; this was generally done by colonial administrators and medical officers with metropolitan advice. But the regulation of pharmacy was usually linked to control of the distribution and supply of medicines through Pharmacy and Poisons Acts, the result of active lobbying by the PSGB.

A key marker of pharmacy professionalization—the examination and registration of pharmacists by pharmacists by an independent Pharmacy Board—proved elusive throughout the period of British rule. Although the practice of pharmacy bore many of the hallmarks of British pharmacy, some features remained absent. Economic circumstances offered little incentive to those contemplating the establishment of chain pharmacies in the West Indies. By 1914, pharmacy was British only in relation to the European enclave. Elsewhere pharmacy remained undeveloped; the development of pharmacy institutions, education and regulation were all greatly hampered by the legacy of slavery.

Pharmacy developments elsewhere in the region appear to have had little impact in the British West Indies. Other Caribbean islands were parts of the empires of France, Spain, the Netherlands and Portugal, countries where pharmacy had developed differently and was a separate and autonomous profession from the early eighteenth century. Legislation concerning pharmacy and medicine supply privileged European health over that of the Black population. It strengthened the position of British chemists and druggists in the medical marketplace, but had a marked negative impact on the health of the local population. If the delay of more than half a century in the professionalization of pharmacy was a legacy of centuries of British colonialism and slavery in the West Indies, so too was greatly reduced access to medicines by much of the population.

Notes

1. Anderson, Stuart. 'Pharmacy and Slavery: Apothecaries, Medicines and the Slave Trade, 1650 to 1807', *Pharmaceutical Historian*, 39 (2009), pp. 11–6.
2. Walvin, James. *A Short History of Slavery* (London, 2007), p. 50.
3. Davies, K.G. *The Royal African Company* (London, 1999), p. 64.
4. Richardson, D. 'Liverpool and the English Slave Trade', in Tibbles, A. (ed.) *Transatlantic Slavery: Against Human Dignity* (Liverpool, 1994), p. 67.
5. Lloyd, T.O. *The British Empire 1558–1995*, second edition (Oxford, 1996), p. 19.
6. Walvin, James. *The Slave Trade* (Stroud, 1999), pp. 48–9.

7. Lloyd, *British Empire*, p. 20.
8. Cumper, G.E. 'Population Movements in Jamaica, 1830–1950', *Social and Economic Studies*, 5 (1956), pp. 261–80.
9. Ward, J.R. 'The British West Indies in the Age of Abolition, 1748–1815', in Marshall, P.J. (ed.) *The Eighteenth Century: The Oxford History of the British Empire* (Oxford, 1998), pp. 432–5.
10. Morgan, K. *Slavery and the British Empire: From Africa to America* (Oxford, 2007), p. 17.
11. Ibid., p. 15.
12. Ibid., p. 17.
13. Bissessar, A. et al. *The Historical Development of the Health System in Trinidad and Tobago* (London, 2001), LSHTM PHP Publication No. 34.
14. Craton, M. 'Death, Disease and Medicine on a Jamaican Slave Plantation: The Example of Worthy Park 1767–1838', in Beckles, H. and Shepherd, V. (eds) *Caribbean Slave Society and Economy* (Kingston, 1991), pp. 183–96.
15. Handler, Jerome S. 'Slave Medicine and Obeah in Barbados, circa 1650 to 1834', *New West Indian Guide*, 74 (2000), pp. 82–3.
16. Sheridan, R.B. *Doctors and Slaves: A Medical and Demographic History of Slavery in the British West Indies, 1680–1834* (Cambridge, 2010), p. 336.
17. Fergusson, William. *Reports on Hospital Establishments in the West Indies*, Royal Army Medical Corps Muniment Collection, 210, Wellcome Library.
18. Harrison, M. *Medicine in an Age of Commerce and Empire: Britain and Its Tropical Colonies, 1660–1830* (Oxford, 2010), p. 19.
19. Sturge, J. and Harvey, T. *The West Indies in 1837* (London, 1838), p. 48.
20. Ibid., p. 216.
21. Sheridan, *Doctors and Slaves*, p. 336.
22. Ibid., p. 70.
23. Schiebinger, Londa. *Secret Cures of Slaves* (Stanford, 2017), p. 84.
24. *Pharmaceutical Journal and Transactions*, 46 (1886), p. 416.
25. Anderson, Stuart. 'Liverpool Apothecary in the Slave Trade', *Pharmaceutical Journal*, 285 (2010), pp. 732–3.
26. Dunn, Richard S. *Sugar and Slaves: The Rise of the Planter Class in the English West Indies, 1624–1713* (Chappell Hill, 1972), p. 250.

27. Burnby, J.G.L. *A Study of the English Apothecary from 1660 to 1760* (London, 1983), p. 9. See also Haycock, D.B. and Wallis, P. (eds). *Quackery and Commerce in Seventeenth Century London: The Proprietary Medicine Business of Anthony Daffy* (London, 2005).
28. Magee, Gary B. and Thompson, Andrew S. *Empire and Globalisation: Networks of People, Goods and Capital in the British World, c.1850–1914* (Cambridge, 2010), p. 135.
29. Porter, R. and Porter, D. 'The Rise of the English Drugs Industry: The Role of Thomas Corbyn', *Medical History*, 33 (1990) p. 292. Liebenau, J., Higby, G.J. and Stroud, E.C. (eds) *Pill Peddlers: Essays on the History of the Pharmaceutical Industry* (Madison, 1990), p. 21.
30. Porter and Porter. 'Rise of the English Drugs Industry', p. 19.
31. Trease, G.E. *Pharmacy in History* (London, 1964), p. 159.
32. Cripps, E.C. *Plough Court: The Story of a Notable Pharmacy 1715–1927* (London, 1927), p. 10.
33. Chapman-Huston, D. and Cripps, E.C. *Through a City Archway: The Story of Allen and Hanburys 1715–1954* (London, 1954), p. 37.
34. Tweedale, Geoffrey. *At the Sign of the Plough: Allen and Hanbury's and the British Pharmaceutical Industry 1715–1990* (London, 1990), p. 30.
35. Tweedale, *Sign of the Plough*, p. 26.
36. Heuman, Gad. 'The British West Indies', in Porter, Andrew (ed.) *The Nineteenth Century: The Oxford History of the British Empire* (Oxford, 1999), p. 476.
37. Ibid., p. 477.
38. Sheridan, Richard B. 'The Formation of Caribbean Plantation Society, 1689–1748', in Marshall, P.J. (ed.) *The Eighteenth Century: The Oxford History of the British Empire* (Oxford, 1998), p. 394.
39. Heuman, 'British West Indies', p. 490.
40. *Chemist and Druggist*, 26 (1884), p. 206.
41. *Chemist and Druggist*, 27 (1885), p. 12.
42. *Chemist and Druggist*, 36 (1890), p. 19.
43. *Chemist and Druggist*, 22 (1880), p. 32.
44. *Chemist and Druggist*, 26 (1884), p. 206.
45. *Chemist and Druggist*, 27 (1885), p. 12.
46. *Chemist and Druggist*, 22 (1880), p. 494.

47. Ferris and Co. of Bristol were large exporters of surgical equipment and appliances in addition to drugs and pharmaceutical preparations. See *British Medical Journal*, 2 (1885), pp. 449–54.
48. The National Archive (TNA), London, CO-137/584/12; Number 449, 30 October 1897, folios 68–78.
49. Heuman, 'British West Indies', p. 470.
50. Anderson, Stuart, 'From "Bespoke" to "Off-the-Peg": Community Pharmacists and the Retailing of Medicines in Great Britain 1900 to 1970', in Curth, Louise Hill (ed.) *From Physick to Pharmacology: Five Hundred Years of British Drug Retailing* (Aldershot, 2006), pp. 105–42.
51. *Chemist and Druggist*, 22 (1880), p. 32.
52. Ibid.
53. TNA, London, CO-137/502/43; Jamaica Office to Pharmaceutical Society, 11 June 1881, folios 295–301.
54. TNA, London, CO-137/500/3; Pharmaceutical Society to Jamaica Office, 1881.
55. *Chemist and Druggist*, 23 (1881), p. 183.
56. Ibid., pp. 209–10.
57. TNA, London, CO-321/92; Pharmaceutical Society to St. Lucia Office, 1885.
58. *Chemist and Druggist*, 27 (1885), p. 5.
59. TNA, London, CO-137/563; Pharmaceutical Society to Jamaica Office, 1894.
60. TNA, London, CO-28/236/85; Pharmaceutical Society to Barbados Office, 19 December 1894.
61. *Chemist and Druggist*, 58 (1901), p. 212.
62. *Chemist and Druggist*, 61 (1902), p. 93.
63. Ibid., p. 98.
64. *Chemist and Druggist*, 33 (1888), p. 793.
65. Trinidad and Tobago Pharmaceutical Society. https://pharmaceuticalsocietytt.org/about-us/ (Bridgetown, 2020). Accessed 12 October 2020.
66. Its Memorandum and Articles of Association were registered under the Companies Ordinance.
67. Pharmacists in Trinidad_and_Tobago. https://rgd.legalaffairs.gov.tt/laws2/Alphabetical_List/lawspdfs/29.52.pdf (Port of Spain, 2020). Accessed 12 October 2020.
68. Rayner, T.C. *The Laws of British Guiana*, volume 4 (Waterlow, 1905), p. 489.

69. Jamaica Pharmaceutical Society. http://pharmasocietyjamaica.com/about-pharmaceutical-society-ja/ (Kingston, 2020). Accessed 12 October 2020.
70. Pharmacy Council of Jamaica. https://jis.gov.jm/government/agencies/pharmacy-council-of-jamaica/ (Kingston, 2020). Accessed 12 October 2020.
71. Pharmacy in Barbados. https://www.who.int/medicines/areas/coordination/pscp.barbados_en.pdf?ua=1. Barbados Ministry of Health (Bridgetown, 2011), p. 17. Accessed 12 October 2020.
72. *Chemist and Druggist*, 27 (1885), p. 12.
73. TNA, CO-295/358/22, Fees for Pharmaceutical Lectures, 27 November (1894).
74. 'An Ordinance to Establish a Government Medical Service, and to Enforce the Registration of Practitioners in Medicine and Surgery', No. 5 of 1886, in Rayner, *Laws of British Guiana*. See also De Barros, J. 'Infant Welfare in Post-Slavery British Guiana', in De Barros, J., Palmer, S. and Wright, D. (eds) *Health and Medicine in the Circum-Caribbean, 1800–1968* (Abingdon, 2009), p. 189.
75. *Report of the West Indies Committee of the Commission on Higher Education in the Colonies* (London, 1945), p. 63.
76. *Pharmaceutical Journal*, 39 (1915), p. 268.
77. 'British Pharmacopoeia 1898, Indian and Colonial Addendum: A Report of Progress, 1899. Appendix XX', *GMC Pharmacopoeia Committee Minutes* (London, 1899), p. 753.
78. *Chemist and Druggist*, 97 (1922), p. 335.
79. Ibid.
80. Ibid.
81. Greene, Jack P. 'Social and Cultural Capital in Colonial British America: A Case Study', *Journal of Interdisciplinary History*, xxix (1999), p. 492.
82. Ibid., p. 501.
83. Hudson, B. *The School of Pharmacy, University of London: Medicines, Science and Society, 1842–2012* (London, 2013), p. 161.
84. Deeney, Colin. 'Tropical Pharmacy: 25 years of the Caribbean Association of Pharmacists', *Pharmaceutical Journal*, 267 (2001), pp. 934–5.
85. Ibid., p. 934.
86. Ibid., p. 935.

CHAPTER 5

Mediterranean Colonies: The Legacy of Palermo

From the early eighteenth century Britain acquired a string of posses sions across the Mediterranean from west to east. It did so not to exploit their natural resources but to secure naval and military bases in support of its political and economic interests in India and the Far East. Over a period of 170 years, it established bases in Gibraltar, Minorca, Malta, Corfu, and Cyprus (Table 5.1). In 1914 Gibraltar, Malta, and Cyprus were still part of the British Empire. The situations met on first arrival by the British authorities in these three territories could hardly have been more different. They were under different systems of political control, had different cultures, and followed different religions.

The Mediterranean islands presented the British medical authorities with many varied challenges. Whilst supplying and using medicines was universal, the arrangements in place for doing so differed greatly. In Malta, the British inherited a system where pharmacy and medicine had long been separate and distinct, in accordance with the 1240 Edict of Palermo and subsequent Italian practice. In Cyprus they found a back-water of the Ottoman Empire where developments in medicine and pharmacy occurring in the main cities of Constantinople, Beirut and Damascus had not yet reached. But one thing they had in common was the initial role of naval and military medicine in shaping services for both service personnel and civilians. Its military origins also played a

S. Anderson, *Pharmacy and Professionalization in the British Empire, 1780–1970*, Medicine and Biomedical Sciences in Modern History,
https://doi.org/10.1007/978-3-030-78980-0_5

Table 5.1 Professionalizing pharmacy in British Mediterranean Colonies

Colony	*Year occupied by British*	*Year occupation ended*	*Pharmaceutical Society or Association*	*Early pharmacy legislation*	*Early poisons legislation*	*Pharmacy education*
Minorca	1708–1756 1763–1782	Restored to Spain 1802	None during British ule	None during British rule	None during British rule	None during British rule
Gibraltar	Occupied 1704: Colony 1713	British Overseas Territory	None	Medical Register 1790	Pharmacy and Poisons Act 1868	Apprenticeship
Malta	Occupied 1802: Colony 1814	Independent 1964	Chamber of Pharmacists 1900	1903	1624	1729
Sicily	Occupied 1806	Left 1815	–	–	–	–
Corfu (Ionian Islands)	Occupied 1809	Incorporated into Greece 1864	None during British rule	None during British rule	None during British rule	None during British rule
Cyprus	Occupied 1878	Independent 1960	Pharmaceutical Association 1954	1879	1900	None locally during British rule

part in shaping the early development of pharmacy in the Mediterranean colonies.

This chapter explores the development of pharmacy in these colonies. It considers the early practice of pharmacy in Gibraltar and its origins in military and naval medicine. It contrasts the impact on pharmacy of existing arrangements encountered by British medical and pharmaceutical practitioners in Malta and Cyprus. It reviews pharmacy in the Ottoman Empire during the time it included Cyprus, and notes the lack of diffusion to the periphery. It explores the professionalization of pharmacy in each territory, describing collective action through the formation of associations, the role of the state in achieving legal recognition, and the development of education. It also considers the extent to which pharmacy in the Mediterranean colonies was 'British'. But it begins with a review of the role of British military and naval medicine and its relationship with pharmacy.

Medical Services in the Armed Forces

British occupation in the Mediterranean began in 1704 when Gibraltar was captured during the War of the Spanish Succession; it was followed shortly by Minorca in 1708. Britain took formal possession of both in 1715 under the terms of the Treaty of Utrecht. Eastern Mediterranean islands were much less affected by the wars of the eighteenth century than those in the west, but Napoleon's actions to block British access to India and the 1798 Egyptian expedition led to a new British policy for the Mediterranean.[1] Once Minorca was returned to Spain under the terms of the 1802 Treaty of Amiens, Britain had no bases between Gibraltar and Alexandria, but the situation was rectified by the occupation of Malta in the same year and of Corfu in 1809. Under the 1815 Treaty of Paris the islands around Corfu became the United States of the Ionian Islands under British protection; Corfu was later ceded to Greece in 1864. Britain also occupied Sicily for a short whilst between 1806 and 1814. By the end of the Napoleonic Wars Britain had an empire of bases in the central Mediterranean.[2] But continuing security concerns emphasized the need for military bases in the eastern Mediterranean; in 1878 Cyprus was occupied and a base established (Fig. 5.1).

The military presence dominated during the early years of British occupation; by the middle of the eighteenth-century Gibraltar was home to a large garrison and naval station. Efficient medical services were essential to keep fighting forces fit for action, and by then military medicine was well established in Britain and her colonies. But doctors based in the Mediterranean did not have to deal with many of the diseases encountered by their colleagues in tropical colonies.[3] In Gibraltar each regiment had its own medical officers and individual sick bays, but there was no central medical facility.[4] Initially the hospital of St. John of God was used as a regimental sick bay, but much larger medical facilities were needed, and a new Naval Hospital was built in 1740 to cater for the needs of both the army and the navy. Experience in Gibraltar greatly influenced the development of medicine and pharmacy in the other British Mediterranean colonies.

Fig. 5.1 Map of the Mediterranean division of the British Empire, 1914

Military medical authorities were responsible not only for the health and fitness of the armed forces but also the welfare of the civilian population. In 1800 a major epidemic of yellow fever spread across Gibraltar, reappearing in 1804 and again in 1810 and 1813–14. Early in 1814 the Governor Designate, Sir George Don (1756–1832), reorganized the medical facilities of the town, including the building of a new hospital for civilians. It was funded by the Colonial Government and opened in 1816.[5] Its first medical officer, Roman Amiel, had been a surgeon in Wellington's army during the Peninsula War.[6] Pressure from the army to have its own hospital led to the Naval Hospital being designated an Army Hospital in 1814, with one ward reserved for sailors. With further expansion of British forces this became too small, and a new British Military Hospital was built in 1903. The old Naval Hospital was handed back to the Navy, and eventually closed in 1927.

Medicine Supplies to Army and Navy Hospitals

The Navy obtained most of its medicines from Apothecaries' Hall in London.[7] At the start of the eighteenth century its Navy Stock—founded in 1703—also secured the contract for the supply of medicines to the Army.[8] In 1704 the British Fleet in the Mediterranean was desperately short of medicines. The Navy Stock was asked to supply medicine chests to be transported to Lisbon. From there they were distributed to thirteen ships operating in the Mediterranean.[9] Apothecaries' Hall then held the contract to fit out surgeons' chests for both hospital ships and the garrisons at Gibraltar and Lisbon.[10] Demand for medicines grew rapidly with successive wars in the eighteenth century. With such large volumes of business involved, the Navy Stock's monopoly was challenged at the outbreak of the Seven Years' War in 1756 by William Cookworthy (1707–1780). But by the end of the century Apothecaries' Hall were sending naval surgeons' chests not only to Gibraltar, Minorca and Lisbon, but to the East and West Indies and to Australia.[11]

The costs of drugs and medicines required for the army hospitals became a matter of great concern, as they more than doubled during the Seven Years' War to cope with up to 3000 patients at a time. Procurement and supply were the responsibility of the Apothecary General, George Garnier (1703–1763),[12] but there were constant complaints about the cost from civil servants, and about the quality of the medicines supplied from the army. These issues came to the attention of John Pringle (1707–1782), who in 1742 was appointed physician to the British Army, a post he held for eight years before being appointed Physician-General.[13] Pringle saw the need to modernize and standardize the drugs used; he drafted the first army dispensatory.[14] He had been a member of the Edinburgh Royal College of Physicians' pharmacopoeia committee, and had a major influence over the content of the *Edinburgh Pharmacopoeia*.[15] He also wrote about army diseases, studied hospital cross-infection, and promoted the welfare of troops, becoming known as the 'father of military medicine'.[16] As Joint Physician-General of overseas forces in 1744, he had oversight of the drugs supplied. Building on his work, further British works on military and surgical pharmacy appeared, with some also published in France and Spain.[17]

Military pharmacopoeias were regularly revised reflecting changes in British practice. In the 1860s and 1870s a drastic decline occurred in the use by the army of 'heroic' therapies, such as purgatives and bloodletting.[18] Venereal diseases amongst soldiers meant that substantial quantities of mercury were used, although this slowly changed. Whilst some regimental surgeons continued to use mercury in the form of calomel (mercurous chloride), one surgeon reported: 'We have had occasion to see so much of the poison of syphilis that I am glad to think that I have not made matters worse by adding that other poison, mercury'.[19] By 1889 mercury was limited to topical application, and chemical therapies including salicylates, morphine and quinine were common.[20] Uniformity in prescribing practices was further facilitated by the merger in 1898 of the Army Hospital Corps and the Medical Staff Corps to form the Royal Army Medical Corps.[21]

Pharmacy in the Army and Navy

Military and naval surgeons were responsible for the dispensing of the medicines they prescribed, usually helped by untrained staff. The role of supporting ships' surgeons by making up medicines slowly evolved into the office of 'dispenser'. Henry Blakely was noted in 1713 as the dispenser at Greenwich Hospital. Initially, dispensers and surgeon's mates were required to undertake each other's duties, but by 1794 this arrangement had ended, with the Royal Navy declaring that

> the subordinate part of the Medical establishment is divided into three branches, namely that employed in assisting the physicians, that employed in assisting the surgeons, and that employed in compounding medicines.[22]

The pharmacies of naval hospitals were managed by the dispenser, and as they grew in size he was assisted by one or more assistants. By 1805 assistant dispensers were termed 'hospital mates' and they were required to take the assistant's examination of the Society of Apothecaries, regardless of whether they served at home, abroad or on ships serving overseas.[23] In 1872 an Order-in-Council directed that persons in charge of naval pharmacies must hold the higher level pharmaceutical chemist qualification, although in 1897 this was relaxed such that the lower-level chemist and druggist qualification was deemed acceptable.[24]

In the army, the regiment's medical officer, or 'surgeon', combined the duties of physician, surgeon, and apothecary. The army abolished the title 'apothecary' in 1820, although it was revived during the Crimean War (1853–1856) and finally abandoned after 1873.[25] Although military doctors were responsible for the supply and issue of medicines, the army employed sergeants and other unqualified military personnel to assist in the dispensing of prescriptions. This arrangement came under criticism during the Crimean War. A hospital dispenser complained to the *Pharmaceutical Journal* in 1855, reporting the engagement of unqualified men working as 'dispensers of medicine' whose duties were no more than 'those of a drug commissariat clerk or a Thames Street druggist's warehouseman'.[26]

There was no prospect of advancement for dispensers, although they had the nominal rank of 'assistant surgeon'. The journal declared that the office of 'pharmaceutical assistant or dispenser' was distinct from that of 'apothecary' and should be formally introduced in the army.

In 1862 the army—like the navy ten years later—specified that a person seeking appointment as a dispenser must have passed the Major examination of the PSGB and be registered as a pharmaceutical chemist. This gave state backing to the PSGB's attempts to raise the standard of pharmacy education.[27] But its implementation was delayed indefinitely; the demand for dispensers was such that in 1905 any man could still join the army as a private and become a compounder. Being already qualified offered no benefit, although promotion to a non-commissioned rank later became possible.[28]

Health and Pharmacy Services in Gibraltar, 1713

Control of the Rock of Gibraltar changed hands many times over the centuries. The Christian Crown of Castile annexed it in 1309, lost it again to the Moors in 1333, and finally regained it in 1462. Gibraltar became part of the unified Kingdom of Spain and remained under Spanish rule until 1704, when it was captured during the War of the Spanish Succession by an Anglo-Dutch fleet. At the war's end, Spain ceded Gibraltar to Britain under the terms of the Treaty of Utrecht of 1713.[29] When British forces landed in 1704, most of the population fled, and the original hospital run by the brothers of St. John of God closed, becoming the barracks for the Royal Artificer's Company. Britain built up its army and navy bases in Gibraltar during the course of the eighteenth century.

To meet the growing needs of the large garrison and naval station, immigrants came from Italy, Portugal, Spain and Malta, and a new civilian town slowly formed.[30] Medical services for the inhabitants were initially supplied by an assortment of practitioners, both civilian and military, with diverse qualification. Pharmacy licenses to trade were granted and drug stores opened from 1809.[31] A new civilian hospital supported by central government funds was opened in 1816 and services were largely free. But patients paid for general medical services, which became very lucrative for practitioners. When ships returning from the Crimean War put into Gibraltar to land their sick and injured, the latter were neglected by the military surgeons at the hospital, 'all their time being taken attending to rich civilians and Jews'.[32] Military and civilian doctors from Britain, Spain, France, and Ireland competed with each other, until eventually civilian practice was restricted to civilian practitioners. The civilian hospital was extensively enlarged and rebuilt from the 1880s and renamed the Colonial Hospital. In the 1960s it was renamed St. Bernard's Hospital.[33]

The first group of health practitioners to be registered in Gibraltar were the doctors. By 1790 a medical registration list had been drawn up, and an official committee was established to examine the abilities and qualifications of individuals seeking admission to the register. The committee consisted entirely of military practitioners, and by 1814 there were ten registered physicians—three British and seven Spanish.[34] But there was no register of chemists and druggists; those wishing to practise pharmacy normally undertook an apprenticeship with one of the doctors. During his visit to Gibraltar in 1894 Silas Burroughs noted that this had been the case for some years:

> Mr Roberts has, as a family relic, the indentures of his father, who served his apprenticeship to a Gibraltar doctor. The doctor promises to teach his pupil all the knowledge and art which he possesses, and his pupil is bound, among other things, not to be out later than nine o'clock in the evening.[35]

This remained the normal route by which pharmacists were trained in Gibraltar until the 1870s.

In the early 1880s, Ordinances were passed implementing British legislation relating to the regulation of pharmacy and the control of poisons, in accordance with the 1868 Pharmacy and Poisons Act. Those wishing to open pharmacies needed to hold the PSGB qualification. There was no school of pharmacy in Gibraltar, and the only option available for taking the PSGB's Minor examination and for joining the register of chemists and druggists was to travel to Britain and attend a school of pharmacy or one of the crammer colleges in London, such as Wills' Westminster College of Chemistry and Pharmacy.[36] Wills should be deleted from the Index entry but not the text Pharmacy in Gibraltar was British; it was subject to British laws and was treated in the same way as other parts of Britain. Professionalization distinct from that in Britain was neither wanted nor necessary.

Pharmacy Practice in Gibraltar

On his visit in 1894 Silas Burroughs was surprised to see so many pharmacies. He wrote to the *Chemist and Druggist:*

> Pharmacy is well represented at Gibraltar, and I doubt if there is any town of 23,000 inhabitants in Europe which has so many well stocked pharmacies, and which do so extensive a trade…The main street, near Mr Roberts's pharmacy, is barely wide enough for two carriages to pass… The pharmacy occupies the entire ground floor of one of the best buildings in Gibraltar, erected on Government land by the founder of the pharmacy, Dr. Roberts.

Gibraltar was popular for chemists and druggists migrating from Britain. There was little to distinguish pharmacies there from those in more affluent areas of England:

> The pharmacy is as thoroughly English as if built and conducted in the 'home' country. There are large ornamental jars in the window, the long counter and show cases filled with all the good things to be found in a chemist's shop in England, with handsome tablets and show cards of the leading English manufacturers. At the side of the pharmacy is a fine garden, now blossoming out with roses.[37]

Gibraltar pharmacies were almost more British than those in Britain. As elsewhere in the Empire they served as potent symbols of colonial social and cultural capital. The customers of the Gibraltar pharmacies, however, were a much more diverse group than the typical clientele of an English pharmacy. Burroughs observed that

> large and varied stocks of goods are necessarily kept, for Gibraltar is an important trade center, not only for Europeans but for the world. Every important nation has its consulate at Gibraltar, and Norway and Sweden, though under the same government, have separate consuls…the Gibraltar pharmacist has customers, therefore, from the ends of the earth, as well as among his fellow townsmen, of varied nationalities.[38]

Burroughs reflected on the level of drug imports, noting that 'although there is a large Spanish population in Gibraltar, but few Spanish chemical and pharmaceutical goods are sold'. He noted that 'the high cost of materials in Spain, owing to an exorbitant tariff, makes goods so dear that manufacturers cannot complete with English goods abroad, and barely do so in their own country'. The Roberts pharmacy nevertheless faced stiff competition: 'Among other popular pharmacies in Gibraltar are those of Patron, Garibaldi's, Dr. Triay, and others'.[39] But pharmacies remained in the hands of individual proprietors; by 1900 there was no sign of company chemists in Gibraltar.

The British in Malta, 1802

More than anywhere else, the distinct status of pharmacy in Malta when the British arrived was the consequence of its long history. Their location in the middle of the Mediterranean meant that the Maltese islands were subject to repeated attack. They were conquered by the Kingdom of Sicily in 1091, and later by the Spanish. But with the defeat of the Byzantine Empire in 1453,[40] much of the eastern Mediterranean area came under Ottoman control, and remained so for over four centuries. In 1522 the Ottomans drove the Knights of the Order of St. John (the Hospitallers) out of Rhodes, dispersing them to their bases across Europe. Fearing that if Rome fell to the Turks Europe would cease to be Christian, the Spanish King handed Malta to the Knights in 1530 in order to protect it from invasion from the south. In 1565 an Ottoman attempt to take the islands was repelled, and for the next 275 years the Knights operated as an all-powerful and autocratic regime. As a vassal state of Sicily, Italian was the official language, and Malta was subject to Sicilian laws and decrees.

In 1798 French troops invaded the island; this time the Knights were expelled and a brief period of French occupation began. But the Maltese rebelled, and the French were themselves expelled in 1800 with British, Neapolitan, and Portuguese assistance. Britain had considered negotiating for the acquisition of Malta from the Knights in 1793, but its capture from the French led to its retention.[41] Its occupation was at first illegal, but Malta subsequently became a British protectorate and a de facto colony in 1813. This was ratified a year later under the terms of the Treaty of Paris in 1814. Malta became an important naval base for Britain, serving as the headquarters of its Mediterranean Fleet. The islands became a self-governing Crown Colony between 1921–1933, between 1947–1958, and between 1962–1964. They became the independent State of Malta in 1964 and a republic in 1974.[42]

Pharmacy Regulation and the State in Malta

Medicine and pharmacy in the hospitals founded by the military Knights were strictly regulated under the Code of Laws laid down by the Grand Master of the Knights Hospitaller.[43] Frederick II's 1240 Edict of Palermo had established clear boundaries between medicine and pharmacy (Chapter 2). With their takeover in 1530 the Knights retained the existing laws regarding the regulation of pharmacy and the control of drugs. Under these, anyone wishing to practise as an apothecary had to be examined by the judicial authorities, consisting of the infirmary physicians and the apothecaries on the island.[44] Prior to examination he had to undertake six months training at the pharmacy of the Holy Infirmary, and to obtain a certificate of proficiency from the Master Apothecary.[45] The first hospital to be established in Malta was the Santo Spirito Hospital in Rabat, where in 1708 Michele Alfort was the apothecary in charge of the pharmacy (Fig. 5.2).[46]

The business of pharmacy was tightly regulated. In order to open a pharmacy, an application for a licence had to be made to the Grand Master, who only gave his assent on the advice of the chief physician, the Protomedicus. The same applied when an established pharmacy was bought or sold, or if a pharmacy was relocated from one district to another.[47] Before 1500, virtually all the pharmacists in Malta were

Fig. 5.2 Restored pharmacy at Santo Spirito Hospital, Rabat, Malta (Courtesy of Heritage Malta)

Sicilian, but with the arrival of the Knights they were increasingly replaced by Maltese nationals.[48] By November 1772 there were thirty apothecaries including those at the Holy Infirmary.[49] In 1782 Valletta had six pharmacies, and many villages had their own.[50] Most opened between 1760 and 1797, before the French occupation.[51] The inhabitants of Gozo were served by the pharmacy of the St. John the Baptist Hospital and by a private pharmacy set up in 1772.[52] Very often the pharmacies passed from father to son or from owner to assistant.[53]

All imported drugs were inspected by the Protomedicus at the Custom House before they could be withdrawn.[54] The apothecary was not allowed to sell medicines without a doctor's prescription, and poisons such as arsenic and mercury were to be kept in a locked cupboard.[55] A Grand Master's decree in 1624 prohibited the sale of 'aromatic substances' and medicines by anyone not in possession of a licence. Stock mixtures and medicines had to be marked on the lid of their container with the date on which they were compounded.[56] Once a year, every pharmacy on the island was inspected by the Protomedicus. Deteriorated drugs were destroyed and the apothecary fined. In 1638 punishment was extended to customers who were unfortunate enough to have bought their items from an apothecary whose drugs were bad.[57]

Responsibility for the quality of medicines passed from medicine to pharmacy in the mid-seventeenth century. Decrees in 1640 permitted trading in drugs and medicaments only after approval had been given by the chief pharmacist of the Holy Infirmary. But the selling price of remedies was fixed by the Protomedicus, and overcharging by apothecaries was made a punishable offence.[58] Pharmacy regulations issued prior to 1724 were eventually embodied in a Code of Laws issued by the Grand Master; they were confirmed in 1784 and by an Edict in 1797. When the French arrived in 1798 'the commander whom Napoleon left behind immediately started the task of introducing the new institutions which were the result of the French Revolution, abolishing all legislative enactments which had been gradually introduced by the Grand masters and to which the Maltese had long adapted themselves'.[59] However, no laws directly affecting the practice of pharmacy in Malta were passed during the period of French occupation.[60]

With its status as a colony confirmed in 1814, English pharmacists were soon attracted to Malta to open pharmacy businesses. The 'Original English Dispensary' was established in 1816 in Valletta, and the 'English Dispensary' opened in 1833.[61] Others followed in subsequent years;

Woolley's 'New British Pharmacy' was active in 1843, and Kingston's 'English Dispensary' in 1859; both were owned and managed by British pharmacists.[62] Kingston targeted British servicemen and their families through newspaper advertisements in which he emphasized that he was an associate of the PSGB.[63] By 1899 Kingston's dispensary had its own dentistry department.[64] But the opening of English pharmacies seriously impacted those run by Maltese pharmacists. By 1850 the number of pharmacies had dropped to twenty-eight, mainly in Valletta, Senglea, Bormla, and Borgu; the private pharmacy in Gozo had closed. By mid-nineteenth century local pharmacies were advertising in newspapers for items including toiletries, cosmetics, and photographic requisites.[65]

Yet more English pharmacies opened in the later decades of the nineteenth century, including the 'Original English Pharmacy' in 1876, Berg's 'English Dispensary' in 1881, and the 'British Central Pharmacy' in 1899. Pharmacies opened elsewhere in Malta, some targeting mixed populations. An 'Anglo-Maltese Dispensary' opened in Sliema in 1881, and the 'English Pharmacy' appeared in 1896 at Bormla, followed in 1914 by the 'Cottonera English Dispensary'.[66] Continuing emigration from Britain and Ireland boosted business; in 1911 there were about 1000 Irish nationals in the Maltese islands.[67] 'Englishness' rather than 'Britishness' was important; in a newspaper advertisement the proprietor of the Collis and Williams Pharmacy in Valletta 'respectfully informed newcomers to Malta' that the pharmacy was run by English chemists and that its clientele would be 'attended by Englishmen only'.[68] The pharmacy also claimed to be the 'cheapest place in Malta compatible with good quality'. As in Gibraltar, English pharmacies were little different from those in England.

The Erosion of Pharmacy Professionalism, 1814

Following the British occupation in 1802, Physicians for the Poor were appointed to pay domiciliary visits to poor sick women living in Valletta.[69] The early years of British rule witnessed a rise in the cost of living. The wages of all hospital employees were increased, and the salaries of medical officers in the Valletta hospitals were doubled.[70] By the early 1830s English military medical officers had become heavily engaged with the health of the wider community. A suggestion to establish a Public Dispensary to serve the local poor was made by John Davy, the Inspector General of Army Hospitals, in 1831. Davy's offer to donate surplus equipment from the army's general pharmacy to the new dispensary was gratefully accepted by the Permanent Committee.[71] The dispensary was staffed by a paid apothecary, two physicians, and two surgeons who gave their services free. By 1833 some 22,000 people had been seen at the dispensary.[72]

Once British occupation had been legalized, steps were taken to bring local regulations regarding the control of medical and pharmaceutical practice into line with British ones. A Proclamation in 1814 stated that requests to be licensed as a physician, surgeon, apothecary, or midwife were, in future, to be made to the Governor, or in his absence, the Lieutenant Governor—not the protomedicus, although applications were to be accompanied by a certificate signed by the protomedicus attesting to the individual's good character.[73] Transfer of responsibility for the registration of practitioners from Maltese to British authorities was completed in 1821. A Medical Board was established, and the certificate previously signed by the protomedicus was in future to be issued by the Board. Some changes to pharmacy practice were made; in 1816 an Order was passed to ensure that uniform weights and measures were used in making medicines across the islands.[74]

For local Maltese pharmacists, British occupation heralded a steady erosion of long-held privileges and status. The gradual shift towards the British model of pharmacy favoured British pharmacists over Maltese ones. In 1827 a law was passed stating that no pharmacist could be appointed to the hospital dispensary unless he could read, write, and speak the English language.[75] In 1829 a new law announced in the *Malta Government Gazette* established a modified trial by jury in certain cases. In the schedule listing exemptions from jury service, no mention was made of pharmacists, although other accepted professional groups (including advocates, priests, physicians, and surgeons) were exempt.[76] This was a reversal of the previous situation in Malta, when pharmacists were recognized as independent professionals, and exempted from both military and jury service.[77]

Education, Examination, and Registration

The British arrival marked a significant shift in the education, examination, and registration of pharmacists in Malta. Under the Knights, pharmacy education had been centred on the hospital. A course in pharmacy started in 1676 soon after the School of Anatomy and Surgery was founded at the Holy Infirmary. The first director gave lectures on chemical composition and pharmaceutical preparations.[78] In 1729 the Infirmary's chief pharmacist was appointed as Director of the School, reflecting the equal status of pharmacy with medicine.[79] Pharmacy education was strengthened: candidates had to pass the final examination, work for six months in the pharmacy of the Holy Infirmary, and obtain the approval of its chief pharmacist (not the protomedicus). In 1771 the School of Anatomy and Surgery became part of the University of Malta, and the pharmacy course was held mainly in the Holy Infirmary dispensary. Successful candidates were granted a diploma in pharmacy by the university, signed by the protomedicus in the name of the Grand Master.[80]

When the British authorities established a Medical Council in 1821 the duty of issuing certificates was devolved to them. But pharmacy education remained the responsibility of the Faculty of Medicine and Surgery at the University. In 1835 the Medical Council specified the subjects to be examined, and in 1838 the university included the pharmacy course in its statutes.[81] To be admitted, students had to meet the same educational standards required for entry to the Faculty of Medicine and Surgery. But in 1887 the entrance requirement for pharmacy was downgraded; applicants had only to pass the entrance examination in subjects required for admission to the Faculty of Arts. The course was reduced from three years to two, and in 1889 training in practical pharmacy was extended across both years. Teaching was undertaken in English. The course was, however, still closer in content and duration to the British pharmaceutical chemist qualification than the chemist and druggist diploma. The Maltese qualification was accepted for pharmacy practice in Malta.

For those wishing to practise pharmacy, another option was to qualify in Britain. Between 1894 and 1906 several Maltese citizens studied at the PSGB's school of pharmacy in London or attended one of the crammer schools that specialized in preparing people to take the Society's qualifying examinations.[82] Of the 73 Maltese pharmacists in business on the island in 1910, twelve were members of the PSGB. Aspiring pharmacists continued to go to Britain to be trained well into the twentieth century.[83] At Malta University the pharmacy department became part of the Faculty of Medicine and Surgery, and the pharmacy course was integrated with the medical course during the 1930s. Pharmacy and medical students studied the same basic subjects during the first three years of the course.[84] Teachers were mainly recruited from Britain; in 1947 the Colonial Office was involved in the appointment of a temporary demonstrator in pharmacy at the University.[85] The course leading to the pharmaceutical chemist diploma was finally abolished in 1949, and replaced by the bachelor of pharmacy degree, with the course extended to four years.[86]

Further steps were taken to bring pharmacy practice in Malta into line with that in Britain at the start of the twentieth century. A new framework of legislation for health and medicine was set out in the Second Sanitary Ordinance; Chapter 4 laid out the basic for regulating the practice of pharmacy.[87] In 1900 Maltese pharmacists took initial steps in working together to protect their professional rights and privileges. In that year a Malta Chamber of Pharmacists was founded 'to maintain the dignity and rights of the profession, and to protect the common interests of apothecaries'.[88] In 1903 the Chamber was given formal legal status following an Ordinance issued by the Council of Government. Over the course of one hundred years the legacy of Palermo in Malta had been replaced by the British model of pharmacy.

Health Services in Cyprus

Before its occupation by Britain, Cyprus had been part of the Ottoman Empire for over four hundred years. The Ottomans initially raided it soon after the defeat of the Byzantine Empire in 1453; it became part of the Ottoman Empire in 1571.[89] But in 1878, as a result of the Cyprus Convention, the United Kingdom took over the government of Cyprus as a protectorate from the Ottoman Empire. In 1914, at the beginning of the First World War, Cyprus was formally annexed by the United Kingdom; and in 1925, following dissolution of the Ottoman Empire, it was made a Crown Colony. The British presence was to last over eighty years; the independent Republic of Cyprus was declared in 1960.[90]

Although Ottoman cities like Constantinople, Beirut and Damascus were developing rapidly during the nineteenth century, the same was not true of Cyprus. On arrival in 1878 the British authorities were immediately concerned about the situation that confronted them.[91] They painted a depressing and consistent picture. Sir Garnet Wolseley, the British High Commissioner in Cyprus, noted that Larnaca 'resembled a quarantine full of dirt and fever', stating that if cholera hit the inhabitants it would decimate the population.[92] Physician William Steele, in charge of medical services in Limassol, described the situation in 1879: 'The streets were filled with human waste and other impurities, and wells were not separated from adjacent toilets'.[93] The provincial medical officer in Famagusta, W.H. McNamara, described sanitary conditions as 'deplorable', noting that 'malaria decimated the people'.[94] Leprosy was also a problem.[95] The chief medical officer, Frederick Heidenstam, noted that sanitary conditions in most private homes were 'not at all satisfactory'.[96]

In Turkey at the time medical and pharmaceutical services were well developed, but such developments did not reach outposts of the Ottoman Empire such as Cyprus. Dimitrakopoulos notes the total lack of adequate health institutions and health professionals on the island during Ottoman rule; there were no organized health services and no effective medical or nursing care available for the residents.[97] A traveller in Cyprus noted the 'total lack of state medical infrastructure during the Turkish occupation'.[98] The only health care facilities were the infirmary of the monastery of nuns in Larnaca, the Leper Farm, and the quarantine facilities operated at the two ports. But during the British period health services were developed rapidly.[99] Military physicians were appointed to each of the six provinces, and the colonial government established outpatient dispensaries and pharmacies for the examination of patients and the supply of medicines. The first of these opened in 1878 in Nicosia. Within a year, dispensaries had opened in Limassol, Kyernia, Larnaca, Famagusta, and Paphos.[100] In 1892 British nurses were appointed at Nicosia General Hospital, and by 1901 inpatient hospitals existed in all six provinces.

Pharmacy in the Ottoman Empire, 1820

The lack of development in Cyprus was little different to that elsewhere in the Ottoman Empire. Throughout the Empire there was an almost total lack of diffusion of innovation from centre to periphery.[101] Yet the process of modernizing medical and pharmaceutical education in the Empire had begun in the 1820s and 1830s. Surgical education began in 1827, and an Imperial Military School of Medicine was founded in Constantinople in 1839.[102] The new school was established to supply doctors for the army. Most of the students as well as the instructors were non-Muslims, and included Armenians, Arab Christians, Bulgarians, and Greeks.[103] Pharmacy education began when pharmacy classes were offered at the School from 1839.[104] In the absence of Turks qualified to teach pharmacy, foreign teachers were recruited. The first was a French chemist who was brought from Paris to run a three-year pharmacy course. With no books or journals in Turkish, the teaching was undertaken in

French, a pattern that continued until the 1870s.[105] Further expansion of medical and pharmacy education occurred in the second half of the nineteenth century; a Civil School of Medicine followed in 1867, and a Military Health School in 1876.[106]

Legislation to regulate civilian pharmacy in the Ottoman Empire was first passed in 1852. But in the absence of adequate inspection and enforcement, the practice of pharmacy was effectively open to anyone, including large numbers of quacks and empirics. As a result of numerous accidents and deaths, in 1861 the Imperial Medical School was asked to formulate regulations for the admission of persons to medicine and pharmacy, which were later approved by Imperial Decrees.[107] The pharmacy decree concerned the general practice of 'the art of pharmacy' and applied to the whole of the Ottoman Empire.[108] It provided several routes by which a pharmacist could be judged qualified. Applicants wanting to set themselves up as pharmacists anywhere in the Empire could obtain the 'master in pharmacy' degree of the Imperial School of Medicine, or they could obtain such a qualification from a European university. Alternatively, candidates could submit themselves to oral examination; those fulfilling one of the criteria could then seek the permission of the medical school to open a pharmacy. Pharmacy courses were later run at universities elsewhere in the Empire; in 1903 the diploma awarded by the Faculty of Medicine and Pharmacy in Beirut was recognized, and in 1908 a School of Pharmacy was founded in Damascus. The teaching at both was given in French and delivered by French nationals. A pharmacy diploma was also awarded by the American Faculty of Medicine in Beirut after two years study, without requiring any practical experience of pharmacy, but this was not recognized by the Ottoman government.[109]

The Ottoman pharmacy model was to establish municipal drug stores under the supervision of a qualified pharmacist in most towns where a governor or vice-governor resided. The costs of these and of prescriptions for the poor were certified by the city physician. The stores also sold drugs direct to the public. Pharmaceutical preparations were mainly made according to the French pharmacopoeia, *Pharmacopée Française*.[110] But enforcement was lax; a French pharmacist reported that

> the authorities are not strict in enforcing the pharmacy law, and on payment of certain sums periodically, they tolerate the sale of all sorts of drugs – indigenous or exotic, but often dangerous – in the 'drug shops', and particularly in the bazaars.[111]

European pharmacists who settled in the Ottoman Empire faced formidable obstacles. The French correspondent noted that

> in these countries the Mohammedan population prefers to have recourse to the art of the [indigenous] druggists rather than to the Turkish or Arab pharmacists...They are superstitious in the extreme, and prefer their own Arab remedies to all others.[112]

Halil Tekiner notes that 'public opinion in those days was that the pharmacy profession was only suitable for either non-Muslim or minority background pharmacists'.[113] As a result many Ottoman pharmacies were owned by pharmacists from European countries. A 'British' pharmacy was opened in Istanbul in 1833 by an Italian pharmacist, Noel Canzuch, in an area of commercial shops and eleven embassies, including the British Embassy. It was later run by his sons.[114]

Pharmacy in British Cyprus, 1878

Whilst Western medicine was established in state-run facilities, the same was not true elsewhere. When the British arrived in Cyprus there were no Western pharmacies, whether run by Europeans or Turkish pharmacists. British pharmaceutical qualifications were recognized in Cyprus from the start of the occupation, although few British chemists and druggists chose to move there, despite the *Pharmaceutical Journal* reporting in 1915 that 'the certificate of the Pharmaceutical Society of Great Britain was accepted for registration in most colonies including Cyprus'.[115] For the great majority of the local population, spice sellers were the main source of remedies.

By 1907 a Turkish physician was examining patients in a Greek-owned pharmacy in Nicosia. Turkish pharmacists only began opening pharmacies in Cyprus after 1912, the first being in Nicosia and Paphos. Many pharmacies were owned by Ottoman subjects other than Turks including Greeks and Armenians. A second Greek pharmacy opened in Nicosia in 1909. In 1912 in Lefka the colonial authorities employed a Turkish pharmacist who had graduated in Cyprus, having completed the 'Special Pharmacy Education Programme of the Cyprus Government', possibly having passed the PSGB's Minor examination.[116] However, most Turkish pharmacists graduated from the school in Istanbul, returning to Cyprus later.

Pharmacy in Cyprus during the period of British rule was thus a very marginal activity. The British brought with them the same expectations they took to other colonies. A great deal of health and medical legislation was passed from 1879 onwards, with new statutes established concerning medicine and public health.[117] New laws regulated the practice of pharmacy in 1900, but they applied to no more than a handful of British pharmacists employed in state institutions.[118] Other laws provided for medical and pharmacy registration in 1907.[119] But few changes were made before passage of a Pharmacy and Poisons Law in 1945.[120] It was 1933 before a British pharmacy was established in Cyprus, and a Cyprus Pharmaceutical Association was founded only in 1954.

The Legacies of Empires

Britain's Mediterranean colonies had all previously been subjected to repeated occupation by a variety of other powers, and in all the initial British presence was a military one. Both these factors played important parts in the subsequent development of pharmacy. In Gibraltar pharmacy was initially shaped by the military presence; in Malta the British found that the medical and pharmacy professions had been separate and distinct for centuries; and in Cyprus they found that pharmacy had not been developed at all. In each colony the actions taken by the British authorities were directed at imposing the British model of pharmacy whilst taking account of local circumstances.

This was largely achieved by the passage of legislation and direct state action through colonial authorities. Yet in none was collective action taken by British pharmacists to escape control by the doctors; in none were Pharmacy Boards established where pharmacists were responsible for their own regulation; and only in Malta were arrangements made for the education of pharmacists locally. When judged against the five markers of professionalization, it is clear that pharmacy in the Mediterranean colonies made little progress during British occupation, and in Malta it took several steps backwards. Whilst pharmacy operated in the shadow of the PSGB, it remained under the overall supervision of the medical authorities; and in Cyprus at least its impact extended little beyond the British and European enclave.

In pursuing the British model in Malta the actions taken undermined an established and functional arrangement and represented a gradual but very real erosion of the professional status of the pharmacist. Pharmacy and medicine had been separate and distinct professions for many years and were accorded equal status, a legacy of the 1240 Edict of Palermo. Medicines too were tightly regulated; the regulation of pharmacy was already linked to the control of poisons. British legislation simply replaced existing legislation. In Cyprus the lack of modern pharmacy was a legacy of Ottoman rule; and in Gibraltar the British model of pharmacy was largely transplanted unchanged. British medicines were exported everywhere across the Empire. For prescribed medicines the *British Pharmacopeia* was adopted as the official pharmacopoeia in all colonies, and the range of drugs available became increasingly uniform.

In Malta 'Britishness' was discernable in a variety of explicit and often subtle ways.[121] Perhaps the most obvious was the requirement that pharmacists needed to be able to read, write and speak English. Until the nineteenth century the pharmacist in Malta was referred to in Italian as 'aromatario' or 'farmacista', or possibly 'chimico-farmacista'. These titles were gradually replaced by English ones, from apothecary and chemist and druggist to dispenser and pharmaceutical chemist.[122] Premises were originally called by their Italian name of 'farmacia', but from about the 1830s they began to be called 'dispensary' or 'pharmacy'. The names of pharmacy shops and the titles used by pharmacists reflected the growing number of British pharmacists on the islands.

There is little evidence of inter-colonial cooperation between pharmacists across the British Mediterranean Division, or that the existence in Malta of pharmacy as a profession separate and distinct from medicine had any impact on thinking in Britain. The difference in status of pharmacists in Britain and Malta is perhaps neatly illustrated by Cassar's observation that well into the twentieth century pharmacies in Malta were

> the rendezvous of small circles of cultured and professional people who gathered there in the evening, so that it was a common sight to see the apothecary, the doctor, the lawyer, the notary and the priest engaged in a dignified discussion of the topics of the day.[123]

Notes

1. Duffy, Michael. 'World-Wide War, 1793–1815', in Marshall, P.J. (ed.) *The Eighteenth Century: The Oxford History of the British Empire* (Oxford, 1998), p. 196.

2. Ibid., p. 197.
3. Hudson, Geoffrey L. *British Military and Naval Medicine,1600–1830* (Amsterdam, 2008). Chakrabarti, Pratik, *Medicine and Empire 1600–1960* (London, 2014), pp. 40–56.
4. Montegriffo, Cecil. 'History of Medicine in Gibraltar', *British Medical Journal*, 2 (1978), p. 553.
5. Martin, Robert Montgomery. *History of the British possessions in the Mediterranean: Comprising Gibraltar, Malta, Gozo, and the Ionian Islands* (London, 1837), p. 95.
6. Montegriffo, 'History of Medicine in Gibraltar', p. 554.
7. Matthews, L.G. *History of Pharmacy in Britain* (London, 1962), pp. 189–205. Cowen, David L., 'Notes on Hospital Pharmacy in the Royal Navy in the Eighteenth Century', *Bulletin of the American Society of Hospital Pharmacists*, 13 (1956), pp. 568–75.
8. Hunting, Penelope. *A History of the Society of Apothecaries* (London, 1998), p. 153. O'Leary, Charles, 'The Elaboratory and Stocks of the Society of Apothecaries', *Pharmaceutical Historian*, 27 (1997), pp. 14–20.
9. A repository for medical supplies from Apothecaries' Hall was established at Lisbon in 1708. Caird Library, Royal Museums Greenwich, 4 November and 21 January (1708), ADM/E/2.
10. Hunting, *Society of Apothecaries*, p. 169.
11. Ibid.
12. Matthews, L.G. 'London's Immigrant Apothecaries, 1600–1800', *Medical History*, 18 (1974), p. 268.
13. Blair, J.S.G. 'Sir John Pringle, 1707–1782', in *Oxford Dictionary of National Biography* (Oxford, 2020). Gordon, Charles. 'Sir John Pringle and the Apothecaries', *Pharmaceutical Historian*, 19 (1989), pp. 5–12.
14. Gordon, Charles, 'Sir John Pringle', p. 8.
15. Cowen, David L. 'The Edinburgh Pharmacopoeia', *Medical History*, 1 (1957), pp. 123–39.
16. Selwyn, Sydney. 'Military and Naval Medicine', *Pharmaceutical Historian*, 3 (1973), p. 5.
17. Cowen, David L. *Pharmacopoeias and Related Literature in Britain and America, 1618–1847* (Aldershot, 2001), pp. 89–90.
18. Padiak, Janet. 'The Role of Morbidity in the Mortality Decline of the Nineteenth Century: Evidence from the Military Population

at Gibraltar 1818–1899', *Journal of the History of Medicine and Allied Sciences*, 60 (2005), p. 83.
19. *Annual Sick Returns and Reports: Hospitals and Convalescent Depots Abroad: Gibraltar, Malta and Ionian Islands* (January–December 1861). The National Archives (TNA), WO 334/26, 17.
20. *Annual Sick Returns and Reports: Hospitals Abroad: Gibraltar, Malta, Cyprus and Egypt* (January–December 1889). TNA, WO 334/91.
21. Lovegrove, Peter. *Not Least in the Crusade: A Short History of the Royal Army Medical Corps* (Aldershot, 1955). Also 'Pharmacy in relation to the RAMC', *Pharmaceutical Journal*, 151 (1943), pp. 182–3.
22. Cowen, 'Notes on Hospital Pharmacy in the Royal Navy', p. 568.
23. Matthews, *History of Pharmacy*, pp. 192–3.
24. Ibid., p. 194.
25. Trease, G.E. *Pharmacy in History* (London, 1964), p. 182.
26. Brewer, John. 'Dispensers in the Crimea', *Pharmaceutical Journal*, 15 (1855) pp. 187–8.
27. Holloway, S.W.F. *Royal Pharmaceutical Society of Great Britain 1841 to 1991: A Political and Social History* (London, 1991), p. 189.
28. Matthews, *History of Pharmacy*, p. 199.
29. Hills, George. *Rock of Contention: A History of Gibraltar* (London, 1974). Jackson, William G.F. *The Rock of the Gibraltarians* (Cranbury, 1986).
30. Montegriffo, 'History of Medicine in Gibraltar', pp. 552–4.
31. Pharmacy: Licences to Trade, 1809, TNA, PC 1/3882.
32. Montegriffo, 'History of Medicine in Gibraltar', p. 555.
33. Benady, Sam G. *Civil Hospital and Epidemics in Gibraltar* (Grendon, 1994), p. 132.
34. Montegriffo, 'History of Medicine in Gibraltar', p. 553.
35. Burroughs, S.M. 'Pharmacy at Gibraltar', *American Druggist and Pharmaceutical Record* (1894), pp. 122–3. *Chemist & Druggist*, 45 (1894), pp. 156–7.
36. Wills, G.S.V. *A Jubilee Souvenir: The Work of GSV Wills and the Westminster College of Chemistry and Pharmacy* (Stratford-upon-Avon, 1899), p. 198.
37. Burroughs, 'Pharmacy at Gibraltar', pp. 122–3.

38. Ibid., p. 123.
39. Ibid.
40. Quataert, Donald. *The Ottoman Empire, 1700–1922* (Cambridge, 2005), p. 4.
41. Duffy, 'World-Wide War, 1793–1815', p. 197.
42. Holland, Rose, J. (ed.) *A History of Malta during the Period of the French and British Occupations 1798–1815* (Malta, 1994), pp. 74–104. Attard, Joseph. *Britain and Malta: The Story of an Era* (Malta, 1988).
43. Cassar, Paul. *Medical History of Malta* (London, 1964), p. 488.
44. National Library of Malta, Valletta (NLM), ms 153, fol.73t.
45. NLM, ms 553, fol.180.
46. Cassar, *Medical History*, p. 28.
47. NLM, ms 1195, fol.19 and ms 1196, fol.115.
48. Borg, John Joseph. 'Pharmacy in Malta from the Late Fifteenth Century Until the Mid-Sixteenth Century', *Pharmaceutical Historian*, 31 (2001), p. 2.
49. Borg, J. 'Pharmacists of Old', *The Pharmacist: Journal of the Chamber of Pharmacists*, 4 (1982), pp. 14–19 and 5 (1983), pp. 16–23.
50. Cassar, P. 'Pharmacies and Apothecaries One Hundred Years ago in Malta', *Journal of the Malta Union of Pharmacists*, 1 (1967), p. 28.
51. Cassar, *Medical History of Malta*, p. 509.
52. NLM, ms 1192, fol.39.
53. NLM, ms 1194, fol.7, ms 1195, fol. 77; ms 1196, fol.23.
54. NLM, ms 1190, fol.12 and 292.
55. NLM, ms 153, fol.73t.
56. NLM, ms 2, fol.601.
57. NLM, ms 149, fol.54–55.
58. NLM, ms 148, fol.80.
59. Borg, G. 'The Influence of the Laws of England on Maltese Legislation', *Scientia*, VIII (1942), p. 51.
60. Hardman, W. 'The French Government of Malta', in Holland, Rose, J. (ed.) *A History of Malta during the Period of the French and British Occupations 1798–1815* (Malta, 1994), pp. 74–104. Also Borg, John Joseph, 'Pharmacy in Malta under the French,' *Pharmaceutical Historian*, 32 (2002), pp. 2–6.

61. Cassar, Paul. 'Impact of British Pharmacy in Malta', *Pharmaceutical Historian*, 8 (1978), p. 2.
62. Ibid., p. 3.
63. Trease, *Pharmacy in History*, p. 191.
64. Cassar, 'Impact of British Pharmacy', p. 3.
65. *il Mediterraneo* (6 April 1853), p. 12; *il Mediterraneo* (20 April 1853), p. 12; *The Malta Times* (28 May 1863), p. 4.
66. Cassar, 'Impact of British Pharmacy', p. 2.
67. Fitzpatrick, David. 'Ireland and the Empire', in Porter, Andrew (ed.) *The Nineteenth century. The Oxford History of the British Empire* (Oxford, 1999), pp. 512–3.
68. *The Malta Times* (26 April 1895), p. 1. Also Evans, G.H., *Guide to Malta and Gozo* (Malta, 1900).
69. Medical and Health Archives, Valletta. *Registro Decreti Lettere*, fol.13. Cited in Cassar, *Medical History of Malta*, p. 348.
70. Cassar, *Medical History of Malta*, p. 490.
71. Resolutions of the Permanent Committee from 3rd March 1831 to 23rd June 1831, fol. 54, Medical and Health Archives, Valletta, Malta.
72. *Malta Government Gazette (MGG)*, 1215 (1832), p. 52.
73. *MGG*, 36 (1814), p. 143.
74. *MGG*, 146 (1816), p. 625.
75. *MGG*, 868 (1827), p. 247.
76. *MGG*, 976 (1829), p. 329.
77. Borg, *Pharmacy in Malta*, p. 74.
78. Borg, John Joseph. 'Teaching Pharmacy in Malta 1676–1990s, Part II', *Pharmaceutical Historian*, 44 (2014), pp. 24–6.
79. Cassar, *Medical History of Malta*, p. 54.
80. Ibid., p. 497.
81. University of Malta Archives. Ms No. 11 (1833–38), fol. 4.
82. Trease, *Pharmacy in History*, pp. 190–1.
83. Cassar, 'Impact of British Pharmacy', p. 3.
84. Borg, John Joseph. 'Teaching Pharmacy in Malta 1676-1990s, Part I', *Pharmaceutical Historian*, 44 (2014), p. 68.
85. 'Demonstrator in Pharmacy', TNA, FCO 141/9968.
86. Borg, 'Teaching Pharmacy Part I', p. 69.
87. *Malta Government Gazette Supplement*, 30 May and 2 July (1901). See also 'Medical and Kindred Professions Ordinance', in *Laws of Malta* (Valetta, 1942).

88. *Malta Government Gazette*, 22 May 1902, p. 511.
89. Quataert, *Ottoman Empire*, p. 4.
90. Mallinson, William. *Cyprus: A Modern History* (London, 2005). Tofallis, Kypros, *History of Cyprus: From Ancient Times to the Present* (Cambridge, 2002).
91. Georgiades, A. *The History of Cyprus Medicine and Nursing during the British Colonization 1878–1960* (Nicosia, 2001). Dimitrakopoulos, Ioannis et al., 'Healthcare Services in Cyprus during the Ottoman Period 1571–1878', *International Journal of Caring Sciences*, 8 (2015), pp. 747–55.
92. Cavendish, A. *Cyprus 1878: The Journal of Sir Garnet Wolseley* (Nicosia, 1991), p. 749. Cited in Dimitrakopoulos et al., 'Healthcare Services in Cyprus'.
93. Georgiades, *Cyprus Medicine*, p. 201.
94. McNamara, W.K. *Topographical and Medical Report of the District of Famagusta of the Year 1878–1879* (Nicosia, 1879). Cited in Dimitrakopoulos et al., 'Healthcare Services in Cyprus'.
95. Heidenstam, F. *Report on Leprosy in Cyprus* (London, 1890).
96. Dimitrakopoulos et al., 'Healthcare Services in Cyprus', p. 747.
97. Ibid., p. 755.
98. Cited in Ohnefalsch-Righter, Magda. *Greek Customs in Cyprus* (Nicosia, 2012).
99. Dimitrakopoulos, I. and Sapountzi-Krepia, D. 'Historical Review of Nursing in Cyprus during the Ottoman Period (1571–1878) and the British Period (1878–1960)', *Cyprus Nursing Chronicles Journal*, 17 (2017), p. 29.
100. Ibid., p. 27.
101. Kasaba, Resat. *The Ottoman Empire and the World Economy: The Nineteenth Century* (New York, 1988).
102. Trompoukis, C. and Lascaratos, J. 'Greek Professors of the Medical School of Constantinople during a Period of Reformation, 1839–76'. *Journal of Medical Biography*, 11 (2003), pp. 226–31.
103. Strauss, J. 'Language and Power in the Late Ottoman Empire', in Murphey, R. (ed.) *Imperial Lineages and Legacies in the Eastern Mediterranean: Recording the Imprint of Roman, Byzantine and Ottoman Rule* (London, 2016), p. 196.
104. Tekiner, Halil. 'A Brief History of the British Pharmacy in Istanbul', *Pharmaceutical Historian*, 36 (2006), p. 64. See also Yavuz,

Mustafa, *Medicine and Pharmacy Education in the 19th Century Ottoman Empire* (Istanbul, 2017).

105. Soyalan, M. et al. *Pharmacy and Pharmacy Profile in the Turkish Republic of Northern Cyprus* (Cyprus, 2013), pp. 5–8.
106. Tekiner, 'British Pharmacy in Istanbul', p. 66.
107. Marcailhou-Daymeric, H. 'Pharmacy in Asiatic Turkey', *American Druggist and Pharmaceutical Record*, 42 (1903), p. 143.
108. Tekiner, H. and Ulu, A. 'The Historic Evolution of Turkish Pharmaceutical Legislation from 1852 to the Present', *Farmacia*, 63 (2015), pp. 619–22.
109. Marcailhou-Daymeric, p. 143.
110. Baytop, T. *Turk Eczacilik Tarihi Arastirmalari* (Istanbul, 2000), p. VIII.
111. Marcailhou-Daymeric, p. 143.
112. Ibid., p. 143.
113. Tekiner, 'British Pharmacy in Istanbul', p. 64.
114. Ibid.
115. *Pharmaceutical Journal*, 39 (1915), p. 268.
116. Soyalan, *Northern Cyprus*, pp. 5–8.
117. Dimitrakopoulos and Sapountzi-Krepia, pp. 25–30.
118. *The Cyprus Gazette*, 48 (1925), p. 90.
119. Ibid., 30 (1907), p. 6118.
120. 'Pharmacy and Poisons Law 1945', TNA, CO 67/330/14.
121. Cassar, 'Impact of British Pharmacy', pp. 2–4.
122. Ibid., p. 2.
123. Cassar, *Medical History of Malta*, p. 512.

CHAPTER 6

West Africa: The Scramble for Professionalization

Post-independence studies of professionalization in West Africa have suggested that the model is of limited value because of the very different circumstances compared to those in Europe and North America. Johnson, for example, suggests that the professions were too dependent on the state to be able to form effective autonomous power bases. Dependence, he argues, was a consequence of the scale of patronage of colonial authorities for professional services: 'in the absence of a heterogeneous middle class providing sources of demand for professional services, the conditions for professional autonomy are also absent'.[1] Robin Luckham in his study of the Ghanaian legal profession highlights the role of British laws and the transplantation of the British legal system on developments there.[2]

In his analysis of the professionalization of pharmacy in Ghana, Paul Bennell demonstrates the value of taking a historical perspective, and the importance of focussing on the nature of a profession's relationships with other occupational groups and their role in the educational process.[3] For pharmacy, the role of the state, the passing of laws, the relationship with the medical profession, the working together in professional associations, the development of education, and the establishment of hierarchies associated with differences in status and remuneration, are so closely interconnected in West Africa that they are best considered as a chronological progression.

S. Anderson, *Pharmacy and Professionalization in the British Empire, 1780–1970*, Medicine and Biomedical Sciences in Modern History,
https://doi.org/10.1007/978-3-030-78980-0_6

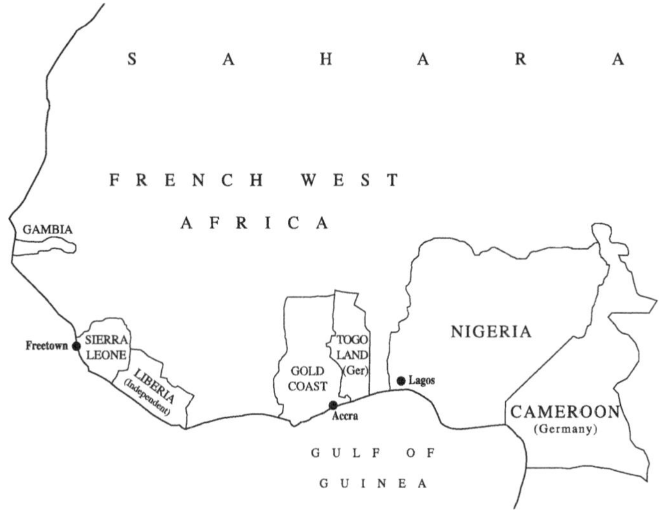

Fig. 6.1 Map of the West African Division of the British Empire, 1914

This chapter explores the process of professionalization of pharmacy in West Africa with particular reference to Nigeria and the Gold Coast (later Ghana). In the early twentieth century Britain had four colonies in West Africa, the others being Gambia and Sierra Leone (Fig. 6.1). Developments in the two colonies closely paralleled each other, although there were differences (Table 6.1). These colonies had relatively small expatriate communities, and they offered few attractions for British chemists and druggists hoping to establish profitable businesses in one of the colonies. British settler chemists and druggists played virtually no part in the evolution of pharmacy in West Africa. The British model of pharmacy could be transplanted unhindered by ex-patriate demands.

The colonial authorities introduced a succession of Pharmacy and Poisons Ordinances to control the issue of poisons and to regulate pharmacy practitioners. But pharmacists remained firmly under the control of the doctors until after independence; pressure for autonomy came not from the PSGB or British pharmacists but from locally trained practitioners. And the escalation of qualifications encouraged by British advisors led to severe shortages of pharmaceutical practitioners and to the reintroduction of lower-level qualifications, and the creation of new

Table 6.1 Professionalizing pharmacy in West Africa

Colony	*Year British colony*	*Year Independent*	*Pharmaceutical society or association*	*Early pharmacy legislation*	*Early poisons legislation*	*Pharmacy education*
Gambia	1766	1965	None during British period	Ordinance	Ordinance	None during British period
Gold Coast	1874	1957	Gold Coast Pharmacist and Druggist Union 1929. Chemist Defence Union 1929 Pharmaceutical Society of the Gold Coast (PSGC) 1935	Ordinance	Drug and Poisons Ordinance 1892	Dispensing School at Korle-Bu Hospital, Accra, 1927 Relocated to Kumasi College of Technology, 1953
Nigeria	1865	1963	Association of Dispensers 1927 Nigerian Union of Pharmacists 1947 Pharmaceutical Society of Nigeria (PSN) 1956	Ordinance	Pharmacy and Poisons Ordinance 1902	Dispensers' training school, Yaba, Lagos, 1927 Dispensers' training school, Zaria, 1930 (closed 1940);
Sierra Leone	1808	1961	None during British period	Ordinance	Ordinance	None during British period

boundaries between pharmacists and subordinate staff. The legacy of British pharmacy in West Africa was distinctly ambivalent.

"From Prayers to Pills": Pharmacy, Medicines, and Missionaries

The first exposure to Western medicine for most of the indigenous population came at mission stations.

Before the arrival of the first European missionaries, few white men ventured to West Africa. In Britain the region had a formidable and entirely justified reputation as the 'White man's grave'.[4] This meant that in the early nineteenth century few Europeans willingly went there. The deadly pathological environment meant that people were wary of going to West Africa to start businesses such as pharmacies, which usually relied heavily on a substantial European expatriate community. This only developed in the last decades of the nineteenth century.

The main practitioners involved in healing the sick were traditional healers, who used clay, plants, and animal parts, or supernatural means (or combinations of these) to bring about the healing of their patients. Nigerian medical knowledge was largely kept by the traditional healers.[5] The arrival of the missionaries represented a serious challenge to them, and they were quick to take any opportunity to discredit the new medicines and to spread rumours about them. Poisons were labelled 'Number 1 medicines'; if a patient was admitted and later died, it was believed that he must have been given a 'number one' medicine. The rumour was very successful in scaring people from attending hospitals, and in encouraging them to go to the traditional healer instead.[6]

Although Africans were usually reluctant to embrace Western medicines, they made exceptions for dramatic results. A chance discovery by missionary doctors was that the external symptoms of yaws—a disfiguring tropical disease—could be quickly removed by injections of bismuth sodium tartrate, led to mass pilgrimages in East Africa.[7] However, Ben Breen cautions against setting a binary between 'African healing' and 'European medicine', noting that the amulets and pouches associated with cures in West and West-Central Africa had close correlation with the use of amulets for medicinal purposes in Europe, as did the use of animal parts believed to have supernatural powers.[8]

A mission was established at Calabar in 1882, and Methodist and Baptist missionaries arrived in Lagos between 1884 and 1888. Independent missions were founded in Lagos and Freetown between 1886 and 1889.[9] More opened in the interior as British commerce and colonial government advanced.[10] The relationship between colonial administrators and missionaries was often a difficult one; in parts of Nigeria Christian missionaries were barred by the colonial authorities.[11] They often brought health services to areas where the colonial government had no effective presence, providing basic medical services even when no health professionals were available. With expansion and increased coverage came improved arrangements for the distribution of medicines through mission facilities.[12] Etherington notes that people who were indifferent or hostile to Christian evangelism often proved to be avid consumers of missionary medical services, which ranged 'from prayers to pills'.[13]

The early twentieth century saw a rapid increase in the provision of medical services in West Africa. Colonial governments built increasing numbers of hospitals and ambulatory clinics. Mission hospitals also increased in number, particularly in the coastal south, and increasing numbers of health professionals—including doctors, nurses, and a few pharmacists—were employed.[14] Many missionaries were also doctors, substantial numbers being trained at the Livingston Medical College in London, founded in 1896.[15] The missionaries developed a system of modern drug distribution that operated extremely efficiently. The system was characterized by a steady supply of drugs and reasonable charges to patients. As a result, there was a consistently high demand for these services.[16]

In 1827 the Church Missionary Society established the Fourah Bay College at Freetown, Sierra Leone, to train missionary teachers.[17] Mission schools played a formative role in the education of Nigerians who went on to play important parts in the development of pharmacy in Nigeria; teaching was delivered in English and schools developed along British lines.[18] Students acquired European political and economic aspirations, and gained leadership and organizational skills that easily transferred to secular roles.[19] The influence of missionary organizations continued into the twentieth century. In June 1958 Jack Rowson, then head of the pharmacy department at Ibadan College of Technology, was asked by the Nigerian Pharmacy Board to visit missionary institutions in Britain in order to 'better advise the Board in deciding whether to grant missionaries permits to practise pharmacy in Mission Hospitals'.[20]

Western Pharmacy in Nigeria, 1887

In 1887 a locally trained dispenser, Richard Zaccheus Bailey (1829–1917), opened a retail pharmacy in Balogun Street, Lagos (Fig. 6.2); it was the first Western retail pharmacy in Nigeria.[21] Bailey had been apprenticed to a British doctor, and obtained his licence from the Governor in Council. He spoke many African languages, and ran a successful business serving European expatriates and a few leading Africans; he was known locally as 'the doctor'. He later trained many dispensers himself through apprenticeship; most were the sons of the Nigerian elite, and some later trained as physicians.[22]

That a local dispenser was the first to open a Western pharmacy in Nigeria illustrates the fact that few if any British chemists and druggists were attracted to the region; prospects were much rosier elsewhere in the Empire. Yet by then Britain already had an established presence. Following the invasion of Lagos in 1851 the British occupation extended further east, west, and north, with the colony being formally annexed in

Fig. 6.2 Richard Zacheus Bailey (1829–1911), 'father' of Nigerian pharmacy (Courtesy of Pharmaceutical Society of Nigeria)

1865. But the new colony offered few attractions for British chemists and druggists, and dispensing medicines was left to the doctors. For colonial authorities advised by the PSGB, Nigeria was an obvious candidate for introduction of the British model of pharmacy. Pharmacy laws and practices developed stepwise in Nigeria; as successive Poisons and Pharmacy Ordinances were passed, pharmacy became ever more aligned with that in Britain.[23] But the British model proved entirely inappropriate, and an autonomous profession was achieved only after independence.

Training Dispensers

The regulation of pharmacy and the control of poisons was not an immediate priority for the colonial authorities; it was only after Nigeria had been a British Colony for 22 years that the first Pharmacy and Poisons Ordinance was passed in 1887.[24] This was intended to control the supply of medicines by issuing licences to those authorized to sell Western medicines; those eligible for a licence were holders of the PSGB certificate and locally trained 'dispensers'. Dispensers were trained through informal apprenticeship provided by European medical doctors residing in Nigeria.[25] There was no dispensing school and no lectures were available; students chose the doctor under whom they wished to serve as apprentice, and went to their house or shop every morning, in some cases living with the master for some years.[26] The training of dispensers was at a considerably lower level than that of chemists and druggists trained in Britain, but in 1899 the authorities reviewed the level of training needed for service in government institutions. Training was formalized, with students sponsored but required to work in government hospitals and dispensaries for a period of five years after qualifying. Candidates undertook a three-year apprenticeship under the direction of medical officers, and were required to learn the composition of mixtures, extracts, decoctions, pills, and tablets then in use.

The first batch of indigenous dispensers under the new arrangements successfully completed their training in 1902.The first Nigeria-born dispenser to be registered with the Colonial authorities on 1 September 1902 for service in government institutions was Emmanuel Caulcrick, who completed his apprenticeship with a British doctor.[27] Once qualified, dispensers often found themselves acting as hospital administrators, laboratory technicians or anaesthetists as well as dispensers, and they were frequently left in charge of the hospital for long periods whilst the medical officer was on tour in the villages and hamlets.[28] On completion of their five years of government service they were free to start their own business; Caulcrick opened a Western pharmacy in Lagos. A new Poisons and Pharmacy Ordinance was passed in 1902 to further regulate pharmacy practice and control the sale and distribution of drugs and poisons. Control was placed in the hands of the Chief Medical Officer, and the training of dispensers was further specified. The Ordinance applied only to the colony of Lagos, the towns of Calabar, Opobo, Warri, Forcados, and a few other places named by the Governor.[29]

Upgrading to the British Qualification, 1930

These arrangements continued until the 1920s, when legislation became necessary for the control of Dangerous Drugs,[30] and there was recognition by the colonial authorities that the training of dispensers needed to be raised. Upgrading of qualifications was ratified with passage of the Poisons and Pharmacy Ordinance in 1927.[31] Two levels of pharmacy qualification were to be established; the existing lower-level Dispenser's Certificate; and a new higher-level Chemist and Druggist Diploma which would more closely mirror the British qualification. The Chemist and Druggist Diploma would entail satisfactory completion of two years full time service and passing the dispenser's examination, followed by a one-year course at a dispensers' training school and the taking of the PSGB's examination. The 1927 Ordinance detailed the standard required by those wishing to register as either chemists and druggists or dispensers.[32] Chemists and druggists would henceforth be described as 'pharmacists', and the dispensers' training school would be designated a school of pharmacy.

Plans were made to open a school of pharmacy in the mid-1920s offering the chemist and druggists course; it was opened at the Yaba Higher College, Lagos in 1927. The superintendent medical officer, Dr. Gordon Taylor, was appointed as its first director; he had served as a sergeant-dispenser during the First World War. He asked the Education Department to provide him with an assistant from Britain, and in September 1929 Edward Arthur arrived to take up the role. Arthur was highly qualified; he had a pharmacy degree, was a pharmaceutical chemist, and a member of the PSGB. He had completed a five-year apprenticeship in a busy chemist's shop in Glasgow, and had been a lecturer at the school of pharmacy there. In Lagos he was assisted by two Nigerian dispensing instructors.[33]

The Ordinance established a Board of Medical Examiners, which in 1930 became responsible for the training and examination of 'dispensers, chemists and druggists, and assistant medical officers' employed by the Government. The Board initially consisted of five doctors, with the government's chief medical officer, Dr. A. Robertson as its chairman. In 1929 it received a request for the reciprocal recognition of pharmacy qualifications awarded by other West African colonies. It refused the request, agreeing that only those holding the PSGB qualification would be allowed to practise pharmacy, an arrangement that remained in force until after independence. With its new pharmacy responsibilities, the Board added Edward Arthur—who was a lecturer in chemistry and an examiner in science as well as being assistant director at the school—to its membership. It registered its first two chemists and druggists—E.E. Moma and A.D. Brown—in 1930.[34]

Nigeria was then producing pharmacists to a level comparable with the British chemist and druggist diploma.[35] But the country's need was for large numbers of lesser trained dispensers. A decision was taken to open a second dispensers' training school in Northern Nigeria, and this was eventually opened at Zaria in 1930, although it closed again in 1940 due to a lack of qualified teachers.[36] Those with the chemist and druggist qualification increasingly opened retail pharmacies. New businesses sprang up, including the Phillips Medicine Store in Lagos. This was founded by Thomas King Ekundayo Phillips (1884–1969), the son of a bishop, who had trained at the school of pharmacy, qualifying as a chemist and druggist. He later opened other branches in Lagos and elsewhere, and went on to become the first president of the Pharmaceutical Society of Nigeria.[37]

Aligning the Curriculum

Amendments were made to the 1927 Poisons and Pharmacy Ordinance in 1929, 1930, and 1934.[38] They were consolidated through new provisions in the Poisons and Pharmacy Ordinance 1936, the main purpose of which was to align more clearly pharmacy education and training in Nigeria with that in Britain.[39] It made provision for the registration of pharmaceutical premises, and empowered Government Medical Officers and police officers to inspect them. Nigeria then had two levels of pharmacy qualification; 'pharmacists' holding the Chemists and Druggists Diploma, and 'dispensers'. In Nigeria a dispenser was defined as

> a person holding a certificate granted under the provisions of the Poisons and Pharmacy Ordinance 1936, or possessing other prescribed qualifications, who by virtue of holding such certificate or other qualifications has been granted a licence under this Ordinance to mix, compound, prepare and dispense drugs and poisons where licence is still in force.[40]

The Ordinance standardized the curriculum of the Chemists and Druggists Diploma and formalized the issue of Dispensers Certificates.

A step towards the separation of pharmacy from medical control occurred in 1936 when the School of Pharmacy at Yaba became an independent institution. It had previously been under the Education Department, with the superintendent reporting to the Director of Medical and Sanitary Services and to the Board of Medical Examiners. But Nigerian pharmacy remained firmly British; in 1940 the Colonial Office appointed Mr. K. S. Feltham as Superintendent of Pharmacy and Head of the School of Pharmacy; Messrs B. Cole, N. O. A. Morgan and A. B. Dikko, all of whom were chemists and druggists, were appointed as assistant lecturers, and in 1941 L. C. Bell was appointed inspecting pharmacist.[41]

A Pharmacy Board, 1945

Within four years of the implementation of the 1936 Poisons and Pharmacy Ordinance a committee was established to review it. Its recommendations were wide-ranging but focused on the level of educational attainment of those practising pharmacy. By the 1940s the number of well-qualified applicants for pharmacy courses was increasing rapidly, and the committee recommended that the entry qualification should be raised. Places for pharmacy students should be increased; the Yaba dispensing school in Lagos should be expanded, and the school in the north at Zaria should be re-opened. The proposed revisions to the Ordinance were considered by the Board of Medical Examiners in June 1941.

The Board approved the revisions, and they were passed as the Poisons and Pharmacy Ordinance 1945. Its main aim was to regulate pharmacy in the same way as it was in Britain.[42]

The most significant change was the establishment of a separate Pharmacy Board, with its own pharmaceutical registrar. The person appointed was K. Feltham, then head of the School of Pharmacy, although he was appointed as the Government Chemist soon afterwards and was replaced as head of the School by A. J. (Jack) Bearman. The number of members of the Board was increased to ten, but the Director of Medical Services continued as chairman, a situation that was to continue until independence in 1963. The 1945 Ordinance also provided for the regulation of the sale and distribution of drugs and poisons in registered premises and the inspection of such premises. It defined 'Packed Goods' and 'Patent and/or Proprietary Medicine' and provided for unqualified practitioners wishing to continue selling packed goods to apply for a Patent and Proprietary Medicine Vendor's licence, due to a shortage of pharmacies in many areas. The Ordinance also authorized the issue of permits to some missionaries, enabling them to practise pharmacy only in mission-owned hospitals.[43]

In 1948 an Education Committee was appointed to 'consider the future education policy of pharmacists in Nigeria and to make recommendations'. One of its concerns was the uneven distribution of trained pharmacy personnel between the north and south of the country. It was chaired by Dr. Stanley Merrett, the Principal of Yaba Medical School. It proposed that, at least for the moment, the two levels of pharmacy education should be retained; the course at the southern school (at Yaba, Lagos) should be of three years duration and lead to the Chemist and Druggist Diploma; the course at the northern school (at Zaria) should also be of three years duration but should be at a lower standard. The northern course would lead to a Dispensers Certificate and would only be available to those born and educated in the north. But as soon as sufficient students with the necessary basic education were available, the Dispenser's Certificate should be discontinued, and replaced by the Chemist and Druggist Diploma, for which the entrance qualification was to be London Matriculation or exemption from it.[44]

The Committee recommended that, for the Chemist and Druggist Diploma, 'the syllabus for these examinations should be substantially the same as is at present in force in the United Kingdom'. The only significant changes necessary for the course at Yaba were the addition of zoology and physiology.[45] They suggested that both be introduced at Zaria 'as soon as possible' to enable holders of the Dispenser's Certificate to progress to the Chemist and Druggist Diploma in due course.[46] With few exceptions the recommendations were accepted by the Pharmacy Board and implemented under the Ordinance, aligning arrangements in Nigeria with those in Britain.[47] No account was taken of the number of pharmaceutical practitioners the country needed, and Nigeria was soon producing a small number of well-qualified pharmacists, but no larger body of practitioners with lesser qualifications.

Associations and Collective Action

From the beginning dispensers working in government hospitals felt that they were not accorded the status and remuneration that they thought their responsibilities and duties merited. Rising levels of qualification were not reflected in the salaries or grading of those employed in government establishments. Widespread discontent led to collective action, and an Association of Dispensers was registered in 1927 under the 1922 Company Ordinance. It agitated for better wages and higher status for dispensers in the civil service, but was unable to achieve any improvement in conditions, and faded away. But discontent continued, and the association was reformed in 1947 as the Nigerian Union of Pharmacists. Two years later it represented 200 government employees and was registered under the Trade Union Ordinance.[48]

In January 1951, the Nigerian Union of Pharmacists called a national strike, and a meeting was hastily called between the Pharmacy Board and union representatives.[49] The Board then persuaded the Director of Medical Services to appoint a Committee of Enquiry under an independent chairman to examine the conditions of service of government pharmacists and to make recommendations. A distinguished Nigerian ophthalmologist and politician, Sir Kofoworola A. (Kofo) Abayomi, was appointed as its chair, and the committee reported later that year. It recommended a grading structure for pharmacists working in Government hospitals along similar lines to that in Britain, with much improved salaries.[50] The proposals were fully supported by colonial medical officers; Andrew Egboh, later the Pharmacy Board's Registrar, noted that progress and 'promotion to higher status followed [in] leaps and bounds' after the appointment of Sir Samuel Manuwa as Director of Medical Services in 1952.

In 1956 retail pharmacists came together to form the Pharmaceutical Society of Nigeria (PSN) organized along similar lines to that of the

PSGB.[51] It organized branches across Nigeria and developed educational, social, and professional activities, and had achieved considerable influence in government circles, despite being an informal and voluntary organization. The Nigerian Union of Pharmacists was later dissolved in 1977, when Nigeria's Central Labour Organization restructured and took over all other unions. The PSN was left as the only professional umbrella body of pharmacists in Nigeria.[52]

Designation as a Profession, 1956

Pharmacy education in Nigeria gained a measure of autonomy from medicine in 1952 when the Pharmacy Board established its own Examination Committee, although it remained under ultimate medical supervision. It was chaired by Dr. G. H. M. Menzies, the Deputy Inspector-General of Medical Services, and included representatives from the retail, hospital, industry, and government sectors in Nigeria. Grading and pay levels in the Civil Service continued to be a source of discontent amongst pharmacists.[53] On 20 January 1956 pharmacy in Nigeria was formally designated as a profession by the Colonial Government. At the same time a committee was appointed to review the operation of the 1945 Pharmacy Ordinance, including the relationship between doctors and pharmacists. Doctors were still in overall charge of pharmacy in Nigeria; whilst there was a separate Pharmacy Board, its chair was not a pharmacist, and its membership did not consist solely of pharmacists.

The authorities in Nigeria followed events not only in Britain but also in its neighbouring colonies. In 1955 Jack Bearman was sent to the Kumasi School of Pharmacy in the Gold Coast to compare pharmacy education in Nigeria with that there. He concluded that the Nigerian chemist and druggist diploma was a more advanced qualification than that in the Gold Coast. But by 1956 it was recognized that there was an acute shortage of pharmacists. The Pharmacy Board recognized the need to train a lower grade of pharmacy practitioner than the dispenser, to be called dispensing assistants (later designated pharmacy technicians). A conference was held at Ibadan to compare the pharmacy courses in Nigeria with those in Britain. It was attended by local British expatriates plus invited participants from Britain, including the president and registrar of the PSGB and the Dean of the School of Pharmacy at the University of London.[54]

The conference proposed that the chemist and druggist diploma become the minimum qualification for practise in Nigeria, and in 1958 the Zaria School of Pharmacy applied to the Northern Ministry of Health for their course to be upgraded. Approval was given in 1961, and in 1962 the Pharmacy Board agreed that holders of the Dispensers' Certificate awarded at Zaria could upgrade their qualification to the chemist and druggist diploma after a 9-month refresher course at the School of Pharmacy and passing the Board's examinations. The following year they were allowed to practise as pharmacists anywhere in the country after two years' experience in community pharmacy.[55]

Transfer of Power

British influence on pharmacy in Nigeria continued up to and beyond independence in 1963. The composition of the Examination Committee was changed in 1957: Jack Bearman was appointed its chair, but most of the British members were replaced by Nigerian nationals. The Yaba School of Pharmacy was moved to Ibadan in 1958, firstly as part of the new Nigerian College of Arts, Science and Technology, and later as part of the University of Ife, founded in 1963. In June 1960 Professors Linnell and Shotton from the London School of Pharmacy were appointed as external examiners.[56]

Despite the efforts of the PSGB key elements in the professionalization of pharmacy were delayed until after independence in 1963. A Pharmacy Board of Nigeria—with only pharmacists as members—was finally established in 1964 following enactment of the Pharmacy Act 1964. A further Poisons and Pharmacy Act followed in 1990, and a wholesale review of laws governing pharmacy in Nigeria occurred in 2004. This represented a shift away from the British model, in which the regulation of pharmacy was linked with the control of poisons, towards the American model in which food and drugs were considered together.[57] Independence saw a transfer of responsibilities from British to Nigerian nationals. Bearman was replaced as Principal at the Yaba School of Pharmacy by J. P. Marquis, and as Registrar by S. O. Ogunsakin, and later by Andrew Egboh in 1964.[58] Bearman returned to England to work for the Crown Agents.[59] The final step in the removal of pharmacy from medical authority in Nigeria came in 1965. The last Chief Medical Adviser to serve as chair of the Pharmacy Board was Dr. S. O. Awoloyi; he formally handed over chairmanship to the Chief Pharmacist of the Federal Government, Mr. A. A. Oluwole, on 27 March 1965.[60]

Pharmacy in the Gold Coast, 1874

In 1900 the colonial authorities in the Gold Coast brought charges against an assistant colonial surgeon, Dr. B. W. Q. Papafio. He was charged with 'keeping a drug store in Accra without permission and with permitting his apprentice to sell a poison contrary to the law'.[61] He was fined £20 by the Executive Council, although they recommended remission of the fine as he had already been fined by the District Commissioner. Thus by 1900, there were Western drug stores in the Gold Coast, there were laws controlling the sale of poisons, and colonial medical staff took on local apprentices and trained them in dispensing.

When Britain proclaimed the Gold Coast as a Crown Colony in 1874, medical officers of the Colonial Medical Service provided for the needs of colonial administrators and expatriates. As in Nigeria, Africans generally relied on traditional healers for medical care, although they became increasingly exposed to Western medicines following the arrival of missionaries. The first civilian hospital was built in Accra in 1878, and Western medical services developed after 1880.[62] Initially medical officers themselves undertook the dispensing of medicines, but this was increasingly delegated to assistants under their supervision. Informal pharmacy training started in the early 1880s.[63] Individuals with a basic level of education were trained in a variety of supporting roles as dispensers, dressers, and nurse-dispensers, to assist medical officers. The nurse-dispenser scheme was adapted from a similar one in Sierra Leone, but it had to be abandoned because on completion of training most chose to practise as druggists, creating a serious shortage of nurses.[64] Dispensers manned dispensaries and small hospitals where resident medical officers could not be stationed; they diagnosed conditions such as malaria, diarrhoea, and yaws, as well as dispensing medicines.[65] In due course some opened drug stores and began selling medicines.

Under Medical Control

The first steps in regulating the practice of pharmacy in the Gold Coast were taken in 1892, with passage of a Drugs and Poisons Ordinance (The Druggists' Ordinance).[66] Its purpose was to regulate the dispensing and sale of drugs and poisons in the Colony; it was enacted by the Governor with the advice and consent of the Legislative Council, who in turn obtained guidance from the PSGB.[67] Henceforth, persons retailing, dispensing, or compounding drugs and poisons would need to have a competent practical knowledge of their business, and 'ignorant and incompetent persons' were to be prevented from assuming the title 'druggist'. The Druggists' Ordinance established a body empowered to regulate the practice of pharmacy and the qualifications required in order to practise as a druggist. This was the Board of Examiners, and it consisted entirely of doctors. The colony's Chief Medical Officer (CMO) was its chair, and he was assisted by other medical practitioners appointed by the governor. In the absence of the CMO a senior government medical officer would chair the Board; the chairman with two other members formed a quorum. No pharmacists were included on the Board.[68]

The Board examined all persons who presented themselves for examination under the Ordinance and granted or refused certificates accordingly. The examinations were held at Accra twice a year, and at other places and times determined by the CMO.[69] Doctors in the Gold Coast therefore had total control over pharmacy at the end of the nineteenth century; medical practitioners trained dispensers in government hospitals, they examined them as the Board of Examiners, and they registered them as druggists. Licences to act as druggists were granted to persons who met one of several conditions. Duly qualified chemists and druggists from Great Britain and Ireland were licenced automatically.[70]

But licences were also be awarded to those who produced 'a certificate that he has been employed in the compounding and dispensing of prescriptions in a colonial hospital in the colony, or any other of Her Majesty's colonies or dependencies for three years', and subsequently passed an examination.[71] Those trained locally needed to have been employed as an assistant to a duly qualified medical practitioner, apothecary, or chemist and druggist, for a period of five years' and to have been examined by the Board. This period was reduced to three years if the applicant had 'been engaged in the colony in the selling, compounding and dispensing of prescriptions in some house or shop kept by him for the purpose'. By 1907 there were 18 locally trained dispensers and 16 dispenser pupils in the Gold Coast.[72] But dispensers did not attend lectures, nor did they sit the PSGB's Minor examination.

Pharmacy Education, 1927

Measures were taken in the 1920s to put pharmacy education in the Gold Coast on a more formal and organized basis. A Dispensing School was opened in 1927 at the Korle-Bu Hospital in Accra, which had opened two years earlier. The entrance qualifications for dispensers were the basic School Certificate, and the training course lasted two years. The colonial authorities considered that only a basic level of training was required for dispensers to meet the needs of a predominantly rural population. The number of trained dispensers in the colony steadily increased following the opening of the dispensing school, although demand continued to outstrip supply. In 1930 a new course was introduced with a raised entrance requirement to produce a new class of practitioner, the nurse-dispensing assistant. But the dispensing qualification was soon upgraded: in 1940 a further adjustment was made, and the title changed to 'Second Division Dispenser'.[73]

Although most of those qualifying at the Dispensing School went on to work in civilian hospitals, by the late 1920s a small number had ventured into retail pharmacy. Newly qualified practitioners soon became extremely dissatisfied with arrangements for the regulation of pharmacy practice—including the absence of druggists on the Board of Examiners—and began working together to lobby for change. Two professional associations, the Gold Coast Pharmacist and Druggist Union (GCPDU), and the Chemist Defence Union, were established in 1929. The members of both were predominately former students at the Dispensary School. The two unions merged in 1935 to form the Pharmaceutical Society of the Gold Coast (PSGC).[74] Membership was entirely voluntary, and it had no legal status. But the Society sought the support of the Minister for Health and members of the Legislative Assembly to obtain changes to the laws regulating pharmacy, and its continuing advocacy eventually persuaded the authorities to act.

Recognition of Pharmacists

By 1946 it was clear that reform of legislation relation to pharmacy and medicines was long overdue. The 1892 Druggist Ordinance was repealed and replaced by the Pharmacy and Poisons Ordinance 1946. The new Ordinance introduced the term 'pharmacist' in the Gold Coast;

the previous distinction between locally qualified 'dispensers' and 'druggists' holding the PSGB's Minor certificate was removed. The Ordinance sought to 'make better provision for regulation of the profession of pharmacy and control of the trade in drugs and poisons'.[75] It followed the British model of combining the control of poisons with the regulation of the pharmacy profession. A new governing body, the Pharmacy and Poisons Board, was appointed. It was to have no more than eight members, and its chair would be the Director of Medical Services (DMS) who also acted as its Registrar.[76] Two members were to be medical practitioners, but now three were to be registered pharmacists; the others were to be appointed by the Minister. The chair and three other members formed a quorum. Pharmacy was now represented on the Board, but it was still firmly under the control of the doctors.

The role of the new Board was extended beyond the examination of candidates, the issuing of certificates of competence, and the keeping of a register, to prescribing the courses of instruction to be undertaken by pharmacy students. It approved and supervised 'private institutions for the training of pupil pharmacists' and it provided for the registration of such institutions and 'for all matters relating to such registration'.[77] In order to register as a pharmacist a candidate now needed to meet one of four conditions; they could be a chemist and druggist duly registered in the United Kingdom; they could hold a valid certificate or diploma of competence as a chemist or druggist from any college, society, council, board or body of persons recognized by the PSGB; they could hold a valid certificate of competence issued by the Gold Coast Pharmacy and Poisons Board; or they could have been licensed and registered under the 1892 Druggist Ordinance before 1 November 1946.[78] This echoed the PSGB's attempts to bring together those with disparate qualifications through the 1868 Pharmacy and Poisons Act (Chapter 2).

There was widespread recognition by the colonial authorities of the need to rationalize pharmacy education. In the mid-1940s a British expatriate pharmacist previously employed by a large drug company was recruited by the government to completely reorganize pharmaceutical training in the Gold Coast.[79] A College of Technology was established at Kumasi in the middle belt in January 1952, and twelve months later the School of Pharmacy previously at Korle-Bu Hospital in Accra transferred to Kumasi. The new pharmacy department initially ran the two-year course in pharmacy which lead to a certificate awarded by the Pharmacy Board. But the 1946 Ordinance was considered unacceptable in several respects; pharmacy was still under the firm control of the doctors; and pharmacists remained in the minority on the Pharmacy and Poisons Board, since the chair with three medical practitioner colleagues could make decisions affecting the profession. The Pharmaceutical Society of Ghana (PSGH) lobbied the government for it to be given control over the regulation of pharmacists and pharmacy businesses. On 16 February 1955, a new Pharmacy and Poisons Bill was read for the first time in the National Assembly, and notice given for a second reading.[80] But several areas of the Bill were still considered unacceptable, including the fact that the CMO would continue to chair the Pharmacy Board.[81] On 21 February 1955 the second reading was deferred indefinitely after discussion with pharmacy-supporting backbenchers.

Professionalization After Independence, 1957

It was at this point, in 1957, that the Gold Coast gained its independence and became Ghana. The British authorities departed, and matters were left in the hands of locally elected representatives. The process of transition delayed progress by over five years. After consultation with the Ghana Medical Association and the PSGH a revised Pharmacy and Drugs Bill was read in Parliament on 24 May 1961.[82] In the new Bill, the government would relieve the CMO of his position, and the chief pharmacist would be appointed chair of the Pharmacy and Poisons Board. A pharmacist would be appointed as Registrar, and the membership would be enlarged to provide more satisfactory representation of all interests.[83] With these changes the National Assembly passed the Bill as the Pharmacy and Drugs Act 1961.[84] The new Act sought to both regulate the pharmacy profession and control the supply, manufacture, storage and transportation of drugs.[85] Members of the Board would in future include a lawyer, the head of the School of Pharmacy, three pharmacists nominated by the Pharmaceutical Society, and three medical practitioners. As well as registering pharmacists, the Board now prescribed courses of instruction and practical training for pharmacy students, arranged for their examination, and awarded diplomas to successful candidates.[86]

As before, the 1961 Act specified the conditions under which a candidate might be admitted to the Register, but there was one significant change. They included holding a diploma awarded by the Pharmacy

Board and 'holding a degree, diploma, or other qualification conferred in a country outside Ghana', but specific mention of a British or colonial qualification had disappeared. At the same time, like the British 1933 Pharmacy and Poisons Act on which it was based, membership of the PSGH was made compulsory for registration by the Board. If a pharmacist ceased to be a member of the PSGH, the Pharmacy Board was notified and the pharmacist's certificate of registration cancelled. The Act affirmed the PSGH as a professional body, but it had no power to terminate a pharmacist's membership except at their own request, or if membership fees were more than one year in arrears. In 1961, a total of 133 pharmacists were registered by the Pharmacy Board. In 1975 the PSGH had 480 members, although only 355 pharmacists were registered with the Board in 1976.[87]

With independence efforts were made to advance science and technology education in Ghana, including pharmacy education, which quickly led to qualification inflation. In 1961 a three-year degree programme was introduced at the now University of Science and Technology at Kumasi.[88] The certificate course ended in 1963, making pharmacy a graduate only entry profession from 1966, four years before it was in Britain. This provided the basis of the campaign by the Pharmaceutical Society of Ghana for pharmacy to be fully recognized as a profession, having equal status and remuneration to other professions. Other symbols of professionalization appeared after independence; a quarterly journal was published from 1959 under the title *West African Pharmacist*. This replaced an earlier publication, the *West African Druggist*. In 1971, the *West African Pharmacist* was itself superseded by an expanded journal, the *African Journal of Pharmacy and Pharmaceutical Sciences*.[89]

The Mills-Odoi Commission

The move to professionalize pharmacy suffered a major reversal in the 1960s. In 1967 the government appointed a committee to investigate the health needs of Ghana. As part of its enquiries the Mills-Odoi Salary Review Commission examined the position of pharmacists within the public sector. It was highly critical of the decisions made during the British period, especially that of stopping the diploma course. 'It is apparent now that the abolition of the Diploma Course in Pharmacy was a mistake' it declared. 'There is a need to train two classes of pharmacists, both of them registrable in the professional register. Professional graduates should continue to be trained'.[90] The problem was not so much the over-professionalization of pharmacy as the neglect of the provision of adequate numbers of lesser qualified staff. Clearly only relatively small numbers could be trained to graduate level. The report concluded that

> the majority of hospital and retail pharmacists will not require this level of training, and a shorter diploma course in pharmacy will be most appropriate. The large majority of pharmacists will be in this category. It is only thus that any impression can be made on the dire shortage of pharmacists in this country.[91]

But the report was also highly critical of the training programmes for pharmacists themselves. They observed that

> it is a sad and costly reflection on the lack of coordination between the investment of funds on education and occupational needs of the country that the University of Science and Technology (UST) [at Kumasi] should insist on providing a degree course in pharmacy which the Government's Chief Medical Adviser does not consider necessary.[92]

Meetings were held between the Government, the University, and the PSGH to determine the content and duration of the new diploma course. But implementation of the Report's recommendations would create sizeable income differentials between pharmacists and other professions.[93] The PSGH therefore mounted a campaign to prevent the reintroduction of the diploma course, culminating in its refusal to register those completing the course as pharmacists. In September 1968 it placed an advertisement in a local newspaper criticizing the university. It noted that they planned to introduce such a course

> despite the fact that the Pharmacy Board—which is the statutory authority in regulating and controlling pharmacy education—has disapproved of mounting such a sub-standard course and has made it categorically clear to the university that it would not register the products of such a course to practise in Ghana. The Society, as guardian of the pharmacy profession in Ghana, shares the same view with the Pharmacy Board, since the trend in pharmacy education the world over is towards higher standards.[94]

The arguments put forward by the Society in support of its case hark back to Jacob Bell's vision of pharmacy as a branch of medicine where pharmacists worked as equal partners with others. 'The surgeon, the physician, the general practitioner, the gynaecologist and the pharmacist are all specialists in their own right' it argued.

> A mistake in the execution of his duties may result in death. The question of status and remuneration should therefore be examined with an open mind. No one member of the group should be permitted to moderate the other, because such a move will not be in the interests of the community.[95]

But despite their objections a diploma course was introduced. Until the early 1970s the UST at Kumasi was the only school of pharmacy in the country. By 1975 they had produced 219 graduates and 66 pharmacists with diplomas. The number of expatriate students dwindled rapidly; in 1960 there were over 20 to 300 Ghanaians, by 1973 there were only 2 expatriates to 282 Ghanaians. This was in striking contrast to other professions such as doctors, engineers, and architects where the representation of overseas personnel was still over 80 per cent of total employment in the early 1960s.[96]

But the plan to train two classes of pharmacists backfired. By 1972 the Ministry of Health was dissatisfied with the calibre of pharmacists completing the Diploma Course. Restrictions were placed on what they could do without the supervision of a pharmacist, and they were redesignated 'pharmacy technologists'. The occupational space between pharmacists and a subordinate group was re-established. In this respect pharmacy practice during this period closely mirrored that in Britain where the role of a subordinate group of pharmacy technicians was well established.[97] The victory of the PSGH was sealed in 1973 when pharmacists were placed on the professional range of salaries in the public sector. The Issifu-Ali Salary Review Committee reaffirmed this decision the following year.[98]

Autonomy from the Doctors, 1994

The presence of the three medical practitioners on the Ghana Pharmacy Board rankled throughout the 1970s and 1980s. The pharmacists on the Board lobbied hard to replace them, since—as they pointed out—there were no pharmacists on the Ghana Medical Board.[99] In 1992 the PSGH called for a review of the Pharmacy and Drugs Act 1961 and the exclusion of doctors from the Pharmacy Board. A draft Pharmacy Bill was submitted to Parliament in 1994.[100] Its provisions regarding regulation were similar to those applicable to the medical and dental professions, and sought to replace the Pharmacy Board with a Pharmacy Council consisting solely of pharmacists.[101] It received careful scrutiny in Parliament, which eventually agreed that pharmacy should be independent of doctors, and that only pharmacists should be on the Pharmacy Council. But the requirement that compulsory membership of the PSGH would be a condition for registration as a pharmacist was deleted, as this was judged to be an infringement of Ghana's Constitutional guarantee of freedom of association. With passage of the Pharmacy Act 1994 a Pharmacy Council was established, and all previous laws relating to the regulation of pharmacy practice in Ghana wererepealed.[102]

Professionalizing Pharmacy in West Africa

This account of how developments unfolded in Nigeria and the Gold Coast illustrates the interdependence of state support, the passage of legislation, the emergence of professional organizations and the role of education. The West African colonies typified the wholesale transplantation of the British model of pharmacy. In both cases the PSGB played an active role, advising on the legislation that should be passed, linking the regulation of pharmacy to the sale of poisons, and creating monopolies for pharmacy practitioners in the supply of medicines. For the PSGB these issues were more important than whether registers should be controlled by a Medical or Pharmacy Board. Pharmaceutical associations were formed by locally trained practitioners, but they appear to have found it more difficult to make an impact than ones formed elsewhere by British expatriate chemists and druggists.

If a key measure of a profession is that it is autonomous and completely independent from others, then pharmacy in West Africa could not be described as fully professionalized during British rule. In both Nigeria and the Gold Coast senior doctors remained as the chairs of pharmacy's regulatory bodies. But British and later African doctors were generally supportive of the escalation of pharmaceutical qualifications. This was driven by educational developments not only in Britain but elsewhere in the world. By the 1950s great emphasis was placed on making local qualifications internationally acceptable for reciprocal recognition, allowing Ghanaians and Nigerians to work overseas. In his analysis of what happened in pharmacy in the Gold Coast after independence, Bennell suggests that it was not so much an active process of professionalization as a defence by professional associations of status and patterns of remuneration which had been inherited from the colonial period.[103]

The two-tier nature of pharmacy qualifications in Britain was reflected in the development of pharmacy education in West Africa, although it operated at a later stage at a lower level. Although there were key differences, the patterns of pharmacy education development were similar in Nigeria and the Gold Coast. Philip Olurinola identifies three stages in the training of pharmacy practitioners prior to independence in Nigeria; the training of 'dispensers' through apprenticeship (1887–1923); the training of 'dispensers' through Schools of Dispensing (1923–1927); and the training of 'chemists and druggists' (1927–1972) in Schools of Pharmacy.[104] Augustina Koduah suggests similar stages for the Gold Coast; a first stage when a certificate of dispensing was offered (1880s–1942); a second dual stage when both the certificate of competency and the diploma were available (1943–1960); and a third stage after 1961 when the bachelor of pharmacy degree was awarded.[105]

These stages represented a gradual alignment of pharmacy education in both Nigeria and the Gold Coast with that in Britain. But there was no equivalent of the British higher-level qualification—the pharmaceutical chemist diploma—prior to independence. Alignment was achieved through the passage of Ordinances, the appointment of British nationals to key pharmacy positions, and the direct involvement of pharmacy authorities in Britain. Together they were able to convince colonial and medical authorities of the merits of the British model of pharmacy and develop it without posing any threat to the doctors, who retained overall control. Disputes between doctors and pharmacists over the rights to prescribe and dispense, respectively, were largely absent in West Africa.

Whilst in both Nigeria and the Gold Coast the final steps in the professionalization of pharmacy were delayed until after independence, the raising of educational standards to British levels was pushed through too quickly in the later stages. The scramble to do so resulted in low numbers of highly qualified individuals, and both countries experiencing severe shortages of the lesser qualified individuals needed to ensure the safe and efficient distribution of effective medicines to widely scattered populations. The legacy of British pharmacy in West Africa was distinctly mixed.

Notes

1. Johnson, T.A. 'Imperialism and the Professions: Notes on the Development of Professional Occupations in British Colonies and the New States', *Sociology Review Monograph No. 20* (1973), p. 295.
2. Luckham, Robin. 'Imperialism, Law and Structural Dependence: The Ghana Legal System', *Development and Change*, 9 (1978), pp. 201–43.
3. Bennell, Paul. 'Professionalization: The Case of Pharmacy in Ghana', *Social Science and Medicine*, 16 (1982), p. 604.
4. Curtin, Philip D. 'The End of the 'White Man's Grave'? Nineteenth-Century Mortality in West Africa', *The Journal of Interdisciplinary History*, 21 (1990), p. 63.
5. *West African Pharmacist*, 3 (1961), p. 101; 9 (1967), p. 8; 9 (1967), p. 67.
6. Egboh, Andrew. *History of Pharmacy in Nigeria: A Guide and Survey of the Past and Present, 1887–1980* (Lagos, 1982), p. 1.
7. Etherington, Norman. *Missions and Empire* (Oxford, 2005), p. 281.
8. Breen, Benjamin. 'Sub-Saharan African Medicines', in Crawford, M.J. and Gabriel, J.M. (eds) *Drugs on the Page: Pharmacopoeias and Healing Knowledge in the Early Modern Atlantic World* (Pittsburgh, 2019), p. 156.
9. Porter, Andrew. 'An Overview, 1700–1914', in Etherington, *Missions and Empire*, pp. 58–60.
10. Ade Ajayi, J.F. *Christian Missions in Nigeria 1841–1891: The Making of a New Elite* (London, 1965), pp. 159–62.
11. Etherington, *Missions and Empire*, p. 2.
12. Carlson, Dennis G. 'Drug Supply Systems in West Africa', *Pharmacy in History*, 24 (1982), p. 75.
13. Etherington, *Missions and Empire*, p. 280.
14. Carlson, 'Drug Supply Systems', p. 76.
15. 'The David Livingstone Medical College in London', *Journal of the American Medical Association*, XXI (1893), p. 938.
16. Carlson, 'Drug Supply Systems', p. 75.
17. McCaskie, T.C. 'Cultural Encounters: Britain and Africa in the Nineteenth Century', in Porter, A. (ed.), *Nineteenth Century, Oxford History of the British Empire* (Oxford, 1999), p. 669.

18. Etherington, *Missions and Empire*, p. 272.
19. Ayandele, A.E. *The Missionary Impact on Modern Nigeria, 1842–1914: A Political and Social Analysis* (London, 1966).
20. Egboh, *History of Pharmacy in Nigeria*, p. 86.
21. Ibid., p. 2.
22. Nigerian Pharmacy Board Archives. Cited in https://www.pharmapproach.com/history-of-pharmacy-in-nigeria-2/. Accessed 5 October 2020.
23. Egboh, *History of Pharmacy in Nigeria*, p. 6.
24. Ibid., p. vii.
25. Ibid., p. 1.
26. Adenika, F.B. *Pharmacy in Nigeria: Historical Glimpses from 1887 to 1997* (Lagos, 1998), p. 16.
27. 'Emmanuel Caulcrick, 1902', Pharmacists Council of Nigeria Archives.
28. Egboh, *History of Pharmacy in Nigeria*, pp. 1–3.
29. Ibid., p. 7.
30. 'Dangerous Drugs Ordinance 1927 and Associated Regulations 1927–28', The National Archives (TNA), CO 583/152/10.
31. 'Poisons and Pharmacy Ordinance 1927', TNA, CO 583/146/3.
32. Egboh, *History of Pharmacy in Nigeria*, p. 62.
33. Ibid., p. 4.
34. Ibid., p. 61.
35. Egboh, A.A. *History of Pharmacy Regulation in Nigeria* (Lagos, 2001), p. 10.
36. Egboh, *History of Pharmacy in Nigeria*, p. 15.
37. Ibid., pp. 166–7.
38. 'Poisons and Pharmacy (Amendment) Ordinance 1934', TNA, CO 583/199/17.
39. 'Poisons and Pharmacy Ordinance 1936'. TNA, CO 583/213/16.
40. Ibid.
41. Egboh, *History of Pharmacy in Nigeria*, p. 62.
42. Ibid., p. 5.
43. Ibid., p. 29.
44. 'Report of the Education Committee, 1948', Pharmacists Council of Nigeria Archives.
45. Egboh, *History of Pharmacy in Nigeria*, p. 5.
46. 'Report of the Education Committee, 1948'.

47. Egboh, *History of Pharmacy in Nigeria*, p. 12.
48. Egboh, A. 'The Nigerian Union of Pharmacists' Accomplishments', *West African Pharmacist*, 2 (1960), pp. 63–4.
49. 'Minutes of Meeting between Pharmacy Board and Nigerian Union of Pharmacists, 15 January 1951', Pharmacists Council of Nigeria Archives.
50. Egboh, *History of Pharmacy in Nigeria*, p. 43.
51. Carlson, 'Drug Supply Systems', p. 77.
52. Ibid., pp. 80–4.
53. Ibid., pp. 83–4.
54. Carlson, 'Drug Supply Systems', p. 77.
55. Egboh, *History of Pharmacy in Nigeria*, p. 88.
56. Ibid.
57. Egboh, Andrew A. *Guide on Product Advertisement of Food, Drugs and Cosmetics in Nigeria* (Lagos, 1994), p. 23.
58. Egboh, *History of Pharmacy in Nigeria*, p. 30.
59. 'Bearman, Jack'. Interviewed by Stuart Anderson and Alan Yoshioka, 17.05.95. British Library Sound, C816/02/01/1.
60. Egboh, *History of Pharmacy in Nigeria*, p. 30.
61. 'Charges against Dr. Papafio', TNA, CO 96/364/13.
62. Addae, S. *History of Western Medicine in Ghana, 1880–1960* (Edinburgh, 1997).
63. Koduah, A., Sekyi-Brown, R. and Kretchy, I. 'The Evolution of Pharmacy Practice Regulation in Ghana 1892–2013', *Pharmaceutical Historian*, 50 (2020), pp. 97–108.
64. Koduah, A., Kretchy, I., Sekyi-Brown, R. et al. 'Education of Pharmacists in Ghana: Evolving Curriculum, Context and Practice in the Journey from Dispensing Certificate to Doctor of Pharmacy Certificate', *BMC Medical Education*, 20 (2020), pp. 6–8.
65. Addae, S. *The Evolution of Modern Medicine in a Developing Country: Ghana 1880–1960* (Durham, 1997), p. 243.
66. *Gold Coast Gazette* (1892), pp. 756–66.
67. 'Poisons law 1894', TNA, CO137/563.
68. *Gold Coast Gazette* (1892), pp. 756–66.
69. Ibid., p. 758.
70. Koduah, 'Evolution', p. 100.
71. Ibid.
72. Addae, *Evolution of Modern Medicine*, p. 244.

73. Bennell, 'Professionalization', pp. 604–5.
74. Ibid., p. 604.
75. Koduah, 'Evolution', p. 102.
76. *Gold Coast Gazette* (1946), pp. 530–85.
77. Ibid., p. 545.
78. Ibid., pp. 540–41.
79. Bennell, 'Professionalization', p. 605.
80. 'Legislative Assembly Debates, Accra', 1 (1955), p. 27. References to Legislative Assembly and Parliamentary Debates cited in Koduah, 'Evolution', pp. 97–108.
81. 'Legislative Assembly Debates' (1955), p. 213.
82. 'Parliamentary Debates', 23 (1961), p. 861.
83. Ibid., p. 862.
84. Ibid., 22 (1961), p. 1032.
85. *Pharmacy and Drugs Act, 1961* (Accra, 1961).
86. Ibid., p. 6.
87. Bennell, 'Professionalization', p. 604.
88. Ibid., p. 605.
89. Carlson, 'Drug Supply Systems', p. 75.
90. *Report of the Committee to Investigate the Health Needs of Ghana* (Accra, 1967).
91. Ibid.
92. *Report of the Commission on the Structure and Remuneration of the Public Services in Ghana* (Accra, 1967).
93. Bennell, 'Professionalization', p. 606.
94. *Ghanaian Times*, 10 September (1968), p. 2.
95. Letter from Pharmacy Faculty to Committee Reviewing Mills-Odoi Salary Commission Report, University of Kumasi Archives (1968).
96. Bennell, 'Professionalization', p. 605.
97. Tackie, A.N. 'The Practice of Pharmacy in Ghana', *Journal of the American Pharmaceutical Association*, 11 (1971), pp. 655–7.
98. Bennell, 'Professionalization', p. 607.
99. Koduah, 'Evolution', p. 103.
100. Parliamentary Debates, Accra, Fourth Series, 6 (1994), p. 28.
101. Koduah, 'Evolution', p. 103.
102. *The Pharmacy Act, 1994*. Accra, Ghana: Publishing Corporation, 1994: 1–15.
103. Bennell, 'Professionalization', pp. 603–4.

104. Olurinola, Philip F. *The Pharmacy Profession: A Focus on Nigeria* (Zaria, 2003).
105. Koduah, 'Education', pp. 4–7.

CHAPTER 7

Southern Africa: Networks and Connections

Britain's expanding interests in Southern Africa during the late nineteenth century attracted increasing immigration by British pharmacists keen to exploit the business opportunities. But they were not the first pharmacy practitioners to establish themselves there: the Portuguese, the Dutch, and the Germans had been there before. As the country opened up with the development of farms and the discovery of gold and diamonds, they spread steadily eastwards from Cape Colony. But their progress was not unopposed: The Second Boer War ended only in 1902. For pharmacy, events were dictated more by the actions of a small number of British-trained individuals than the actions of the PSGB. Pharmacy reform became part of overall medical reform, unlike in Britain: and inter-colonial networks—both within country and across continents—played important parts in the professionalization of pharmacy.

Professional networks performed crucial roles in the development of occupations across the British Empire, as the demand for qualified individuals grew rapidly from the 1850s as part of the process of social and economic development in the colonies.[1] In places with substantial settler communities, those carrying out the same occupation came together to form associations for the purpose of securing recognition by the state and advancing their interests.[2] But for British expatriates they served

S. Anderson, *Pharmacy and Professionalization in the British Empire, 1780–1970*, Medicine and Biomedical Sciences in Modern History,
https://doi.org/10.1007/978-3-030-78980-0_7

Fig. 7.1 Map of the South African Division of the British Empire, 1914

another important function; Magee and Thompson note that for participants, networks 'reinforced a sense of belonging to a worldwide British community'.[3] For pharmacy practitioners in Southern Africa professional networks were central to progress. They were the conduit through which information, opinions and attitudes were shared not only between neighbouring colonies but also between colleagues in Britain and on other continents. As a result, they were able to see how things worked out and to avoid repeating mistakes made elsewhere. They could avoid some aspects of the British model of pharmacy, such as the linking of representative and regulatory functions, by having a Pharmacy Board independent of the professional association, and they could prevent the emergence of company pharmacy.

Although pharmacy in South Africa was to emerge as rather less 'British' than elsewhere, the process of professionalization was no less tortuous. Medicine and pharmacy were already well established when Britain invaded Cape Colony in 1795, but it was another fifty years before British occupation of Natal and the Orange Free State, and not until the end of the Boer War in 1902 that Transvaal came under British control (Fig. 7.1). The four colonies became Provinces and were brought together as the Union of South Africa in 1910, with the status of a Dominion—a semi-independent territory under the British Crown. The Union became the Republic of South Africa in 1961.

This chapter considers the professionalization of pharmacy in the four colonies that became the Union of South Africa, highlighting the local, regional, and international networks that guided it. Pharmacy practice had to a large extent been shaped by other European practitioners before the British arrived. Pharmacists had to adapt to pre-existing local circumstances, and their leaders had to accommodate diverse and often conflicting interests. They had the same issues to address as their colleagues elsewhere in the Empire: obtaining the support of the state, achieving legal recognition, establishing clear boundaries with other groups, raising educational levels, and securing autonomy from the doctors. Developments occurred first in the Cape, and these largely shaped subsequent developments elsewhere in South Africa and across the border into Rhodesia, where British pharmacy thrived.

Cape Colony, 1795

A consequence for pharmacy of the Dutch and German legacy in South Africa was that many of the principles espoused in the 1240 Edict of Palermo (Chapter 2) were in force when the Cape of Good Hope became a British possession for the second time in 1806. A year later the first Medical Proclamation was made by the British Governor; it stated that in future all medical personnel, whether physicians, surgeons, or venders of medicines, must appear before the Supreme Medical Committee (later the Colonial Medical Committee) with their diplomas proving their qualifications. The committee consisted of several doctors and one apothecary. If practitioners were unable to produce diplomas their fitness to practise was examined. The names of nine apothecaries appeared on the List of Licensed Practitioners in 1807. The Committee was also authorized to inspect chemists' shops at least quarterly to examine all medicines and drugs, and to destroy any found to be unfit for use. Only the *London Pharmacopoeia* of 1788 was to be used when compounding medicines. Poisons such as arsenic and opium were to be kept in a secure place, and none were to be sold in dangerous quantities. These conditions greatly exceeded those in Britain at the time.[4]

A few months later the Governor made a second Proclamation: 'I do hereby most strictly forbid any physicians or surgeons from keeping any medicine whatsoever for the use of his patients'.[5] This confirmed European over British practice in recognizing the separation of medicine and pharmacy, and legally established the dispensing of prescriptions as the work of an apothecary whilst forbidding physicians and surgeons from selling or preparing medicines. But the Proclamation was widely flouted with impunity; as Hilary Deacon points out, in the first half of the nineteenth century, doctors in the Cape had to be self-sufficient.[6] Another attempt was made to impose separation through a Third Proclamation in 1823:

> It having been a prevailing custom in this Colony for the physician and surgeon not only to prescribe but to prepare medicines themselves for their patients to the manifest injury of such patients, who in most cases are thereby neglected, only receiving their medicines at night, such custom is in future to be discontinued.[7]

The businesses of the apothecaries were suffering. Behind the Proclamation was a difference between the role of Dutch and British apothecaries; the Dutch apotheker was a pharmacist, a maker of medicines, but not a medical practitioner as the British apothecary was by 1824. As Burnby points out, by then the titles apothecary and chemist and druggist were not interchangeable.[8]

The first locally trained apothecary to be licensed to practise at the Cape was Carl Friedrich Liesching in 1824, although this depended on the acceptance of his credentials by the Colonial Medical Inspector, Dr. James Barry. The Governor instructed Barry to set up a Medical Board to examine Liesching, saying that 'the apothecary, chemist and druggist, taken either individually or collectively, were the most important branches of the profession because the physician, surgeon and patient were wholly at their mercy'.[9] Liesching was a chemist and druggist rather than a general medical practitioner and questioned the competence of the Medical Board to examine him. The pharmacists had good reason to claim that the doctors were not qualified to examine them or to approve their registration. The view of the president of the Medical Board, Dr. Ebden, was that 'we take it for granted that a man who is competent to prescribe for disease is also competent to prepare the medicine he requires for the treatment of it'. All doctors were examined in practical pharmacy, botany and materia medica, he pointed out, 'and in branches of the apothecary's calling'. But he also acknowledged that many doctors lacked any practical experience, and many had learnt such skills 'for examination purposes only'.[10]

But the separation of prescribing and dispensing functions proved unworkable. Doctors continued to dispense for their patients and British apothecaries continued to both treat patients and supply medicines from their shops. By 1840 there were twelve pharmacies in the Colony and up to fifteen pharmacists including those based in the hospitals.[11] For most of the nineteenth century pharmacy was seen as a branch of the medical profession by the colonial authorities in South Africa, and the role of the apothecary as both doctor and pharmacist persisted.[12] But the boundary was porous; in 1860 F.B. Spencer, a British-trained chemist and druggist, was appointed as Acting District Surgeon at Hanover, Cape Colony, and licensed by the Colonial Medical Committee as an 'apothecary'. But in 1887 he issued a death certificate which the Town Clerk refused to accept.[13] A year later the Medical Committee reprimanded the Government for continuing to appoint men with chemists' qualifications as district surgeons. As elsewhere in the Empire, frontier conditions made European distinctions between physician, surgeon and apothecary inappropriate.[14] The difference in meaning of 'apothecary' in England compared to the rest of Europe was overlooked.[15] In remote rural areas, pharmacists would frequently perform the role of 'general practitioner' if there was no doctor in reasonable travelling distance.[16] But by 1884 the number of pharmacists in Cape Colony had risen to 236, and by 1891 it had risen by another thirty to 266.[17]

Expansion Eastwards

If Cape Colony drew on experience from across the Empire, the other three South African colonies based their pharmacy legislation entirely on that of Cape Colony. British occupation spread eastwards in the second half of the nineteenth century, and by 1902 Britain had four South African colonies. Natal in the south east was proclaimed a British colony in 1843 after the Boer Republic of Natalia was annexed. The Orange Free State was an independent Boer republic; its borders were determined by Britain in 1848 when the region was proclaimed as the Orange River Sovereignty. It surrendered to the British at the end of the Boer War in 1902, having been annexed as the Orange River Colony in 1900. Following a period of direct rule by the British, it attained self-government in 1907. The Transvaal (or South African) Republic was also an independent Boer state in the late nineteenth century. Britain annexed it in 1877, although it regained self-government following a Boer revolt in 1881; it became the

Transvaal Colony after the Boer War. In 1910 the four British colonies were brought together to form the Union of South Africa, and in 1961 the union became the Republic of South Africa.

The starting point for medical and pharmaceutical arrangements in these colonies was the 1830 Cape Medical Ordinance. In Natal, the British authorities copied it with only a few amendments, passing their own Ordinance in 1856. In Transvaal, the authorities based their regulations on the Natal Ordinance, establishing Law No. 8 in 1881, with all medical and pharmaceutical matters controlled by the Transvaal Medical Committee.[18] The economy of the Transvaal was transformed with the discovery of gold in 1886, and the opportunities presented by its rapid development attracted pharmacists from Britain and elsewhere in southern Africa (Fig. 7.2). Pharmacists centred on Johannesburg formed a Transvaal Pharmaceutical Society in 1894.[19] Its first president was Victor Brown, who had qualified in Britain in 1886; its secretary was Edwin Adcock, who had moved from the Cape.[20] Together they brought experience of pharmacy politics and administration from two other places.

Fig. 7.2 Golden Mortar Dispensary, Johannesburg, 1886 (Courtesy of Pharmaceutical Society of South Africa)

Pharmacists in Natal established a Natal Pharmaceutical Society, based on Durban, in 1892.[21] Finally, an Orange Free State Pharmaceutical Society, based in Bloemfontein, was founded after the Boer War in 1903.

A Pharmaceutical Association

By the early 1880s it was widely recognized in Cape Colony that the 1830 Medical Ordinance was in urgent need of review. In 1883 a Select Committee on Medical Reform was appointed to investigate its operation. It proposed a new law dealing with the regulation of both medicine and pharmacy and the control of poisons. No changes were made to laws governing pharmacy in the Colony between 1830 and 1891, and no chemist and druggist served on the Medical Committee between 1855 and 1891. It was discontent with this situation and concern about impeding new regulations that prompted chemists and druggists to work together to demand improvements in their circumstances. In echoes of the meeting that led to the founding of the PSGB (Chapter 2), seven pharmacists met in the public library at King William's Town in the Eastern Cape on 13 June 1885 to form the 'South African Pharmaceutical Association' (SAPA). A more accurate description would have been the 'Eastern Provincial Pharmaceutical Association', but its founders had hopes of attracting Cape Town pharmacists and wanted a more inclusive title.[22] It became the 'Eastern Districts Pharmaceutical Association' in 1911.

At the same time a shortage of qualified pharmacists occurred as a result of poor planning and training facilities, and in the early 1890s a number were recruited from England.[23] They received free passage and a salary of £12 per month, increasing to £13 in the second year and £14 thereafter. In some cases accommodation was also provided.[24] Once established in their own businesses they recruited qualified assistants from Britain. William Mager emigrated from England in 1881 and settled at Queenstown in the Eastern Cape Province, which had been established in 1853. He recruited several others from England, including A. Lomax, T. Wardley, J. Choat, all of whom went on to play important roles in the development of South African pharmacy.[25] New immigrants, including ones from Germany and Holland, opened pharmacies in emerging parts of the country, but their customers were mainly European expatriates. They spoke a variety of languages and had different cultures: in King

William's Town German settlers who had remained after the British occupation formed the backbone of the farming community.[26] But the real obstacle to progress was the level of animosity between pharmacists in the Eastern Province and the Western Province. It was not helped by the distances involved; between Cape Town in the west and East London in the east was over 1000 kms by road.

Professional and Scientific Networks

Professional networks played an important part in the development of pharmacy not only in Southern Africa but throughout the Empire. Cooper defines networks as 'forms of affiliation and association that are less defined than structures, but more than just a collection of individuals engaging in transactions'.[27] Magee and Thompson emphasize the British origins of many colonial networks and identify three key features: they were voluntary, with individuals joining by choice; they bound people together through a shared sense of purpose and culture; and they transcended territorial boundaries.[28] For pharmacy, the sense of international community and of being connected was encouraged and promoted by the PSGB and supported by publications including the *Pharmaceutical Journal and Transactions* and the *Chemist and Druggist*. The impact of the former was more limited than the latter since its circulation was restricted to members of the PSGB, whereas the *Chemist and Druggist* was available to any subscriber anywhere in the world. In some colonies it became the official publication of the pharmaceutical association. The pages of both provided a forum for the exchange of information, advice, and criticism.

Another important network was the British Pharmaceutical Conference. This was a network independent of the PSGB formed to advance the scientific foundation of pharmacy. Its first meeting was held in 1863 following a notice in the *Pharmaceutical Journal*. Membership was open to anyone, and the participation of chemists and druggists at home and overseas regardless of qualification was actively encouraged. At the time of its third meeting in 1865 it had over 300 members.[29] For pharmacists based in the colonies, Conference membership became an important badge of professional identity and a show of affinity with British pharmacy aspirations. In 1885 the *Chemist and Druggist* published the names of ninety colonial pharmacists who had been elected to membership of the Conference; many were not obviously British. The list included eight Cape Colony pharmacists.[30]

Individuals were usually members of several networks—some official, some business, some professional—that often worked together hand in hand.[31] Membership of Freemasonry Lodges was important for many pharmacists in South Africa and was usually something that they had taken with them from Britain. Harland-Jacobs observes that the Masonic brotherhood functioned as 'a kind of global patronage network that helped men to find employment, secure promotion, and cope with hardship'.[32] George Cook, described as the 'founder of organized pharmacy in South Africa', was a prominent Freemason.[33] He had been born in Lowestoft in 1857, served an apprenticeship in London, and registered in Cape Town in the late 1870s. He worked for E.J. Mieze in King William's Town before taking over the business. He was the driving force behind the founding of the SAPA, the timing and location of whose meetings were arranged to coincide with Masonic meetings between 1885 and 1892. When the timing of the latter's meetings was changed, a special meeting of the pharmaceutical association was called to decide whether to do the same 'especially as...there were several chemists in the Colony who were also members of the Freemason society'.[34] They changed the date. Cook went on to be a Justice of the Peace and was twice the elected Mayor of King William's Town.[35]

Freemasonry was shared with other leading pharmacists who had emigrated from Britain. William Pocock, another leading figure in Cape Town, was active in Freemasonry and served as Grand Master of the Lodge de Goede Hoop.[36] Pocock trained in Britain and passed both the PSGB's Minor and Major examinations, having studied for them at the South London School of Chemistry and Pharmacy.[37] Like Cook, he was appointed as a Justice of the Peace and served as the elected Mayor of Muizenburg and Kalk Bay. The importance of Freemasonry to pharmacy in South Africa continued well into the twentieth century. In 1909 the *Chemist and Druggist* announced that Mr. J. Sowden, a chemist of Rondesbosch, was 'installed as Grand Master of the Rondesbosch Lodge for the coming year'.[38] The extent to which freemasonry infiltrated pharmacy businesses in South Africa is difficult to assess, but it is clear that it provided fellowship, mutual support and opportunities for personal advancement.[39] A Scottish pharmacist, Alexander McDonald Fyvie, was recruited by Lennon Ltd in 1906 and became a leading figure in Natal;

he was actively involved not only with the Freemasons but also with his local Rotary Club and the Caledonian Society.[40]

Drafting a Pharmacy Bill, 1885

Only a minority of those present at the inaugural meeting of the SAPA were British-trained, and Ryan suggests that it would be wrong to view its founding as the natural consequence of an influx of newly trained pharmacists keen to emulate in the colony the structure and duties of the PSGB.[41] News of the foundation of the SAPA in 1885 was reported in the *Chemist and Druggist* in London. They considered its formation to be an important step towards professionalization, although they noted that the emphasis was on obtaining a monopoly in trade. The correspondent noted that its aim was

> to procure the amendment of the existing Acts affecting the trade, which are useless and absurd, and to endeavour to get a Pharmacy Bill through Parliament based upon that of the old country; and to limit the sale of poisons, which are at present retailed, both in large and small quantities, by every storekeeper in the colony.[42]

The proposal was to link the regulation of pharmacy with the sale of poisons in accordance with the British model, one fully supported by the writer. 'If any trade ever asked bare justice, surely it is the claim of the pharmacists to a monopoly in poison'.[43] But problems emerged immediately: if the sale of poisons was restricted to pharmacists, many farmers would be seriously disadvantaged, as they would have to travel great distances to collect them. The parcel post was suggested as an alternative.

An attempt was made to see if control of pharmacy could be wrested from the doctors by voluntary means. It was a forlorn hope. The president of the association wrote to the Medical Committee asking if they would be willing to relinquish the right to examine candidates for the chemist and druggist licence if a Pharmacy Bill was introduced in Parliament the following year. The committee replied that they could not legally decline carrying out what the 1830 Act instructed them to do. They suggested that the association contact the government direct.[44] In October 1885 its executive committee met and agreed a draft Pharmacy Bill. All chemists in the Colony were asked to support it by making 'petitions during the coming session of Parliament'.[45] It received a first reading on 27 May

1886. Four days later pharmacists in Cape Town met to discuss it at a hastily arranged meeting. They had many objections. William Pocock protested via the pages of the *Cape Times* that 'the Bill was absurd in its provisions and incongruous in its details…it was unnecessary and uncalled for legislation…and would impose responsibilities that would be irksome and undesirable'.[46] In the face of such opposition the president withdrew the Bill, and the Cape Town pharmacists opted to form their own association.

The animosity between pharmacists in the Eastern Province and Cape Town had deep seated political roots, based in large measure on urban versus rural differences. Pocock and many of his colleagues were wholesale chemists, and they would lose a great deal of their business with rural farmers if the Bill was passed. Cape Town retail chemists were not affected to the same extent as their country colleagues by the sale of poisons by storekeepers and grocers. But the drive for a monopoly in their sale may not have been the sole motive for the stance taken. Pocock claimed that

> no Bill will meet with approval that does not provide for the maximum of security to the public as well as to the profession…any attempt to secure a monopoly in trade at the expense of the public will meet with the fate it deserves.[47]

Once the initial vindictiveness had died down the two sides eventually got together. Pharmacists in Cape Colony's Eastern and Western Provinces shared common interests in reforming the Medical Committee and the system of examinations for pharmacists.

In planning a Cape Town Pharmaceutical Association Pocock wrote to the existing eastern association to enquire whether, if a Cape Society was formed, they would join to make 'one strong representative body'.[48] They replied that they were already 'a society in full work and with a cash balance in hand' and that 'it was not possible to amalgamate with a society not yet in existence'.[49] The eastern association changed its constitution and made other changes to make it more acceptable to Cape Town pharmacists, but the latter were not prepared to play second fiddle to their colleagues in the east and went ahead with the formation of a Cape Pharmaceutical Society. Thus, by 1887, Cape Colony had two bodies representing the professional interests of pharmacists, one in the east and one in the west, but no Pharmacy Act. Professionalization required pharmacists to work together towards common goals. But whilst the lack of

Table 7.1 Professionalizing pharmacy in Southern Africa

Colony	*Year British colony*	*Year joined Union*	*Pharmaceutical society or association*	*Early pharmacy legislation*	*Early poisons legislation*	*Early pharmacy education*
Cape Colony (Western Cape-Cape Town)	1795	Union of South Africa 1910	Cape Pharmaceutical Society 1887	Cape Medical Ordinance 1830	Cape Medical Ordinance 1830	Informal classes 1885
Cape Colony (Eastern Cape-King William's Town)	1795	Union of South Africa 1910	South African Pharmaceutical Association 1885	Cape Medical Ordinance 1830	Cape Medical Ordinance 1830	Correspondence course 1889
Natal	1843	Union of South Africa 1910	Natal Pharmaceutical Society 1892	Natal Ordinance 1881	Natal Ordinance 1881	Pharmacy classes 1908
Orange Free State	1848	Union of South Africa 1910	Orange Free State Pharmaceutical Society 1903	South African Pharmacy Act 1904	South African Pharmacy Act 1904	3-year apprenticeship
Transvaal	1902	Union of South Africa 1910	Transvaal Pharmaceutical Society1894	South African Pharmacy Act 1904	South African Pharmacy Act 1904	Transvaal School of Pharmacy 1904
Southern Rhodesia	1890	n/a	Pharmaceutical Society of Rhodesia 1898	Medical Ordinance	Medical Ordinance	None before 1970

unity might weaken the cause of pharmacy, government lobbying could be carried out by two bodies rather than one. The stage was set for another attempt at reform (Table 7.1).

Inter-colonial Connections

Criticism of the Medical Committee continued both in Parliament and in the press, and demands were made for wider health reform.[50] Changes essential to pharmacy resulted from this process. In 1883 a Select

Committee on Medical Reform made extensive recommendations, which included updating regulations for the control of pharmacy. But it was not until June 1890 that a Medical, Dental and Pharmacy Bill was introduced in Parliament. The remit and membership of the Medical Board would be extended to include pharmacists and dentists as well as doctors. The Bill did not confer legal recognition on a pharmaceutical society empowered to keep a register of chemists and druggists, nor did it establish a separate and autonomous Pharmacy Board to do so. But the link between pharmacy and poisons was retained; the section that dealt with these was taken directly from the British 1868 Pharmacy and Poisons Act. Pocock again argued that it would be unworkable in Cape Colony because of the scattered nature of human settlement; farmers would have to buy them in person, resulting in a substantial loss in revenue for wholesale chemists such as himself.[51]

The medical, dental, and pharmaceutical professions were to be regulated by a single enlarged Medical Committee which would include representatives of all three professions. But in Cape Town the doctors declined to serve on a Medical Board which included a pharmacist and a dentist, even though they could only vote on matters relating to their own profession. 'Nowhere in Europe' declared Dr. Pope, a medical member of the House of Assembly, 'were chemists allowed seats on a Medical Council'.[52] He did not mention that in many places doctors alone decided whether chemists were qualified or not. The matter was referred to a Select Committee, and received considerable attention in the press. The Attorney-General asked whether the doctors considered themselves 'of such high importance' that they could not work on the Board with a dentist and a pharmacist?[53] One medical reader pointed out that, since doctors studied pharmacy as part of their training, this qualified them to regulate the pharmacy profession. But their continuing regulation by doctors was totally unacceptable to the pharmacists, and those in Cape Town submitted a series of recommendations that included the creation of an independent Pharmacy Board.

Neither of the Cape Colony pharmaceutical associations had any legal powers which might be extended to include the registration of pharmacists; the British model was untenable. Instead, Cape Colony pharmacists looked to the experience of other British colonies in Canada, Australia, and New Zealand. But it was politicians who brought these developments to the attention of the pharmacists as part of the work of the Select Committee reviewing the Medical Bill. In 1890 its chairman, Thomas

Fuller, invited the two leading pharmacists, Pocock and Mally, to study the Victoria Pharmacy and Poison Act of 1876 at the library of the House of Assembly. It is likely that they were already familiar with much of its content. They noted that most of the Australian colonies had concluded that the British model, where the PSGB itself acted as the examination and registration authority as well as being pharmacy's representative body, had distinct limitations, and instead delegated examination and registration activities to an independent Pharmacy Board. Cape Colony pharmacists then proposed the creation of a Pharmacy Board on the lines of those that had first been established in Victoria (Chapter 10).

The doctors demanded representation on any Pharmacy Board that might be established. The Board as eventually described in the 1891 Medical, Dental and Pharmacy Act consisted of one doctor nominated by the Medical Council and five pharmacists, two of whom would be nominated by the Governor. Elected members would serve for a period of five years. The sections of the Act relating to the regulation of pharmacy were largely copied from the 1876 Victoria Pharmacy and Poison Act, although the principle that appeals against decisions could be appealed to the Supreme Court was borrowed from the New Zealand Act.[54] The Victoria Act followed the British model in linking the regulation of pharmacy with the control of poisons, and the Cape pharmacists were anxious to maintain the link. Others sought additional controls; the Eastern Province Pharmaceutical Association (as it now was) submitted a clause relating to the wholesale distribution of poisons. As finally agreed, the section of the new Act dealing with poisons was a mixture of the 1868 British and 1876 Victorian Pharmacy and Poison Acts, although certain clauses were copied from the Medical Act passed in Ontario.[55] In South Africa, lessons were drawn from across the Empire, not just from the metropole.

The Pharmacy Board would deal with all questions relating to the practice of pharmacy, including the examination of candidates, and the registration of pharmacists from abroad. The doctors' hard line had largely backfired; instead of seeking to maintain their dominance over the pharmacists, the Select Committee concluded that the only solution was to have separate Medical and Pharmacy Boards. As Ryan notes: 'ironically, the creation of a totally separate and partly-elected Pharmacy Board was due in a large measure to the stubborn refusal of the medical fraternity to have a pharmacist on the Medical Board'. It was a decision that set the future pattern of pharmacy in South Africa, since it led to the creation of colonial Pharmacy Boards in each of the Provinces, and to the eventual creation of a South African Pharmacy Board in 1928.[56]

Pharmacy Education

The training of pharmacists was first laid down by the Cape Colony Medical Committee under powers given to it under the 1830 Ordinance. Settlements were widely scattered and there was a total lack of suitable educational institutions, so they specified a four-month period of apprenticeship followed by a single examination.[57] The Committee readily acknowledged that its examination was far easier than that in Europe, and European trained pharmacists enjoyed greater respect from the public than locally trained ones. But its standard was nevertheless higher than that achieved elsewhere in southern Africa at the time. A subscriber reported on pharmacy qualifications in Cape Colony to the *Chemist and Druggist* in 1885.

> The English certificate and a fee of 50s. is accepted in lieu of the colonial examination, which is conducted by a paper of questions set by the Medical Board of Cape Town, and forwarded by post, to be answered, under the eye of a magistrate, in three hours. The paper covers little ground and consists of questions on chemistry and the preparations of the [British] Pharmacopoeia.[58]

In 1886 and 1887 there was a 50 per cent failure rate: four passed in 1886 and five in 1887. The Medical Committee claimed that the high failure rate was due to candidates underestimating the examination as 'only qualifying them as dispensers' rather than as chemists and druggists.[59]

No formal pharmacy education was available, although classes for chemists' apprentices and assistants were given by British chemists and druggist such as William Pocock, who set up an informal school in Cape Town; in 1885 he had a class of nine students. But most aspiring pharmacists outside of Cape Town were dependent entirely on their apprentice master for help preparing for the examination, although a correspondence course was offered from 1889. This was run by a German pharmacist and botanist, Dr. H.W.R. Marloth who had managed a pharmacy in Kimberley. He opened an analytical laboratory in Cape Town from which he ran his correspondence school of pharmacy. He was later appointed professor of chemistry at Victoria College in Stellenbosch. Apprentices who could afford it could attend a few lectures at the South African College in Cape Town, and some other pharmacists such as Walter Froembling later devoted themselves to the training of young pharmacists. Attempts were made to improve pharmacy education across South Africa in 1899, but these initiatives came to an abrupt halt with the outbreak of the Boer War.[60] Pharmacy classes were inaugurated in Durban in 1908 and became the basis for the Natal Technical College.[61]

Many of the English pharmacists who went to South Africa had attended one of the 'crammer' colleges in London to prepare them for the PSGB's chemist and druggist qualification.[62] Some sent their apprentices there to do the same. In 1899 George Wills, founder of the Westminster College of Chemistry and Pharmacy, indicated that at least four of his former students came from South Africa; two from Cape Town, and one each from East London and Johannesburg.[63] Those who qualified in the Cape hoped that their qualification would give them some exemption from the PSGB examinations. In 1886 a Cape colonist wrote to the *Chemist and Druggist*, reporting that he was qualified by virtue of having been examined by the Medical Board there and had been practising for some years. He hoped to qualify as a member of the PSGB whilst in England. He wanted to know if his Cape Colony qualifications would be of any assistance, what subjects he would have to study further, and how long it would take. The Cape examination had comprised Latin, materia medica and pharmacy and chemistry, and he suggested hopefully that it 'almost' equated to the Minor examination in England.

The *Chemist and Druggist* helpfully spelled out the position of the PSGB for the benefit of pharmacists throughout the Empire.

> The Pharmaceutical Society of Great Britain does not recognise any other examination as a substitute for its own Minor or Major examinations. In lieu of the Preliminary examination certain British matriculation and other examinations are accepted; but except in this respect the English diploma can only be obtained by passing first the preliminary examination, then, after an interval of at least three months, the Minor, which must be passed either in London or Edinburgh. This confers the right to use the title 'chemist and druggist'. After another interval of least three months the Major examination may be taken, either in London or Edinburgh. This would confer the title 'pharmaceutical chemist'.[64]

The unwillingness of the PSGB to recognise any qualification other than its own was an issue for other colonies across the Empire.

The Boer War was followed by a period of reconstruction and a new wave of immigration of British pharmacists and the movement of those already there. J. Reid was a Scottish chemist and druggist who had arrived with British troops in 1900 and was stationed at a military hospital in Bloemfontein. He stayed on managing a retail pharmacy, and later became a lecturer in pharmacy at the Grey University College, and an examiner for the Orange Free State Medical and Pharmacy Council. Another Scot, R. Macintosh, settled in the Transvaal in 1903 having previously worked in Rhodesia for six years.[65] The South African Pharmacy Act of 1904 provided for a Transvaal Pharmacy Board, on the lines of the Cape Pharmacy Board.[66] Macintosh became an examiner for the Transvaal Board and took a leading role in founding a Transvaal school of pharmacy in 1904 He went on to chair the Associated Pharmaceutical Societies of South Africa.[67]

Reciprocal Recognition of Qualifications

By 1889 substantial numbers of pharmacists were moving between the Cape and Natal and the two Boer republics (Orange Free State and Transvaal). For the authorities this raised the issue of reciprocity in the recognition of qualifications.[68] In all the colonies registration of pharmacists was still in the hands of Medical Boards, but not all imposed the same conditions; whilst most required a four-year apprenticeship this was reduced to three years in the Orange Free State. The first move was made by the Transvaal Medical Board; its chairman asked the Cape Medical Committee whether it was necessary for pharmacists who had been examined in the Transvaal to be re-examined in the Cape Colony before being licensed there. They replied that it was necessary 'at present'.[69] Discussion about reciprocal recognition was already taking place elsewhere in the Empire. In 1881 Australian pharmacists had approached the PSGB with a view to recognition of their qualification (Chapter 10). As a first step in building trust between South African colonies the Transvaal Medical Board announced in September 1889 that in future pharmacists licensed in the Cape would be recognized in the Transvaal without further examination.[70]

Initially the Cape Medical Committee agreed to reciprocal recognition, but the question was highly political, both pharmaceutically and more widely. A separate Pharmacy Board was eventually established in 1892. When the Cape Pharmacy Board took over the pharmacy functions of the Cape Medical Committee in January 1892 the decision was reversed; it announced that certificates awarded by other colonies would not be accepted. There were genuine concerns about the standard of pharmacy training in other parts of South Africa, although their stance may have been politically motivated. Certainly, pharmacists elsewhere considered the Boer republics to be substantial obstacles to progress regarding professionalization across a unified South Africa.[71] A first step was to unify these two colonies; there were about twenty pharmacists in the Orange Free State in 1898, and at the end of the war they were subject to the same British administration as the Transvaal. A Medical and Pharmacy Ordinance passed in 1904 established a unified Medical and Pharmacy Council consisting of five doctors, one dentist and two pharmacists. The *Bloemfontein Post* noted that 'owing to the comparatively small number of chemists in the colony it has been thought best to combine the two'.[72] In Orange Free State a Bloemfontein Chemists' Association was formed in 1903; it was quickly embroiled in the discussions about how to deal with the reciprocal recognition of qualifications between South African colonies.

It took until 1920 for agreement to be reached between the Cape, Natal, and Transvaal Pharmacy Boards. All three provinces specified a four-year apprenticeship, but the Orange Free State continued to be bound by the Free State Act of 1904 which specified a three-year apprenticeship. The Free State Medical and Pharmacy Council did not consider that 'there were sufficient men affected to warrant the matter being taken further'.[73] Reciprocity between all four provinces was only finally resolved following passage of a new Medical, Dental and Pharmacy Act in 1928.

Company Chemists

One feature of pharmacy which South Africa shared with Britain was the development of company chemists, as there was no legislation to prevent them operating. Because of the large capital sum required to set up a pharmacy the retail chemist business tended to be limited to family businesses and chains of outlets owned by large companies. As elsewhere, the sons of chemists often followed their fathers into the business and opened branch shops. At the end of the nineteenth century large drug companies developed, including Lennon Limited and Sive Brothers and Kernovsky.[74]

By 1898 Lennon had 10 branches, and between 1884 and 1934 its staff increased from 7 to 800. Newly qualified pharmacists were usually offered posts managing small pharmacies in rural areas before being offered larger branches in the towns after proving themselves. Many then set up on their own account in the larger urban areas. They could also buy a chain store franchise and were then able to negotiate wholesale goods at low bulk prices.[75]

The tendency of large companies to register and trade as pharmacies when only one of the directors was a registered pharmacist was viewed as unfair competition by the pharmaceutical associations. In 1927 Morris Stranack, president of the Natal Pharmaceutical Society, proposed that the managing director—not merely any director—should be a registered pharmacist, and this was accepted by the Senate Select Committee. When it was passed in 1928 the Medicine and Pharmacy Act included the Stranack proposal.[76] This prompted the emergence of new chains of retail pharmacies, such as Publix, which was set up by the large Schlesinger Group. This was based on the Boots model in Britain. Publix opened its first store in 1934, and within three years it owned 34 stores. Unable to keep pace with recruitment, they brought out pharmacists from England to work in their stores. But the chain failed within a few years, at least in part because non-pharmacist managers were appointed to senior posts in the Publix operation, causing great resentment amongst the group's qualified pharmacists.[77]

Inter-colonial Differences, 1910

With the end of the war in 1902 renewed efforts were made to improve pharmacy cooperation across the colonies and to form a national pharmaceutical society. The British authorities took steps to ensure that all four colonies followed similar approaches regarding the regulation of pharmacy and the control of poisons. Until 1902 the role of the PSGB in South Africa was more that of an interested bystander than an active participant and advisor, although British developments were followed closely. In September 1901, the Cape Pharmaceutical Society decided to circulate the British Pharmacy and Poisons Act of 1868 to its members as a basis for a General Pharmacy Act for the whole country. The significance of peace in attracting new immigrants was well recognised, and the president hoped that they could 'reasonably expect considerable additions to [the] membership from newcomers to the Land of Hope'.[78] In

Britain pharmacy in South Africa was reported in an extensive feature in the *Chemist and Druggist* in 1909 under the heading 'South Africa Revisited'.[79]

Most licensed pharmacists working in the former Boer republics of the Transvaal and Orange Free State were English-speaking and many were British subjects. Their businesses suffered greatly during the war, with some being forcibly closed by the Boer governments. But those in Natal fared much better, not least because of a massive influx of foreign soldiers and the demand for medicines to treat the sick and wounded.[80] One such was D.S.B. Anderson, who had arrived in South Africa having been trained in Scotland and worked in London. The pharmacy where he worked in Durban compared favourably with the best in London at the time. It was the largest and best equipped he had seen anywhere, and was the centre for all photographic work, accounting for most of the cameras sold in Durban during the war. The pharmacy had a 'soda fountain with a marble-top slab, about eight feet wide and four feet deep', and the shop employed a special assistant to act as barman. 'It was' he wrote 'the only thing of its kind in Durban and became very much in favour'.[81]

Once the Union of South Africa had been established in 1910 pharmacists sought a new Bill that would bring all the colonies under a single pharmaceutical administration. The Medical and Pharmacy Council of the Orange Free State convened a conference of delegates from all the Provincial Medical Councils and Pharmacy Boards to discuss closer cooperation. The pharmacy delegates met separately from the medical practitioners to discuss the details of a draft Pharmacy Bill.[82] A second conference was held in Johannesburg, after which the Bill was submitted to the Union government 'with a strong and unanimous recommendation for its acceptance and introduction at the earliest date'.[83] But it was never published; disagreements broke out and pharmacists in some parts of the Union pleaded for 'special provisions' in their area. South African pharmacists were not speaking with a single voice. The result was that progress was delayed by eighteen years: the new Medical, Dental and Pharmacy Act was finally passed only in 1928.

Ryan suggests that the pharmacy struggle to get the new Act passed can be considered in two halves: the period 1917–1923 when regional prejudices and suspicions led to provincial pharmaceutical associations appearing divided and disorganized; and the period 1924–1928 when a common goal—opposition to a proposed medicine stamp tax—led to formation of a federation of pharmaceutical societies, the Associated Pharmaceutical Societies of South Africa in 1923, and to a united voice.[84] But they also had to address attempts at interference from the doctors.

The Threat from the Doctors

A threat to the autonomy of pharmacy emerged in 1917 as the South African Medical, Dental and Pharmacy Bill was going through the Senate of the Union Parliament; by June 1917 it had passed its second reading. Anxiety spread rapidly in pharmacy circles as a result of some of the wording, with echoes of the British experience in the mid-nineteenth century (Chapter 2). One section announced that

> any person who, not being registered as a medical practitioner, professes to pursue, or by advertisement, representation of any means whatsoever, holds himself out as pursuing the calling of a medical practitioner, or professes to diagnose disease, bodily defects or injuries, and thereafter to prescribe for or treat the same, or professes to give advice in medicine, surgery or midwifery, or holds himself out as aforesaid as so acting.[85]

would be committing an offence under the Act. For the doctors, Sir Thomas Watt claimed that the Bill was intended 'to protect the public and the profession against improperly qualified practitioners, and also to apply the principle of reciprocity as between the Union of South Africa and other countries granting medical degrees'.[86]

The pharmacists thought this clause would 'take away all privileges of chemists and druggists, in regard to legitimate prescribing for minor ailments'. Great effort was made to change the wording of the Bill, and they were ultimately successful in convincing the authorities that legitimate counter trade had always been one of their privileges. The wording was revised to exclude the phrases 'bodily defects or injuries' and 'or professes to give advice in medicine, surgery or midwifery'. The words 'bodily defects or injuries' would have vitally affected the business of chemists, many of whom undertook optical work, and the pharmacists claimed that the public would have been caused great inconvenience and hardship had this been left unchanged.

The same Bill dealt with the regulation of the importation, sale and use of opium and other habit forming drugs and of articles for preventing conception. It stated that 'no person other than a pharmacist shall import, keep for sale, or sell, barter or exchange or otherwise supply any article or appliance calculated to prevent or intended for the prevention of conception'. The *Chemist and Druggist* noted with satisfaction that 'the above is a new provision, and a step in advance of what obtains in this country [Great Britain]. Moreover, it indicates that the official point of view is one of faith in the pharmacist, who may be trusted to carry out the law with discretion'.[87] Two years later it reported that

> the Ministry of Health has arranged for a conference of delegates of the four Pharmacy Councils to meet at Cape Town… to discuss the question of reciprocity between the various provinces, the formation of a joint examination board and other questions affecting pharmacy in the Union.[88]

The Bill passed its third reading with the amendments listed. It was then passed by the Senate and referred to the House of Assembly for ratification. But this was not proceeded with due to 'lack of time'. The *Chemist and Druggist* declared that this excuse 'seems paltry', noting that there was veiled opposition to the Act in some quarters. Amongst the pharmacists there was great dissatisfaction that much good work would be lost, and that its promoters might have to start all over again.[89] The Act which eventually passed in 1928 established a single set of rules for the profession across the country. It adopted the term 'pharmacy' for referring to both the profession and the place where medicines were dispensed. Persons who performed this function were referred to as 'chemists and druggists', and the terms 'chemist and druggist', 'pharmacist', 'pharmaceutical chemist', 'dispensing chemist or druggist', 'dispenser or compounder of drugs', 'pharmacy', 'chemist's shop' and 'drug store' were all reserved for the exclusive use of registered pharmacists.[90]

Other features of professional identity appeared in the 1920s. Although the provincial pharmaceutical associations came together as a loose federation in 1924 it was not until 1946 that a single unified Pharmaceutical Society of South Africa was inaugurated.[91] In 1921 a new journal, the *African Chemist and Druggist*, appeared.[92] This set out to 'promote and maintain a spirit of fellowship' amongst pharmacists, and hoped to 'knit them in a close bond of union, to elevate their status'.[93] It also provided a forum for discussion of issues such as the Pharmacy Bill, and increased communication between pharmacists in urban and rural areas. In 1934 it joined forces with the *South African Pharmaceutical*

Journal.[94] The Act also created the first National Pharmacy Board of South Africa; reciprocity between Provinces became a reality; and it gave pharmacy professional recognition in law. The pharmaceutical societies had played a vital part in the process of professionalization.[95] The Act formed the basis of pharmacy in South Africa for the next thirty years, until its revision in 1957.[96]

But a key element of the relationship between doctors and pharmacists—the quest by pharmacists for monopoly in the dispensing of prescriptions—remained largely unresolved.[97] Leah Gilbert has shown that pharmacists in South Africa were unable to secure a monopoly over the one activity that constituted their core role.[98] Despite achieving a separate Pharmacy Board and finally coming together in a united South African Pharmaceutical Society, the medical profession still exercised substantial control over pharmacy. They achieved this by ensuring that enabling legislation was retained. As Friedson noted in 1981 'most of medicine's control has not been exercised directly in negotiation with clients or employers, but rather indirectly, through licensing, registering and certifying legislation that establishes constraining limits'.[99] Legislation was effective at controlling the work of others and setting the boundaries between occupational groups. 'Through their influence on regulatory agencies the organized professions are often responsible for writing the job description for their members'.[100]

Pharmacy in Southern Rhodesia

Macintosh was not alone amongst British pharmacists venturing further afield into Rhodesia as the continent opened up.[101] By 1898 a Pharmaceutical Society of Rhodesia had been formed in Bulawayo by six founder members: Messieurs Smart, Copley, Moore, Conrath, Cattel, and Scott. Within a few months a Mr. Byrne had taken over the chair from Moore, who became its president. In 1923 the society moved its headquarters temporarily to Salisbury 'for closer contact with Government during deliberations on the forthcoming Medical, Dental, and Pharmacy Act'.[102] Pharmacy legislation in Rhodesia closely mirrored that in South Africa. Apprenticeship training was the norm, with few pharmacy courses available in Rhodesia before 1970.[103] By 1959 the society had joined with its neighbour to become the Pharmaceutical Society of Rhodesia and Nyasaland,[104] and on independence in 1980 it become the Pharmaceutical Society of Zimbabwe.

In 1912 Norman Harris described the work of the pharmacist in Rhodesia for readers of the *Pharmaceutical Journal and Pharmacist*. He painted a vivid picture of life there and of British pharmacy in the Empire. 'Pharmacy itself is much the same throughout the civilized world, but it is left to the environment in which it is practised to give it that touch of romance and interest which to some extent distinguishes it from the "home" variety'. He also revealed attitudes to the indigenous population held by himself and many of his colleagues. He noted that 'it was but a few years ago that the provinces of Matabeleland and Mashonaland were the homes of the untamed African savage—indeed, it is just twenty-one years ago that the pioneer column of white men arrived at Fort Salisbury...and established what was to become the thriving capital of Southern Rhodesia'.[105] As elsewhere in the Empire, the presence of modern Western pharmacies was a potent symbol of colonial dominance and contributed significantly to colonial cultural capital.

British pharmacy had a presence in Rhodesia almost as soon as it was settled. Harris noted that 'both at Bulawayo and Salisbury there are established pharmacies, which for handsome, up-to-date appearance are not beaten and rarely equalled in London itself'. Pharmacy conditions were similar to those at home. 'It is difficult...to realise that one is really in the wilds of Africa, with lions, baboons, leopards, and many other animals within a ten-mile radius'. The business was geared to the expatriate community. 'A whole department is devoted to mining requisites, including ovens, crucibles, retorts, and moulds for production of the royal metal. Here we see medicine chests for the miners and settlers, and cyanide of potassium by the ton'. The customers were described as 'extremely cosmopolitan':

> ladies (mostly wives of Government officials) attired in the smartest costumes arrive by rickshaw, and side by side with them a raw Mashona [person] clad only in an old, starched shirt...Boer farmers and Hindu coolies jostle one another at the counter. [Indigenous] police boys resplendent in khaki tunics and red tarbush come in, followed by an officer of the B.S.A. Police, who has ridden up on a smart charger.[106]

The business extended far beyond supplying local needs. 'A considerable business is done in supplying prospectors, travellers, and hunters bound for the Congo and other parts of Equatorial Africa beyond the great Zambezi, with quinine and other impediments for use on their journey across the veldt'. Harris ended his account with some words of encouragement for 'those budding pharmacists who have in their blood what Jack London terms the "call of the wild"'. There was plenty of work, he said, but 'it is useless for a London man to expect the pleasures of that city away on the veldt. No doubt he will get them in Durban or Johannesburg, but certainly not in Rhodesia'. He believed that Rhodesia had a promising future, and 'the assistant with the Minor at his back can do much worse, when considering foreign experience, than remember the possibilities of that great land between the Limpopo and the Zambesi, which is as yet in its infancy'.[107] Several more took his advice.

The Britishness of Professionalization

Professionalizing pharmacy occurred piecemeal in South Africa. It began in Cape Colony where the pharmacy that the British encountered had already been strongly influenced by previous colonists.[108] The other colonies came under British jurisdiction only later, with Transvaal the last in 1902. These differences had an important impact on the way in which medicine and pharmacy developed in South Africa, although the factors determining professionalization were the same as those elsewhere. The ability for pharmacists to work together towards common goals proved a great challenge in South Africa. Cape Colony ended up with two associations, and cooperation with neighbouring colonies was beset by obstacles. As elsewhere the role of the state—in the shape of the colonial authorities—was central, particularly in relation to seeking change to the membership and function of the Medical Committees that retained a stronghold over the registration and practice of pharmacists.

There were other substantial differences between Britain and South Africa, where pharmacists were included in medical reform at the end of the nineteenth century, and regulated by means of integrated Medical, Dental and Pharmacy Acts. They established clear boundaries between professional groups, but there was just one level of pharmacy qualification. There was no lower qualification as dispenser, as there was in West Africa, and there was no distinction between 'chemist and druggist' and 'pharmaceutical chemist' as there was in Britain. But pharmacists were eventually at least regulated by a Pharmacy Board rather than a Medical Board, even though such Boards included a medical representative. This was an important departure from the British model of pharmacy. Only in the 1920s was formation of a joint examination board considered by Pharmacy Councils in South Africa.

Some features of the British model were reflected in South Africa, not least the link between the regulation of pharmacy and the control of poisons. Pharmacists were able to a secure a monopoly in the supply of some poisons, despite local difficulties associated with supplying farmers and others in remote areas, possible restrictions in the availability of patent medicines, and concerns about the availability and taxation of Dutch medicines. The emergence of company chemists in South Africa also followed the British model. The explanation for the muted role of the PSGB lies partly in the complexity of local politics and partly in the actions of British pharmacists who went there. Whilst keen to see the development of pharmacy in the British image they were also anxious to learn from the experience of British settlers elsewhere in the world. Professional, educational, scientific, and social networks all played important parts in shaping events. Political differences may have delayed the professionalization of pharmacy in South Africa, but the pharmacy practised by pharmacists there was British in all but name.

Notes

1. Dubow, S. 'The Commonwealth of Science: The British Association in South Africa, 1905 and 1929', in Dubow, S. (ed.) *Science and Society in Southern Africa* (Manchester, 2000), pp. 66–100.
2. Thompson, A.S. *The Empire Strikes Back? The Impact of Imperialism on Britain from the Mid-Nineteenth Century* (Harlow, 2005), pp. 17–20.
3. Magee, Gary B. and Thompson, Andrew S. *Empire and Globalisation: Networks of People, Goods and Capital in the British World, c.1850–1914* (Cambridge, 2010), p. 27.
4. Burnby, J. 'The Three Reislings', *Pharmaceutical Historian*, 23 (1993), p. 2.
5. Ibid.
6. Deacon, Harriet. 'Cape Town and the "Country" Doctors in the Cape Colony During the First Half of the Nineteenth Century', *Social History of Medicine*, 10 (1997), pp. 25–52.
7. McMagh, P. *The Three Lieschings: Their Times and Contribution to Cape Medicine 1800–1843* (Cape Town, 1992).
8. Burnby, 'Three Reislings', p. 2.

9. McMagh, *Three Lieschings*, p. 123.
10. *Report and Minutes of Evidence of Select Committee on Public Health,* S.C.25–'83, pp. 12–3. Cited in Ryan, p. 49.
11. Ryan, Mike. *History of Organized Pharmacy in South Africa 1885–1950* (Cape Town, 1986), p. 2.
12. Deacon, Harriet. 'Medical Gentlemen and the Process of Professionalisation Before 1860', in Deacon, H., Phillips, H. and van Heyningen, E. (eds) *The Cape Doctor in the Nineteenth Century: A Social History* (Amsterdam, 2004), pp. 85–9.
13. Ryan, *Organized Pharmacy*, p. 9.
14. Van Heyningen, E.B. 'Agents of Empire: The Medical Profession in the Cape Colony 1880–1910', *Medical History*, 33 (1989), p. 459.
15. Burnby, J. 'The Reislings', *Pharmaceutical Journal*, 249 (1992), p. 652.
16. Ryan, *Organized Pharmacy*, p. 9.
17. Ibid.
18. Goyns, Dorothy W. *Pharmacy in the Transvaal, 1894–1994* (Pretoria, 1995), p. 6.
19. Goyns, *Transvaal*, p. 23.
20. Ryan, *Organized Pharmacy*, p. 56.
21. Wrinch-Schilz, Joyce. *Carboys, Capsules and Crucibles: The History of Pharmacy in Natal* (Pretoria, 2000).
22. *Cape Times*, 14 June (1886).
23. Ryan, *Organized Pharmacy*, p. 51.
24. Couldridge, J.W. *African Chemist and Druggist*, 6 (1926), pp. 25–33.
25. Ryan, *Organized Pharmacy*, p. 51.
26. Ibid., p. 11.
27. Cooper, F. 'Networks, Moral Discourse and History', in Callaghy, T., Kasimir, R. and Latham, R. (eds), *Intervention and Transnationalism in Africa* (Cambridge, 2001), p. 23.
28. Magee and Thompson, *Empire and Globalisation,* p. 27.
29. Holloway, S.W.F. *Royal Pharmaceutical Society of Great Britain 1841 to 1991: A Political and Social History* (London, 1991), p. 216.
30. *Chemist and Druggist*, 27 (1885), pp. 156–7.
31. Clark, P. *British Clubs and Societies, 1580–1800: The Origins of an Associated World* (Oxford, 2000), pp. 452–3.

32. Harland-Jacobs, J.L. *Builders of Empire: Freemasonry and British Imperialism, 1717–1929* (Chapel Hill, 2007), p. 208.
33. *South African Pharmaceutical Journal*, 2 (1935), p. 12.
34. Ryan, *Organized Pharmacy*, p. 14.
35. *South African Pharmaceutical Journal*, 7 (1940), p. 17.
36. Ryan, *Organized Pharmacy*, p. 48.
37. Ibid., p. 33.
38. *Chemist and Druggist*, 74 (1909), p. 42.
39. Magee and Thompson, *Empire and Globalisation*, p. 136.
40. Ryan, *Organized Pharmacy*, p. 53.
41. Ibid., p. 14.
42. *Chemist and Druggist*, 27 (1885), pp. 600–1.
43. Ibid., p. 601.
44. Ryan, *Organized Pharmacy*, pp. 16–7.
45. Ibid., p. 17.
46. *Cape Times*, 21 June (1886).
47. Ibid., 1 June (1886).
48. *Chemist and Druggist*, 30 (1887), p. 291.
49. Ryan, *Organized Pharmacy*, p. 22.
50. Ibid., p. 41.
51. Ibid., p. 43.
52. *Cape House of Assemblies Debates*, 1890, pp. 48, 79–80. Cited in Ryan, p. 49.
53. Ryan, *Organized Pharmacy*, p. 44.
54. Feehan, H.V. *Bond and Link* (Melbourne, 1978), pp. 2–4.
55. Ibid., p. 46.
56. Ibid., p. 49.
57. Ryan, *Organized Pharmacy*, p. 31.
58. *Chemist and Druggist*, 27 (1885), p. 600.
59. Ryan, *Organized Pharmacy*, p. 33.
60. Goyns, *Transvaal*, pp. 18–9.
61. Ryan, *Organized Pharmacy*, p. 52.
62. Ibid., p. 33.
63. Wills, G.S.V. *A Jubilee Souvenir: The Work of GSV Wills and the Westminster College of Chemistry and Pharmacy* (Stratford-upon-Avon, 1899), p. 198.
64. *Chemist and Druggist*, 28 (1886), p. 640.
65. Ryan, *Organized Pharmacy*, p. 53.
66. Ibid., p. 58.

67. Goyns, *Transvaal*, p. 37.
68. Wrinch-Schilz, *Carboys*, p. 32.
69. Ryan, *Organized Pharmacy*, p. 57.
70. *Medical Committee Minutes*, MC 31, 3 December 1891, cited in Ryan, p. 64.
71. Ryan, *Organized Pharmacy*, p. 57.
72. *Bloemfontein Post*, 23 November (1903).
73. Ryan, *Organized Pharmacy*, p. 65.
74. Goyns, *Transvaal*, p. 10.
75. Ibid., p. 48.
76. Ryan, *Organized Pharmacy*, p. 71.
77. Ibid., p. 99.
78. Ibid., p. 63.
79. *Chemist and Druggist*, 74 (1909), pp. 172–5.
80. Wrinch-Schilz, *Carboys*, p. 6.
81. *South African Pharmaceutical Journal*, 11 (1944), p. 3.
82. Ryan, *Organized Pharmacy*, p. 63.
83. Ibid., p. 65.
84. Ibid., p. 67.
85. *Chemist and Druggist*, 89 (1917), p. 38.
86. Ibid.
87. Ibid., 90 (1918), p. 42.
88. Ibid., 92 (1920), p. 52.
89. Ibid., 90 (1918), p. 45.
90. Ryan, *Organized Pharmacy*, p. 73.
91. Ibid., p. 110.
92. Ibid., p. 67.
93. *African Chemist and Druggist*, 1 (1921), p. 5.
94. Ryan, *Organized Pharmacy*, p. 104.
95. Ibid., p. 74.
96. Matthews, Leslie G. *History of Pharmacy in Britain* (London, 1962), p. 146.
97. Gilbert, Leah. 'Dispensing Doctors and Prescribing Pharmacists: A South African Perspective', *Social Science and Medicine*, 46 (1998), pp. 83–95.
98. Ibid., p. 92.
99. Friedson, E. 'The Future of Occupational Monopoly', in Wertheimer, A.I. and Smith, M.C. (eds) *Pharmacy Practice: Social and Behavioural Aspects* (Baltimore, 1981), pp. 17–26.

100. Ibid.
101. Marks, Shula. 'Southern Africa', in Brown, J.M. and Louis, W.R. (eds) *The Twentieth Century: The Oxford History of the British Empire* (Oxford, 1999), p. 556.
102. 'History of the Society', *Pharmaceutical Society of Zimbabwe*, psz.co.zw. Accessed 31 March 2021.
103. Castiglio, Mary, et al. 'Pharmacy education in Zimbabwe', *American Journal of Pharmaceutical Education*, 60 (1996), pp. 182–5.
104. Anon. 'The Pharmaceutical Society of Rhodesia and Nyasaland', *Central African Journal of Medicine*, 5 (1959), pp. 645–7.
105. *Pharmaceutical Journal and Pharmacist*, 88 (1912), pp. 9–10.
106. Ibid., p. 9.
107. Ibid., p. 10.
108. McMagh. *Three Lieschings* (1992), pp. 80–8.

CHAPTER 8

British India: The Failure of Professionalization

In January 1899, the *Chemist and Druggist* observed that 'the practice of pharmacy in India is centred in Calcutta, Bombay, Madras, Karachi and Rangoon, and is almost identical with good class pharmacy in Britain' (Fig. 8.1).[1] This reflected the narrow view of British interests in India; for those in London 'practice of pharmacy in India' meant pharmacy as practised by British pharmacists for the British ex-patriate community. 'A prescription made up in London' it continued 'must be dispensed in the same manner in Calcutta or Bombay, and the drugs and chemicals in usual demand must also correspond'. Indigenous drugs were best avoided: 'No substitution of one drug for another is practicable unless of identical composition'. Developing pharmacy in India meant persuading the authorities in India to adopt the British way of doing things. 'The steadily increasing stream of Anglo-Indians and visitors passing to and fro between India and Great Britain', concluded the *Chemist and Druggist*, 'makes it incumbent on Anglo-Indian pharmacists to adapt their practice to English methods'.[2]

This statement highlighted the fact that there were at this time a great variety of practitioners involved in the making or supply of medicines in India; there were British pharmacists operating British pharmacies, but there were also large numbers of Indians and Anglo-Indians working in diverse settings. Several types of pharmacy practitioner worked in the

S. Anderson, *Pharmacy and Professionalization in the British Empire, 1780–1970*, Medicine and Biomedical Sciences in Modern History,
https://doi.org/10.1007/978-3-030-78980-0_8

Fig. 8.1 Map of the Indian division of the British Empire, 1914

military and civilian sectors, with their responsibilities clearly laid out in military regulations; divisions were made on strictly racial grounds. Initiatives concerning the development of pharmacy usually occurred at a local level, based around the big cities of Calcutta, Bombay, and Madras. Societies and associations emerged within individual occupational groups and based on narrow sectional interests. There was no common purpose, no common organization, and no common vision for the future. Slowly, as their numbers increased, it was Indian pharmacists who pressed for change during the later years of British India.

Categories of practitioner operating in Britain in the early nineteenth century but not at the end—particularly the surgeon-apothecary—continued to be viewed as relevant and useful in India. This chapter describes the various pharmaceutical practitioners operating in India in the late nineteenth century, and considers initiatives taken in attempts at professionalization. It illustrates the division of opportunities and roles along racial lines, and how pharmacy was kept firmly under the control of doctors in the Indian Medical Service. It was not until a Drugs Enquiry Committee reported in 1931 that formal recommendations were made for the passing of legislation relating to the regulation of the pharmacy profession and control over the sale of medicines. A Drugs Act was finally passed (although not implemented) in 1940, and a Pharmacy Act only in 1948, after independence.[3]

Medical and Pharmaceutical Roles in British India

When the British first arrived in India, medical care was largely in the hands of surgeon-apothecaries—people whose roles encompassed everything from surgery to the making and supply of medicines. There was then no clear distinction between medicine and pharmacy; making medicines was an important part of the surgeon-apothecary's training, and pharmacy was simply an integral part of medicine. Physicians were few in number, surgeon-apothecaries took on Indian nationals to help with compounding medicines, and the Indian Medical Service was subject to strict hierarchical division of labour—a military command structure with power resting firmly in the hands of the doctors.

As the British expanded across India and new medical facilities were opened, the demand for medical officers grew. They soon began employing local Indian helpers to carry out a variety of duties. These were variously named 'native dressers' in Madras, 'country doctors' or 'black doctors' in Bengal, and 'black assistants' and 'apothecaries' in Bombay. 'Native doctors' attached to regiments began as dressers or compounders and could later take examinations for promotion to higher ranks.[4] In Bengal, increasing military action required a distinction to be made between military and civilian surgeons. The first steps in organizing medical services in India in a way which made use of local labour came with the creation of the Bengal Medical Service in 1763. This created fixed grades, rules for promotion, and specified conditions of service.[5] Similar services were established by 1764 in both Madras and Bombay, and this rigid scheme set the framework for the practice of pharmacy in India during the nineteenth century.

With the regularization of medical services, the central government moved to assimilate its 'native doctors' into the lower ranks of its medical organization. In 1812 a Subordinate Military Medical Department was established, and the 'native doctors' were termed 'third class servants'. But the arrival of 'Western' medicine in India risked a potential clash between British and indigenous practitioners. Before the 1857–1858 rebellion a unique form of Anglo-Indian medicine had emerged which resulted from the interaction between Western medicine and indigenous medical traditions.[6] Colonial expansion from the seventeenth century lead to a hybrid medical culture, with new drugs appearing in Asian and European markets, and the emergence of surgeons and apothecaries as the new 'agents of medicine'.[7] By the middle of the eighteenth-century hospitals in the larger towns 'typically employed two surgeons, two or three surgeon's mates, and a host of Indian hospital assistants, dressers and medicine-compounders'.[8] The use of local drugs was encouraged. Mridula Ramanna points out that 'in the early nineteenth century British medical policy had encouraged Indian drugs', noting that 'peaceful co-existence prevailed between the Indian and Western systems of medicine'.[9]

The Indian Army was divided into two branches based on nationality, the British and the Native, and the medical and pharmaceutical arrangements for the two divisions differed. Each British (i.e. non-indigenous) regiment had a surgeon, and the strength of the Medical Service grew during the 1850s. Military and civilian medical services were separated in 1858, with the three military medical services (Bengal, Madras and Bombay) combining to form the Indian Medical Service (IMS) under the direction of a Surgeon General reporting to the Government of India.[10] The IMS was responsible not only for the health of both the British and Native Indian armies, but also for the medical care of the civilian non-indigenous population. It remained in existence until independence in 1947. The medical care of the civilian indigenous population remained in the hands of local practitioners. In the IMS, pharmacy was part of medicine, with medical practitioners being responsible for the making and supply of medicines. Dispensing in both the British and Native Armies was sometimes undertaken by themselves, but was more likely to be delegated to staff of the Indian Subordinate Medical Department.

The 'Lingering Apothecary': Assistant Surgeons and Hospital Assistants

The IMS evolved out of a medical system in which the apothecary-surgeon, working alongside the medical officer, played an essential part. It was a system that worked, was understood by all, and there was little reason to change it. The strict rules and grades meant that a clear distinction existed between indigenous apothecaries, later known as 'assistant surgeons', who worked exclusively with European troops, military hospitals, and depots for the British Indian Army, and 'hospital assistants', who worked exclusively with indigenous troops and civilian hospitals in Native Regiments.[11] The recruitment, training, and medical and pharmaceutical duties of the two groups were very different. Assistant surgeons were recruited only from domiciled Europeans and Eurasians: indigenous Indians were barred from admission to the grade in 1873.[12] Candidates had to be aged 16–18 years, provide evidence of good character, and one or both parents had to be of European birth. If successful in the entrance test, they were sent to Calcutta, Madras or Bombay Medical College to attend a medical course which ran over four years. It included instruction in chemistry, pharmacy and materia medica, anatomy and surgery, physiology, and medicine. They then became 'assistant apothecaries', and after further examination they could progress to 'first class apothecary'.

Assistant surgeons remained subordinate to medical officers but were in charge of medicines. A correspondent to the *British Medical Journal* in 1877 noted that 'he either makes up the medicines with his own hands or superintends the native compounder when preparing them'.[13] Their training and role was closely allied to that of the English apothecary of the early nineteenth century. As a correspondent to the *Chemist and Druggist* noted in 1904, on completion of training, the new recruit was 'exactly similar in attainments to the old licentiates of the apothecaries' societies, prior to the passing of the [British] Pharmacy Acts [in 1852 and 1868]. The only drawback' he concluded 'is that apparently there are no "failures"'.[14] On completion of training they were usually posted to a 'Station Hospital' in their local province. They accompanied medical officers on ward visits, noted any treatment prescribed, administered any medicines ordered, and allocated admissions to the wards. They assisted with surgical dressings and dispensed any medicines required. They also dispensed all medicines ordered for officers and their wives and families entitled to medical attendance. Each hospital normally employed three assistant surgeons working on rotation. The *Chemist and Druggist's* correspondent reported that 'he treats all mild cases of illness himself, only sending for the medical officer in severe cases and accidents' and also fulfilled the role of house surgeon.[15] By employing locally trained assistant surgeons, the IMS could staff large garrisons with a third of the number of medical officers they would need elsewhere.[16]

The medical and pharmaceutical arrangements provided for the Native Indian Army bore little resemblance to those in force for the British troops. Each Native Regiment had its own doctor and its own regimental hospital.[17] The doctors were officers in the IMS, whilst the pharmaceutical duties were performed by 'hospital assistants', a branch of the Indian Subordinate Medical Department. The majority of those recruited were the sons of local tradesmen who had shown interest in medicine but were not sufficiently well-off to afford medical training at one of the medical colleges. The remainder were recruited from amongst 'ward servants' at indigenous regional hospitals who showed a special aptitude and were recommended by the regimental surgeon. They needed to be able to read and write in the vernacular, but they did not need to take an entrance examination, and they did not need to know any English. Their training lasted two years, during which they got 'a good grounding in pharmacy and materia medica, and a good smattering of medical knowledge'.[18]

Hospital assistants signed a contract to work for the Government for five years after leaving college. On completion of training, they were posted to a Native regiment for duty, as were the assistant surgeons. They then worked in the hospital for up to two years under the supervision of a more senior person. Their duties were much lighter than those of the assistant surgeons; whilst they paraded the sick for the medical officer in the morning and accompanied him on his ward visits they were not required to dispense medicines themselves, as they were usually assisted not only by a compounder, but by one or more ward servants. Pharmacies in Native regimental hospitals were like those in the British hospitals but on a much smaller scale. The range and quantity of drugs available was very limited compared to those in 'Station Hospitals', and the supply of instruments and surgical dressings was minimal. If a doctor wanted to use a remedy not listed in the 'Scale of Drugs', he could either purchase it himself or persuade the colonel to obtain it from regimental funds. The hospital assistant was also responsible for all other supplies for the hospital.

Hospital assistants held the same rank in the Native Indian Army as assistant surgeons in the British Army. They had a similar system for promotion through several grades; after at least 25 years' service a few were admitted to the ranks of 'jemadar' and 'sabadar', corresponding to lieutenant and captain. The *Chemist and Druggist's* correspondent reported that the hospital assistant was styled 'doctor babu' by the men of the regiment, but that they were of very mixed ability. All those who went to college eventually graduated. 'Some...know their work better than the average Minor man [one holding the PSGB's chemist and druggist qualification] whereas others could not obtain the assistant's qualification of [Apothecaries'] Hall'.[19] Nevertheless, he concluded, pharmacy in the Indian Army 'offers an excellent career for both young members of the domiciled European community...and intelligent native youths. Both get a cheap education, steady employment, and a pension, with the prospect of attaining to a marketable value in civil life which no other Government department confers'.[20]

Subordinate Staff and Indigenous Practitioners

Although the assistant surgeons had to do their own dispensing they often had unofficial assistance. The Indian Army Hospital Corps included indigenous people—chiefly of low caste—who were divided into ward servants, water carriers, cooks, and sweepers. Ward servants assisted assistant surgeons in taking temperatures, administering medicines, applying dressings, and sometimes helping in the dispensary, where some worked full-time. A correspondent reported that

> in some hospitals a pernicious habit has grown up of allowing an old ward servant who has been long employed in the drug-room to pick up a smattering of pharmacy and do much of the dispensing. To such an extent is this carried out that the ward servant in charge of the drug room frequently styles himself the 'compounder' and is proportionately venerated by his colleagues.[21]
>
> This was against the rules; assistant-surgeons were required to do the dispensing themselves, and the passing of such work to a 'compounder' was strictly forbidden. If discovered by a visiting medical officer, firm action was usually taken.

Both the assistant surgeons and the hospital assistants soon became a source of irritation to the English pharmacists in India. On retirement, assistant surgeons frequently opened 'medical halls' in either a cantonment where they were known, or at a hill station, in competition with the English pharmacies. But they were rarely very successful as—according to a *Chemist and Druggist* correspondent—'the cream of this sort of work is done by English chemists who have settled in India. Hospital assistants do better, as they start as doctors and chemists in the native bazaars, where there is a wide scope for counter prescribing'. They used titles such as 'chemist' and 'apothecary' and gave their businesses impressive-sounding names. They treated minor ailments with Western medicines, and often made a lot of money. The correspondent continued

> I know one retained native pharmacist who has a fine well-stocked chemist's shop in a certain bazaar and does an enormous practice as a doctor as well...he is now a rich man. He even attends the poorer class of Europeans and has a great vogue amongst sergeants' wives and people of that ilk.[22]

Bazaars competed with the English pharmacies, and were places where Western and indigenous medicines existed side by side. They were also important spaces where European merchants met with Indian traders to buy local therapeutic products that were then traded overseas.[23]

For the vast majority of the Indian population medical care was in the hands of local practitioners.[24] These included practitioners of Ayurveda medicine (the Vaids); the physicians or 'wise man' who practised herbal medicine in Unani and Islamic medicine, the Hakīms (or Hakeems); and siddhars who were practitioners of Siddha medicine, a traditional medicine originating in Tamil Nadu in the south of India. Whilst most of the indigenous population relied on vaids and hakims, there were in addition huge numbers of herbalists who practised an amalgamation of Ayurveda, Unani, and Siddha forms of medicine. But the two main traditions slowly strengthened along religious lines; Ayurveda became increasingly associated with Hindu nationalism, and texts were translated into Punjabi, Bengali and Hindustani. Unani writings were translated from Arabic and Galenic texts into Urdu. But by the start of the twentieth century greater emphasis was being placed on differences than similarities.

In the early twentieth century indigenous practitioners began to standardize and codify their practices in an attempt to make it appear more 'scientific'. Indian nationalism started to be used to further their cause, and in 1920 the nationalist party, the Indian National Congress, recommended the adoption of the Ayurveda system of medicine as India's national healthcare system.[25] However, interest in indigenous medicine amongst practitioners of Western medicine continued well into the twentieth century. In 1912 a member of the Bengal Medical Council of Registration, E. Houseman, proposed that practitioners of indigenous systems be trained to help Western practitioners through their knowledge of indigenous drugs and methods of treatment.[26] Local sensibilities were to play an important part in determining the future development of pharmacy in India.

British Pharmacists in India

The first 'English' pharmacy in India was opened in Calcutta in 1811 by a Scottish chemist and druggist called Bathgate. The basis of his business was the dispensing of prescriptions and selling toilet requisites, although later he added a photographic department and began manufacturing products such as aerated waters, galenicals and biologicals. The company later opened several branches, including ones in Calcutta and Ballygunge.[27] Bathgate was soon followed by others; around 1815 an apothecary's shop was opened in another part of Calcutta by two British surgeons. They were joined by other partners and the firm changed its name to Smith, Stanistreet and Co. The business initially supplied apothecaries and surgeons, although it gradually built up a substantial retail pharmacy trade. It was later acquired by a young surgeon, Dr. Charles Noyce Kernot, who turned it into a very successful business. Silas Burroughs, on his travels through India in 1883, noted that Kernot had made 'a very large fortune'.[28] Another early pioneer, E. J. Lazarus, a Welshman from Carmarthen, arrived in India around 1839, gained experience in an English pharmacy, before starting his own firm at Benares, where he remained for forty years. The early British pharmacists in India had to be largely self-sufficient. Initially Lazarus undertook his own printing, but this side of the business grew rapidly, and by 1880 he was employing a large staff and doing everything on the premises except making paper.

With the rush of British expatriates at the start of the Raj in 1857, hundreds of British chemists and druggists went out to India to open English pharmacies and 'make their fortune'.[29] New pharmacies opened not only in Calcutta and the other main centres of European population but also their winter retreats. Some quickly became substantial businesses; R. Scott Thomson & Co. opened for business in Calcutta in 1863; Thomson and Taylor opened English pharmacies in Bombay and Poorna in 1872, followed by a large number of branches. J. L. Lyell set up the first English pharmacy business in Mussoorie in the north east of India in 1862, known as the 'Brighton of India'. Like Kernot, Lyell 'made a large fortune' before selling out to Mr. C. F. Fitch in 1885. Many of these pioneering pharmacists came from Scotland. Tom Bliss had undertaken a four-year apprenticeship with a local chemist in Nairn, before sailing to Calcutta in 1863. He moved to Simla in 1869, eventually also opening branches in Lahore and Delhi. David Skinner Kemp qualified as a pharmaceutical chemist in Scotland in 1855. He went out to India and set up in business in Bombay in 1864. He took on two local partners and later returned home to Scotland. William Treacher built up a manufacturing and wholesale chemists' business in Bombay (Fig. 8.2). By 1889 he had branches in Poona and at a Bombay hill station, Mahableshwar.

Fig. 8.2 Treacher & Co. Pharmacy, Bombay, 1894 (Courtesy of Wellcome Collection)

The typical customer of an English pharmacy was described in 1901 by George Cecil:

> officers, members of the Indian Civil Service comprising judges, 'collectors,' magistrates, commissioners, and 'joint magistrates,' opium wallahs, police superintendents, merchants, engineers of the Canal and Public Works Departments, chaplains, officials of the Education Department, and a certain number of Eurasians and natives.[30]

He noted that 'with the exception of officers in British regiments' English chemists frequently dealt with customers 'endeavouring to force him to reduce his prices by quoting those existing in native and Parsee establishments'. English pharmacies employed only English pharmacists; from the 1860s advertisements appeared in the *Pharmaceutical Journal* and the *Chemist and Druggist* for pharmacists to go to India as 'assistants'. Usually, posts were offered for fixed terms—3, 4 or 5 years—and were 'passage paid' and sometimes 'return passage paid'. In practice many stayed, returning to Britain occasionally as travel became easier and quicker. But things soon became more difficult; with the pound falling against the rupee, items imported from England were more expensive, and the value of earnings paid in rupees was much less when returning home.

English pharmacies faced increasing competition from local practitioners. One correspondent reported that: 'There is one common enemy we all have to fight against, and that is the natives, who open a "Medical Hall" in the native part of the town, and get a native compounder – dismissed from some regimental hospital for some fault or other – to do the dispensing'. They included Butto Kristo Paul, who had started a business in the 1860s 'in a tiny little shop located in a small one-storeyed building in the busiest part of Bara Bazar, Calcutta'. By 1896 it had moved to 'a busy narrow side lane, and occupied two upper flats of the building, with an elegant sign board'.[31] By 1905 it employed 300 assistants in retail outlets in Calcutta, and by 1920 it had over 1500 workers.[32] Affluent Indian clients increasingly shunned the British pharmacies in favour of local businesses; and English customers increasingly complained about the prices charged, and sought cheaper alternatives. One by one the English pharmacy businesses were sold on to Indian owners, and increasing numbers of Western pharmacy businesses were started by Indians and continued under Indian management.[33]

The Petition to the Viceroy, 1880

By the late nineteenth century concerns were being expressed about the lack of regulation of both the sale and supply of medicines and of those engaged in doing so. In 1880 a first attempt was made to bring order to the state of pharmacy practice in India. The initiative was taken by J. T. Bilney, a pharmaceutical chemist who worked for a pharmacy business in Agra. Bilney was not only a member of the PSGB but also of the Victoria Pharmaceutical Society in Australia, where a Pharmacy Act had been passed four years earlier (Chapter 10). In 1880 he raised a petition amongst European chemists and druggists throughout India to the Viceroy and Governor-General of India to pass a Pharmacy Act. The petition resulted from 'the great necessity of having some Act passed prohibiting the indiscriminate sale of poisonous drugs or compounds'.[34] The petitioners claimed that public safety required that persons following the occupation of chemist and druggist 'should have an intimate knowledge of the drugs and medicines that they deal with, as men using such compounds may through their ignorance cause grievance danger to their patients'. The action needed to improve matters was spelled out:

> to that end persons not already engaged in such occupations should, before commencing business, be duly examined as to their practical knowledge, and a register of their names should be kept, and also an Act should be passed, entitled an 'Act for Regulating the Qualifications of Pharmaceutical Chemists', to be hereafter described as the 'Pharmacy Act'.[35]

The petitioners suggested that three pharmacy boards be set up to examine candidates, in Calcutta, in Allahabad and in Lahore. They were keen to see developments in Britain replicated in India. 'Such an Act has been passed at home, and subsequently was extended to the Colonies. Therefore, the undersigned beg that immediate attention be granted to the subject, and that an Act entitled the "Pharmacy Act" be speedily brought into law'.[36] But the petition fell on deaf ears. Existing arrangements for the supply and dispensing of medicines in the IMS worked well, and there was no need for legislation to bring things into line with British pharmacy. The British pharmacies were part of retail trade and outside the control of the IMS.

But civilian doctors had been no more successful in securing a Medical Act in India themselves. In the 1860s European and some Indian doctors qualified in Western medicine campaigned for a Medical Registration Act along the lines of the British Medical Act of 1858. Their efforts were repeatedly thwarted by the government. In the light of the 1857 rebellion it was extremely wary of treading on the sensibilities of its new subjects. Local practitioners of Ayurveda and Unani were opposed to such developments, fearing that registering some practitioners but not others would imply a distinction between legitimate and illegitimate forms of medical practice.[37] A Medical Registration Act was eventually passed in 1912 in the Bombay Presidency; similar legislation for the other two presidencies was enacted in 1914.

The European pharmacy community in India persisted in their efforts, and a second attempt to secure a Pharmacy Act was made in 1894.[38] Although described as an 'Act for Regulating the Qualifications of Pharmaceutical Chemists' it was in fact a pharmacy and poisons Act based on the 1868 British Act. The need for a law to establish a pharmaceutical society authorized to examine candidates and keep a register of practitioners, on the lines of the PSGB, was restated. By limiting the sale of poisons to registered pharmacies it would protect the public from poisoning by unqualified dispensers.[39] But again, no action was taken by the authorities; the British pharmacists were a small group and their views could be easily dismissed. In any case such a scheme would seriously disadvantage much of the population even if it were enforceable. The petitioners claimed that their petition had arisen from a need to prohibit 'the indiscriminate sale of poisonous drugs or compounds'. Yet any such move was likely to receive a hostile reception from indigenous people and traders.

Training Chemists and Druggists in India

In the early 1850s steps were taken to provide training for Indian-born Europeans and Eurasians wishing to qualify as chemists and druggists. In 1852 the Government approved a plan whereby persons wishing to open pharmacies in Madras could be examined by a committee of professors at the Medical College. The college provided a Western education for both Indian nationals and Anglo-Indians. Pharmacy education then became a function of medical colleges, giving doctors overall responsibility for the training of pharmacists.[40] The first pharmacy class at Madras Medical College in 1860 was aimed at medical students rather than aspiring chemists and druggists.[41] One of the first pharmacy students was H. W. Honey, who worked at the Esplanade Dispensary in Madras. He attended courses in chemistry, materia medica and pharmacy, and in June 1866 he was examined by the professors and received a certificate confirming his qualification as a chemist and druggist.

It was subsequently agreed that a course specifically for chemists and druggists should be run annually in Madras. This course started about 1874; initially it was of one year's duration and no entrance examination was required. By the late 1870s only a handful of chemists and druggists were being trained each year. In 1886, the duration of the course was extended to two years, and an entrance requirement was specified; candidates must have passed the middle school examination. Numbers completing the course remained extremely low; a report from 1887 to 1888 recorded that 'the students who had completed the two years' curriculum recently appeared before the Board, and both passed creditably'.[42] In 1893 the entry qualification was raised, to having passed either the matriculation examination or the compulsory subjects of the upper secondary examination.

Some basic pharmacy training was provided for compounders employed by the Indian Medical Service, although arrangements varied from province to province. In Bengal statutory provision was made for the education and training of compounders in 1881. Practical classes were offered at Medical Colleges in Calcutta, Patna, Dacca, and Cuttack in Orissa. Successful candidates received a certificate enabling them to compound and dispense European medicines. Approval was later given to classes given at several Mission Hospitals, and a class for female compounders was established at Campbell Medical School in Calcutta.[43] In Bombay no qualifications were specified for compounders. In Assam, Bihar, and Orissa one year's training was required after 1926. A similar scheme started in the United Provinces in 1928, but there was no system of training for compounders at all in the Punjab.[44]

The state of pharmacy education in India was reported in Britain.[45] In 1894 the *Pharmaceutical Journal* noted that the Madras course had only a handful of students, whilst the great majority of pharmacy practitioners were trained at a lower level in hospitals under government schemes. The *Chemist and Druggist* reported that 'while medical education in India has reached a fairly high status, there has been little corresponding advance as far as pharmacy is concerned…the examiners were medical men in government service'. The examination was little different to that used for 'native compounders'. It noted that 'classes are available at the Medical Schools, but [are] seldom taken advantage of'. It was unfortunate that 'this is the only examination available to apprentices and assistants brought up in the country. There should certainly be a higher examination provided for these'.[46]

Committee of Enquiry, 1895

In May 1895, the British authorities in India appointed a committee to consider the education and training of chemists and druggists and others. The committee was strongly guided by the regulations governing qualifications in Britain. Their report was considered by the Government of India in 1898, by which time the number of student chemists and druggists at the College had shown no sign of increasing. Over the decade 1884–1894, only 28 students had enrolled on the course. The committee recommended that its duration should continue to be two years and made several proposals for changes to its content to bring it more into line with that in Britain. A course in botany was added, along with a 3-month laboratory course in the second year. Candidates would need to complete a one-year apprenticeship in a chemists' shop after their first year of study. Finally, it was recommended that the scheme be brought under the Governments' Technical Examination Scheme.[47]

The Government accepted almost all the committee's recommendations, although the suggestion that the entrance qualification be lowered was rejected; existing requirements would remain. When fully implemented the course would, it was claimed, effectively be on a par with the chemist and druggist qualification in Britain. But the course failed to attract candidates, with less than half a dozen students enrolled per year. Regular attempts were made to abolish the course, and in 1903 the entrance qualification was lowered against the government's wishes, although this decision was reversed in 1907. Continuance of the course was confirmed for three years initially and annually thereafter, with exemptions made to the admission criteria in special cases. The course continued to have Government support, and it was directed that it should continue 'in the Madras Medical College as a permanent arrangement'.[48] Further changes to the curriculum were made, including the addition of organic chemistry and the exclusion of the special laboratory course, and the course carried on running for several decades.

The reasons for the unpopularity of the course were all too clear. This was a British qualification based on Western medicine designed to prepare candidates for work in European pharmacies in India. For most newly qualified chemists and druggists there was little prospect of employment unless they had family connections, and with no Pharmacy Act there was no statutory protection of their professional interests against those with few or no qualifications. These included compounders who had previously worked in hospitals and dispensaries for the IMS, whose own qualifications were gradually improving. It was made compulsory for trainee compounders to be able to read and write English, and later that they must be trained at a major dispensary for six months and pass an examination. With each additional requirement, pay scales improved and differentials between compounders and chemists and druggists narrowed. But with no regulations preventing them doing so, compounders called themselves 'chemists and druggists' and opened up their own shops.

The Committee of Enquiry also examined the education of compounders; it recommended that the course be extended to at least nine months at the major dispensaries. A syllabus in practical dispensing was laid down, a qualifying examination was held once a year under the Technical Examination Scheme, and both oral and practical examinations were set. Its recommendations were accepted in full by the government. India had a means of training chemists and druggists to the level of those in Britain, but no means of providing them with protections such as restrictions on the use of titles and monopoly in the sale of poisons that would be given through a Pharmacy and Poisons Act.

Pharmaceutical Organizations in India

The differences between the various groups engaged in pharmaceutical activities in India—the British pharmacists, the assistant surgeons, and the hospital assistants—were all too apparent to those involved. They had very different concerns, and the different groups took steps to establish their own pharmaceutical organizations. From the beginning, the Indian apothecaries (later called assistant surgeons), felt discriminated against. In the early 1850s they had formed an informal society which lasted until the 1857 Rebellion, but it was revived again in 1864 as the Madras Apothecaries Society.[49] The names of attendees were those of Indian practitioners born to European or Eurasian parents.[50]

British pharmacists outside the big cities were often the first to act. In 1894 the *Chemist and Druggist* noted that 'the pharmacists of the north west of India, though numerically small, prove to be a very active body'.[51] With the failure of attempts to secure a Pharmacy Act, initiatives were taken in 1905 to establish an Indian Pharmaceutical Society, with centres in Calcutta, Bombay and Madras.[52] Progress was extremely slow, but by 1913 the *Chemist and Druggist* was able to report that

> the project of forming a Pharmaceutical Society is being favourably received. Practically every European chemist of standing in India has promised to join. In view of the absence of cohesion between those who practice pharmacy, important developments are expected from the movement.[53]

A Pharmaceutical Society of India (PSI) was founded that year, with Walter T. Grice as its first president.[54] The Society agreed several initiatives, including proposals for anti-adulteration legislation.[55] But within a year, with war looming, things came to an abrupt halt and the Society folded.[56]

Other Indian groups engaged in Western medicine and undertaking pharmaceutical activities set up their own associations (Table 8.1). Hospital assistants (later 'sub-assistant surgeons'), were members of the All-India Sub-Assistant Surgeons Association.[57] Those whose duties were mainly pharmaceutical founded a Pharmaceutical Section at their conference in Madras in 1916.[58] With the demise of the PSI this group proposed representing 'all classes of qualified men, qualified pharmacists, and scientists'. It was a bold and ambitious call to unite the British pharmacists, the Eurasian assistant surgeons, and the Indian hospital assistants. Its aims were vague and wide-ranging, including to 'promote the interests of Indian drugs' and to 'give impetus to the knowledge and practice of pharmaceutical chemistry in general'.[59] Nothing seems to have come of the initiative, and nothing more is heard about it after 1918.

With the end of the war renewed efforts were made to unite pharmaceutical practitioners across India. This time the initiative was taken by Indian rather than British practitioners. In 1921 a Calcutta Chemists and Druggists Association was founded by H. N. Paul, the grandson of B. K. Paul. This had a very narrow remit—lobbying for favourable exchange rates from British banks—and was very short-lived.[60] A Pharmaceutical Society of India re-emerged in Madras in 1923 as the 'Pharmaceutical Association', but two years later it reverted to the name 'Pharmaceutical Society of India'.[61] The key figures behind it were two Indian-born Eurasian pharmacists, A. N. Lazarus and Wilfred Pereira. Membership was

Table 8.1 Pharmacy in British India before 1931

Indian province	*Year boundary set*	*Local and national pharmaceutical society or association*	*Early pharmacy related legislation (all India)*	*Training of compounders and chemists and druggists*
Assam	1874	None before 1931	Sea Customs Act 1878	One year training after 1926 but no minimum entrance qualification
Bengal	1826	Indian Pharmaceutical Society 1923	Opium Act 1878	Training at Dacca Medical School 1881
Bihar and Orissa (from Bengal)	1912	Indian Pharmaceutical Society 1923	Merchandise Marks Act 1889	Pharmacy classes at Orissa and Temple Medical School Patna 1880s
Bombay	1687	Indian Pharmaceutical Society (Calcutta, Bombay and Madras 1923)	Tariff Act 1894	No regular rules for training and qualification of compounders
Calcutta	1793	Calcutta Chemists and Druggists Association 1921	Poisons Act 1919	Pharmacy classes at Calcutta Medical School 1880s
Madras	1640	Madras Apothecaries Society 1864	Cantonments Act 1924	Chemist and Druggist pharmacy classes at Madras Medical College 1860
The Punjab	1849	None before 1931	Sale of Goods Act 1930	No regular system of training compounders
United Provinces of Agra and Oudh	1902	None before 1931	Dangerous Drugs Act 1930	Scheme for training of compounders at six centres 1928

Sources Singh, Harkishan, *Pharmaceutical Education* (Delhi, 1998), pp. 11–25; *Eastern Pharmacist* 42 (1999), pp. 31–7; *Eastern Pharmacist* 44 (2001), pp. 21–8

restricted to those with recognized pharmaceutical qualifications. Thereafter voluntary organizations of chemists and druggists were formed at the city or state levels in all the major metropolitan centres.[62]

Central Indigenous Drugs Committee, 1894–1895

Until the mid-nineteenth century there was no legislation in India which might help prevent drug adulteration or assure the quality of medicines.[63] In the absence of an Act linking the regulation of pharmacy with the control of poisons, separate legislation was required. From the 1870s several statutes were passed which provided a degree of protection in response to rapidly growing imports.[64] Penalties could be applied under the Indian Penal Code for the intentional adulteration of a drug, or the sale of a drug 'not of the nature, quality or substance demanded by the purchaser'. In 1878 a Sea Customs Act prohibited the import of goods with misleading trade descriptions. A Merchandise Marks Act in 1889 provided checks on misbranding, false marketing and trade description, and an Indian Tariff Act followed in 1894. Acts to control the manufacture, import, or sale of certain drugs were passed, including an Opium Act in 1878, a Poisons Act in 1919, and a Dangerous Drugs Act in 1930, but none provided controls over adulteration or potency, and all were widely ignored. A Cantonment Act of 1924 gave powers to the authorities to enter any place and seize any medicine which was adulterated or different from how it was represented, but action was rarely taken against offenders.

Nandini Bhattacharya locates the root of the adulteration problem in the composite nature of the medicines market in India; patent and proprietary medicines were imported in vast quantities from Britain and elsewhere, British Indian agents traded in preparations they had manufactured themselves, and Indian druggists imported on a large scale as well as trading with local Indian firms.[65] Competition was fierce and there was little incentive to cooperate; this only came when the Government of India attempted to regulate the prices charged for drugs and profit margins. A Drugs Control Order in 1943 mandated that no wholesaler or retailer was to sell a drug at a price higher than that specified in the Order, and the entry of newcomers to the trade was restricted. It affected all categories of dealer and served as a rallying call to chemists and druggists across the country. An All-India Chemists and Druggists Federation was founded in 1944, bringing together associations in Bombay,

Calcutta, Madras and Delhi. But it was not until 1960 that an All-India Retail Chemists Association was founded, with aims that bore a close resemblance to those of the PSGB when it was founded in 1841—'to protect the interests of chemists in general and retailers in particular, and to promote improvement and advancement in their scientific and professional knowledge'.[66]

Pratik Chakrabarti notes that the circulation of drugs in India can only really be understood in terms of enriching the intellectual as well as the material wealth of European imperialism.[67] The issue of drug substitution (expensive European drug imports being replaced by cheaper local ones) was a key topic discussed at the first Indian Medical Congress in Calcutta in 1894. Attention was directed to the study of indigenous drugs, their cultivation, and their increased use in medical depots.[68] In response, the government reluctantly agreed to set up a Central Indigenous Drugs Committee in 1894.[69] It consisted of three British doctors; the Inspector-General of Civil Hospitals in Bengal, the Medical Storekeeper in Calcutta, and the Senior Medical Officer, British Troops at Calcutta. They were explicit in denouncing drug substitution; they considered that raw drugs obtained in India were impure and adulterated. Locally produced medicines were of doubtful efficiency and the cost of removing adulterants made them much more expensive than imported ones.[70] For drugs such as quinine and morphine it observed that 'there are no manufactories in India which can turn out such salts at cheaper rates than they can be imported'.[71] The aim of the British authorities was to keep Indian drug requirements largely dependent on imports from Britain.[72] Consumption of allopathic medicines increased enormously in India during the course of the nineteenth century, and was a small but important part of trade.[73]

Drugs Enquiry Committee, 1930–1931

The early twentieth century brought significant political and social changes to British India.[74] Fewer British pharmacists travelled there, and European pharmacies were increasingly sold to Indian-born pharmacists. But the initiative which lead ultimately to the regulation of the pharmacy profession came not from pharmacists but from growing concern about the quality of drugs in use. The country continued to depend heavily on imports for Western drugs, many of which were fraudulent or of inferior quality. A campaign for effective legislation gradually gained momentum; the Indian Merchants' Chamber in Bombay called on the Government of India to take action; Sir P. C. Ray declared that 'it certainly was a reproach to the big cities of Calcutta, Bombay and Madras that no legislation existed as yet to compel the local druggists to obtain the services of duly qualified men'.[75] In March 1927 a resolution was adopted by the Council of State recommending the Governor-General 'to urge all Provincial Governments to take such steps as may be possible to control the indiscriminate use of medicinal drugs and to legislate for the standardization of the preparation and for the sale of such drugs'. The Government of India subsequently appointed a committee 'to explore and define the scope of the problem, and to make recommendations as to the measures that should be taken'.[76]

In August 1930, a second Government resolution set out its terms of reference. The Committee was

(1) to enquire into the extent to which drugs and chemicals of impure quality or defective strength, particularly those recognised by the *British Pharmacopoeia*, are imported, manufactured or sold in British India, and the necessity, in the public interest, of controlling such importation, manufacture and sale, and to make recommendations;
(2) to report how far the recommendations made in (1) may be extended to known and approved medicinal preparations other than those referred to above, and to medicines made from indigenous drugs and chemicals; and
(3) to enquire into the necessity for legislation to restrict the profession of pharmacy to duly qualified persons and to make recommendations.[77]

The terms of reference indicated Government of India recognition of pharmacy as a 'profession' for the first time. The Committee was chaired by Lieutenant Colonel R. N. Chopra, professor of pharmacology at Calcutta,[78] and he was assisted by an Indian analytical chemist, a pharmacologist from Bombay, and Mr. H. C. Cooper, a British-born pharmaceutical chemist with Smith Stanistreet & Co. The committee sought the views of 'all persons, associations or bodies interested in this question' and drafted a preliminary questionnaire to be widely circulated in India. The questionnaire asked different questions of the medical profession, manufacturers of drugs and chemicals, importers and dealers, and dispensing chemists. Doctors were asked if they considered that control of therapeutic agents on the lines enacted in Britain or the United States was desirable in India. Alternative suggestions were invited. The questions for dispensing chemists related solely to the supply of medicines and poisons.

The appointment of the committee produced an indignant response from nationalists; it was thought to be a government response to a campaign by Congress to boycott British drugs. Not only did European drugs still dominate the drugs market, but the government was accused of failing to support the indigenous drug industry.[79] But the *Indian Medical Gazette* welcomed it: its appointment marked 'a new era for medicine in India. We look forward to a day when India will have its own pharmacopoeia, based largely upon minerals and plants obtainable in this country'.[80] The committee received responses from an enormous number of those engaged in the drugs trade, virtually all of whom were Indian.[81] The Chemists and Druggists' Association of Madras was the only drug trade organization to submit written evidence.[82]

The committee sat for over a year and presented its report with appendices in 1931.[83] It recommended that all proprietary medicines with a secret formula, whether manufactured in India or imported, should be registered along the lines of the Patent and Proprietary Medicines Act of Canada.[84] Imported medicines with undisclosed formulas would be subject to additional customs duty of 20 per cent, and the import duty on manufactured drugs increased by 5 per cent.[85] A number of crude drugs not available in India had to be imported, but were subject to 30 per cent import duty. This put the Indian drug industry at a considerable disadvantage. The committee noted that 'in the interest of the indigenous pharmaceutical industry the import of raw materials not available in India deserves special treatment'. It recommended that the import duty be either abolished or significantly reduced. But this was not implemented at the time, and in 1938 Indian drug manufacturing companies approached the Government of India directly seeking the abolition of the duty on raw drugs imported from overseas.[86]

Pharmacy and the Drugs Enquiry Committee

Pharmacy education received considerable attention from the committee. Sir Prafulla Ray had visited several schools of pharmacy in England in 1921 and was greatly impressed with what he saw at the PSGB's School of Pharmacy in London.[87] He regretted that no such facilities were available for aspiring pharmacists in India. He proposed the establishment of separate schools of pharmacy awarding statutorily recognized qualifications in India.[88] However, the teaching of pharmacy remained the responsibility of the medical schools, and—as Anil Kumar has noted—it 'was never taken up seriously in the medical schools and colleges of the country'.[89] Some minor changes were made; the training programme for compounders in Bengal was improved in 1928.[90] But complaints continued; better trained pharmacists were needed both in retail and industry. In 1930 Jyotish Ghosh stated that without systematic pharmaceutical education it would be 'impossible either to produce the drugs of the quality required in order to be of therapeutic value, or to provide the desired machinery for enforcing legislative control of the sale, manufacture, analysis, and dispensing of drugs'.[91]

The PSI submitted evidence emphasizing the need for a Pharmacy and Poisons Act, to restrict dispensing to qualified chemists, and to raise the standard of the examination set for compounders, to make them 'more useful to the pharmaceutical profession'.[92] In London the PSGB followed events in India with interest. An editorial in the *Pharmaceutical Journal* in 1930 noted that the PSI was 'a small but very active and earnest body' and referred to the evidence it had submitted to the Committee. But it remained concerned about the state of pharmacy in India. It began

> Chemists in this country [Britain] already in possession of some of the elementary rights for the attainment of which the PSI is seeking, will sincerely wish them success in their efforts, to put an end to the outrageous anomalies and scandals by which, under existing anachronistic conditions, so much of the drug trade in India is disgraced.[93]

Despite the fact that the majority of the civilian population depended on Ayurveda, Unani or Siddha practitioners, the demand for Western medicines in India continued to grow. F. V. Wells, a British pharmacist in India, wrote that 'there are millions of people in this country suffering from curable complaints. There are thousands of doctors to prescribe for them. There are forty or so pharmacists qualified to carry out the prescriptions. Forty! The position of pharmacy in India is truly a farce'.[94]

Having considered the evidence, the Committee found the state of pharmacy in India to be 'uncontrolled and unsatisfactory'. It recommended the statutory regulation of the profession, although it did not specify how. It suggested that appropriate legislation might consist of either a combined drugs and pharmacy Act or separate drugs and pharmacy Acts.[95] Chopra reflected on the state of pharmacy in a presidential address to the Bengal Pharmaceutical Association in 1937.

> Though at the present time no organized and self-contained profession of pharmacy is in existence in India, as in other civilised countries, during the last two years there has been unmistakable evidence of an urge on the part of the Association to improve the existing conditions.

Evidence submitted to the Committee showed that 'the character and quality of the work turned out by the profession as a whole in this country was far from being complementary'. The Committee concluded that this 'regrettable state of affairs' was due to two main factors; firstly, the lack of educational qualification of those who took up the profession; and secondly, the absence of restrictive laws preventing the practice of the profession by unqualified persons. These were the root of the evil, he thought. If action could be taken to improve these, pharmacy 'would be able to take its proper place by the side of the honourable and dignified profession of medicine in this country'.[96]

Chopra was keen not to apportion blame for the situation, although there was little doubt about where he thought it lay. 'The profession of pharmacy has lagged behind' he said, 'through no fault of its own perhaps'. He was also clear that 'pharmacy is intimately related to medicine, and it may be said without contradiction that the practice of pharmacy is an integral part of the practice of medicine'. Nor could the blame be placed entirely at the feet of the medical profession. In the winter of 1936–1937, Dr. G. C. Anderson, the secretary of the BMA in London, was sent to India to investigate the conditions of medical practice and organization in the country. His report to the BMA Council noted that the medical profession in India 'was disturbed by the fact that there was no organized or self-contained profession of pharmacy, and that there was no proper measure of drugs control'.[97] The report noted that 'there is a lack of adequately qualified men among those who take up the profession of pharmacy, and there is an absence of any restrictive laws preventing the practice of pharmacy by unqualified persons'. Furthermore, 'no attempt is made to control the quality of drugs sold'.[98]

The Government of India was in no hurry to act on the recommendations concerning both the control of drugs and the regulation of pharmacy. It took nine years for it to enact a Drugs Act in 1940, and another five for the Act to be implemented. In 1942 the Government established an Advisory Panel on Drugs and Medicines to advise on which items were essential imports for India.[99] It was again chaired by Chopra who consulted widely.[100] It was 1945 before the Drug Rules needed to implement the Drugs Act were ready; it took even longer to prepare the way for a Pharmacy Act.[101] Some progress was made concerning pharmacy education; a second chemist and druggist course was started in 1937 at Vishakapatnam, and in 1940 the name of the course at Madras Medical College was changed to a 'diploma in pharmacy'. Its duration was extended to two and a half years, but the number of students remained tiny, varying between four and six per year. On independence in 1947 there were just three pharmacy degree awarding institutions and two offering diploma level courses in the whole of India. Pharmacy training at the Madras Medical College finally ended in the 1970s.[102] The statute to regulate the practice of pharmacy in India—the Pharmacy Act of 1948—was passed one year after independence, when the PSI achieved legal status.[103] Pharmacy in India finally received recognition as an autonomous profession, but it had to wait for the end of British rule to do so.

The Failure to Professionalize

Although the Drugs Enquiry Committee 1931 recommended the changes that would eventually lead to the transformation of pharmacy in India, the tardiness of the British authorities to pass the necessary legislation and their unwillingness to implement it before independence constituted the total failure of professionalization. This failure was reflected in all five of the markers identified earlier; collective action, the passage of legislation, autonomy from other professional groups, raising educational standards and above all the role of the state. To emerge as a self-regulating independent profession in India, pharmacy clearly had many hurdles to overcome, not least the legacy of the long-standing British presence. From an early date, medical practitioners with all-round qualifications, including pharmaceutical skills, were greatly valued. The surgeon-apothecary suited the needs of the Indian Medical Service, and the availability of locally trained individuals with medical, surgical, and pharmaceutical skills offered a means of supporting the British and Indian Armies at least cost.

The disparate groups engaged in the sale of drugs and the practice of pharmacy—the assistant surgeons, the hospital assistants, the former 'compounders' and unqualified individuals selling 'Western' medicines in the bazaars, along with the British chemists and druggists—had very little in common, often not even a language. There was no way in which they could come together to attain the degree of unity needed to form societies to promote their collective interests. This lack of a single identity led to groups forming their own associations and societies on a voluntary basis without statutory recognition. Throughout British rule, the medical profession kept pharmacy firmly under its control; it remained an integral part of medicine in India until after independence. Ultimately, the successful professionalization of pharmacy required the support of the Government of India. The Government had more pressing matters to deal with, and pharmacy was never going to be a high priority.

If circumstances precluded the professionalization of pharmacy, a recognizably 'British' pharmacy was nevertheless practised in English pharmacies in India before independence. But its influence did not extend much beyond the British and European community. Yet it was supported by few aspects of the British model of pharmacy; there was no Pharmacy Act providing for its legal separation from medicine, and no division of representative and regulatory functions between a pharmacy association and a Pharmacy Board. Pharmacy was not a dual qualification profession in India; British-trained pharmaceutical chemists were treated no differently to those holding the chemist and druggist qualification. There was no link between the control of medicines and the regulation of pharmacy during British rule. After independence the Indian Government implemented separate legislation for medicines and pharmacy. But small chains of pharmacies did develop in the absence of anything preventing them from doing so.

The failure of pharmacy to emerge as an independent and autonomous profession in India during a hundred years or more of British rule has been the subject of considerable criticism by both historians and pharmacists. Many explanations have been offered, but lingering support for the all-round 'apothecary' of the early nineteenth century and the power and authority vested in the IMS made significant contributions. Harkishan Singh's assertion that pharmacy practice stood neglected throughout the period of British rule is difficult to refute. As he concludes, on independence India 'inherited from the British a profession of pharmacy practice which was unorganized and in a very poor state'.[104] In the end, India was just too large, too diverse, and too politically difficult for a single vision—a British vision—of Western pharmacy to be established while the British themselves were still there.

Notes

1. *Chemist and Druggist*, 54 (1899), p. 22.
2. Ibid.
3. Singh, Harkishan. *Pharmacy Practice* (Delhi, 2002), p. 5.
4. Ibid., p. 18.
5. *Indian Medical Gazette*, 36 (1901), pp. 41–8.
6. Harrison, Mark. 'Medicine and Orientalism', in Pati, B. and Harrison, M. (eds), *Health, Medicine and Empire: Perspectives on Colonial India* (New Delhi, 2001), p. 74.
7. Chakrabarti, Pratik. *Medicine and Empire 1600–1960* (London, 2014), p. 2.
8. Harrison, Mark. *Medicine in an Age of Commerce and Empire: Britain and Its Tropical Colonies, 1660–1830* (Oxford, 2010), p. 22.
9. Ramanna, Mridula. *Health Care in Bombay Presidency, 1896–1930* (Delhi, 2012), p. 158.
10. Harrison, Mark.*Public Health in British India: Anglo-Indian Preventive Medicine 1859–1914* (Cambridge, 1994), p. 7.
11. *British Medical Journal*, 2 (1868), pp. 86–7.
12. Annual Report of the Madras Medical College, Session 1873–1874 (1874), p. 18.
13. *British Medical Journal*, 2 (1877), pp. 396–7.
14. *Chemist and Druggist*, 64 (1904), pp. 690–2.
15. Ibid., p. 690.
16. Singh, Harkishan. 'Apothecaries and Hospital Assistants in Colonial India', *Pharmaceutical Historian*, 32 (2002), pp. 15–6.
17. *Chemist and Druggist*, 64 (1904), p. 690.
18. Ibid., p. 691.

19. In Britain army apothecaries did not need to be examined by the Society of Apothecaries. Matthews, Leslie G. *History of Pharmacy in Britain* (London, 1962), pp. 116–7, footnote.
20. *Chemist and Druggist*, 64 (1904), p. 690.
21. Ibid., p. 691.
22. Ibid., p. 692.
23. Chakrabarti, Pratik. 'Medical Marketplaces beyond the West: Bazaar Medicine, Trade and the English Establishment in Eighteenth Century India', in Wallis, P. and Jenner, M. (eds) *Medicines and the Market in England and its Colonies, c.1450–c.1850* (Basingstoke, 2007), pp. 196–215.
24. Chakrabarti, *Medicine and Empire*, p. 187.
25. Ibid., p. 188.
26. Bala, Poonam. *Imperialism and Medicine in Bengal: A Socio-Historical Perspective* (London, 1991), p. 54.
27. Singh, *Pharmacy Practice*, p. 76.
28. Burroughs, S.M. 'Notes on Travel', *Chemist and Druggist*, 25 (1883), pp. 143–4.
29. Anderson, Stuart. 'Look East Young Man: Tales of British Pharmacists in India during the Raj', *Pharmaceutical Journal*, 281 (2008), pp. 746–8.
30. Cecil, George. 'European Pharmaceutical Establishments in India', *The Pharmaceutical Journal and Pharmacist*, 107 (1921), pp. 88–9.
31. Marshall, John D. 'A Tour of India', *Chemist and Druggist*, 104 (1926), pp. 757–60.
32. Bhattacharya, Nandini. 'Between the Bazaar and the Bench: Making of the Drugs Trade in Colonial India, c.1900–1930', *Bulletin of the History of Medicine*, 90 (2016), p. 76.
33. Singh, *Pharmacy Practice*, p. 92.
34. *Chemist and Druggist*, 22 (1880), pp. 83–4.
35. Ibid., p. 83.
36. Ibid.
37. Harrison, 'Medicine and Orientalism', p. 75.
38. *Chemist and Druggist*, 22 (1880), p. 83.
39. *Indian Journal of Pharmacy*, 1 (1894), pp. 5–6.
40. Singh, Harkishan. *Pharmaceutical Education* (Delhi, 1998), p. 13.

41. Education, Government of Madras, G.O. No. 87, 27 March (1866), Tamil Nadu Archives, Madras.
42. Singh, *Pharmaceutical Education*, p. 13.
43. Ibid., p. 18.
44. Ibid., p. 20.
45. Ibid., p. 14.
46. *Chemist and Druggist*, 45 (1894), p. 476.
47. Singh, *Pharmaceutical Education*, p. 15.
48. Ibid.
49. *Madras Quarterly Journal of Medical Science*, 8 (1865), pp. 466–73 and 9 (1866), pp. 190–3.
50. Singh, *Pharmacy Practice*, p. 31.
51. *Indian Journal of Pharmacy*, 1 (1894), pp. 5–6.
52. *Chemist and Druggist*, 67 (1905), p. 823.
53. Ibid., 83 (1913), p. 41.
54. Singh, *Pharmacy Practice*, p. 75.
55. *British and Colonial Druggist*, 64 (1913), pp. 485–6.
56. Gardner, H.C.T. 'Pharmacy in India', *British and Colonial Druggist*, 65 (1914), p. 547.
57. *Eastern Pharmacist*, 44 (2001), pp. 21–8.
58. *Indian Medical Record*, 37 (1917), p. 58 and *Indian Medical Gazette*, 52 (1917), p. 116.
59. *Indian Medical Record*, 38 (1918), pp. 56–7.
60. *Chemist and Druggist*, 94 (1921), p. 70.
61. *Indian Journal of the History of Science*, 35 (2000), pp. 67–76; *Indian and Eastern Druggist*, 9 (1928), pp. 226–7. *Chemist and Druggist*, 109 (1928), p. 273.
62. Singh, Harkishan. 'Chemists and Druggists' Organizations', *Eastern Pharmacist*, 44 (2001), pp. 21–8.
63. Singh, Harkishan. 'History of Drugs and Pharmacy Statutes', *Eastern Pharmacist*, 42 (1999), pp. 31–7.
64. Ibid., p. 36.
65. Bhattacharya, 'Between the Bazaar and the Bench', pp. 61–91.
66. *Memorandum of Association and Regulations of the All-India Retail Chemists Association* (1961).
67. Chakrabarti, *Materials and Medicine*, pp. 171–204.
68. Kumar, Anil. 'The Indian Drug Industry under the Raj, 1860–1920', in Pati and Harrison, *Health, Medicine and Empire*, p. 356.

69. Home, Medical, 31 October 1895, No. 24, National Archives of India, New Delhi (cited in Kumar, p. 358).
70. *Report of the Central Indigenous Drugs Committee, 1896* (Calcutta, 1901), p. 165.
71. Ibid.
72. Kumar, 'Indian Drug Industry', p. 380.
73. Ibid., p. 364.
74. Brown, Judith M. 'India', in Brown, J.M. and Louis, W.R. (eds), *The Twentieth Century: Oxford History of the British Empire* (Oxford, 1999), pp. 421–47.
75. Singh, Harkishan. 'Drugs Enquiry Committee, 1930–31', *Eastern Pharmacist*, 44 (2001), pp. 21–5.
76. *Indian Medical Gazette*, 65 (1930), p. 640.
77. Singh, 'Drugs Enquiry Committee', p. 22.
78. Singh, Harkishan. *Sir Ram Nath Chopra: Work, Vision and Legacy* (Delhi, 2010).
79. Bala, *Imperialism and Medicine*, p. 56.
80. *Indian Medical Gazette*, 65 (1930), p. 642.
81. Singh, *Pharmacy Practice*, p. 104.
82. Ibid., p. 126.
83. *Report of the Drugs Enquiry Committee 1930–31* (Calcutta, 1931), p. 294.
84. Ibid., pp. 167–9.
85. *Indian and Eastern Druggist*, 13 (1932), p. 33.
86. *Chemist and Druggist*, 128 (1938), p. 526.
87. *Indian and Eastern Druggist*, 2 (1921), pp. 14–6.
88. Singh, *Pharmaceutical Education*, p. 19.
89. Kumar, Anil. 'Progress of Pharmacy and Pharmaceutical Industry', in Kumar, Anil, *Medicine and the Raj: British Medical Policy in India, 1835–1911* (Walnut Creek, 1998), pp. 110–25.
90. Singh, *Pharmaceutical Education*, p. 19.
91. Ghosh, J.C. 'Pharmacy in India', *Pharmaceutical Journal and Pharmacist*, 124 (1930), p. 50.
92. *Report of the Drugs Enquiry Committee 1930–31*, pp. 294–7.
93. *Pharmaceutical Journal*, 125 (1930), pp. 529–30.
94. Ibid., 124 (1930), pp. 409–10.
95. *Report of the Drugs Enquiry Committee 1930–31*, pp. 296–7.
96. Chopra, R.N. 'Needs of the Pharmacy Profession', *Indian Medical Gazette*, 72 (1937), pp. 101–3.

97. Singh, 'History of Drugs and Pharmacy Statutes', pp. 31–7.
98. Ibid., p. 36.
99. *Indian Medical Record*, 62 (1942), p. 326.
100. Ibid., 63 (1943), pp. 95–6.
101. Singh, *Pharmaceutical Education*, p. 16.
102. Ibid., p. 15.
103. Singh, Harkishan. 'The Pharmacy Act and Pharmaceutical Education', *Eastern Pharmacist*, 30 (1987), pp. 47–9.
104. Singh, 'History of Drugs and Pharmacy Statutes', pp. 31–7.

CHAPTER 9

Eastern Colonies: A Melting Pot of Medical Traditions

In 1914 Britain's Eastern Division of Empire was a disparate collection of colonies that included Hong Kong, the Straits Settlements, and Ceylon (later Sri Lanka). They mainly came into British possession during the nineteenth century (Fig. 9.1). Nowhere else in the Empire was the impact of such a diverse range of cultures, languages, and health beliefs more apparent than in this Division. Like other occupations pharmacy was influenced by constraints imposed by religious beliefs and climate differences, and medical and pharmaceutical practices evolved in the light of experience with the other medical approaches they encountered. Medicine and pharmacy were shaped more by developments in India than by those in Britain. Local factors were also important; in the Straits Settlements, pharmaceutical activities were taken on mainly by unqualified assistants following the move by apothecaries to be medical practitioners. In Hong Kong British pharmacy became established as a result of immigration by British pharmacists. Pharmacy was heavily influenced by the social, political, and economic contexts in which it was practised, which included ethnic diversity, geographical challenges, and external threats.

The people of Britain's new eastern territories came from a multitude of ethnic groups, adhered to a diverse range of religions, spoke a variety of languages, and engaged in a plethora of practices. It was a melting pot of medical and pharmaceutical traditions. Malay, Indian,

S. Anderson, *Pharmacy and Professionalization in the British Empire, 1780–1970*, Medicine and Biomedical Sciences in Modern History,
https://doi.org/10.1007/978-3-030-78980-0_9

Fig. 9.1 Map of the Eastern Division of the British Empire, 1914

Chinese, and other forms of traditional medicine were all practised and interacted with each other and with western medicine. Different medical philosophies borrowed from each other, and choices and preferences were sometimes made on racial grounds. Beliefs not uncommon in eighteenth-century West Indian colonies concerning racial susceptibilities to certain diseases—that different treatments were needed for different racial groups—lingered in nineteenth century eastern colonies, although it seems that Sloane's view that, for example, venereal diseases 'had the same symptoms and course among Europeans, Indians and Negroes' was more commonly held, and treatments were often shared across medical systems.[1]

This chapter contrasts the professionalization of pharmacy in the Straits Settlements with that in Hong Kong. In the Straits Settlements the situation more closely mirrored that in India, whilst in Hong Kong the British model of pharmacy was largely adopted unaltered. In Hong Kong British chemists and druggists established pharmacy businesses; in the Straits Settlements pharmacy was largely controlled by medical officers of the Indian Medical Service (IMS). As elsewhere professional networks played an important part in the development of pharmacy in the region, although racial exclusionism and protectionism also played a part, as these were a very common reflex of imperial networks.[2] Even networks that started out relatively open tended to become more restrictive over time. But where numbers were small, as in pharmacy, there was little to be gained by exclusion. Pharmacy developed differently in the two colonies, but by the end of British rule professionalization was incomplete in both.

The Indigenous People of the Eastern Colonies

The Straits Settlements, consisting of Penang, Malacca, and Singapore, were occupied over a 40-year period, and remained British possessions for nearly 130 years. British occupation began in 1786 when Sir Francis Light claimed Penang for Britain.[3] In 1819 Sir Thomas Stamford Raffles concluded an agreement with a local sultan to cede Singapore to Britain.[4] East India Company officials from Penang quickly brought Singapore's administration into line with that elsewhere.[5] Its transformation was extremely rapid; the population grew from around 5000 in 1821 to 10,000 in 1825 and to 81,000 in 1860. In 1824, Malacca was transferred to Britain under the Anglo-Dutch Treaty, and in 1826 the three settlements were united as the Incorporated Settlements of Prince of Wales Island (Penang), Singapore and Malacca, or 'Straits Settlements'.[6] They were initially administered by the East India Company as part of the Bengal Presidency, with Penang as the capital,[7] but in 1832 the capital was transferred to Singapore, and in 1835 the headquarters of the Indian Medical Department followed.[8] The status of the Settlements changed following the Indian mutiny in 1857, and in 1867 control was transferred from the India Office to the Colonial Office, and they became a Crown Colony with their own legislative Council.[9] They retained this status until the Japanese occupation in 1942.

The people of these new territories came from a multitude of ethnic groups, adhered to many different religions, and engaged in a variety of medical practices. Immigration from India into Ceylon, Burma, and the Straits Settlements rose during the last 25 years of the nineteenth century from an average of 300,000 to over 425,000 per year.[10] By the end of the century around half the population of the Straits Settlements and Federated Malay States (FMS) were Chinese. Indian immigrants constituted about 8 per cent of the total population of Malaya as a whole.[11] The Straits Settlements also became a convenient dumping ground for convicts from India. Initially convicts were transported to Penang, but Singapore soon became the main port of disembarkation, and separate Convict Hospitals were established.[12]

Multi-Cultural Medicines

The Chinese and Indian communities were largely self-contained, along with their traditional medical systems, although there was some absorption of indigenous Malaysian medicine and Western medical practice into both. Chinese and Indian medicine flourished alongside each other and with the traditional medicine systems of the indigenous people. In British colonies the Chinese largely ran their own affairs; they had a shared culture and heritage, a common philosophy, and a common written language. Although there were minor differences there was essentially one form of Traditional Chinese Medicine (TCM). In the Straits Settlements their treatments incorporated western and traditional Malay practices. Chinese practitioners embraced the use of anaesthetics and the techniques of acupuncture, vaccination, moxibustion, and hydrotherapy. Nevertheless, more affluent Chinese increasingly chose to put their trust in western medicine. Chinese benefactors made donations to western medicine institutions rather than TCM ones, where most Chinese people were treated. The Tan Tock Seng Hospital in Singapore was funded with mainly Chinese money, but provided only western medicine to its patients.[13]

For the most part the practice of western medicine was limited to Europeans. British practitioners felt that there was much to learn from the medical traditions they encountered. Charles Hose noted that

> After all, the medicine man acts according to his lights and his experience, even if he does not realize the value of careful diagnosis; and as for faith-healing, mesmerism, homeopathy, and (if one wishes) psycho-analysis, the West has a good deal to learn from the East.[14]

When the TCM Tung Shin Hospital was opened by the Acting British Resident he declared that

> it is very possible that the knowledge of medical science...may be appreciably increased by a study of Chinese methods of treatment, and [of] the effects of Chinese drugs, which have been in vogue amongst your countrymen for many hundreds of years.[15]

But he added 'your own Sin Sangs, on the other hand, will doubtless find much to learn from the methods of treatment, especially from the point of view of general hygiene, pursued in the Government Hospitals'. The British authorities were nevertheless happy for health facilities for a variety of medical traditions to be provided through local initiatives.

Unlike the Chinese, Indians in the region did not come from a single tradition but from one that was divided by race, religion, and culture.[16] The Indian community was also divided into two mutually exclusive groups; a small middle and professional class which included merchants, shopkeepers, teachers, and priests; and the vast number of Indian labourers working on the sugar and coffee plantations or building roads and railways for the colonial authorities. As a result, Indian immigrants lacked unity and organization, an issue that extended to medicine.[17] Frederick Colley reported that 'For the most part Indian medicine has been confined to rural estates and small towns in the past and has never been firmly organized'.[18] The Ayurvedic medicine of the Hindus was just as elaborate and extensive as TCM, he noted, and the range of its materia medica was just as diverse. But it was only one of several medical traditions followed in India. Early Muslim conquests had brought Unani medicine to the sub-continent; and the traditional medicine of South India, Siddha medicine, had its origins with a Tamil mystic and sage.[19] Ayurvedic and Siddha materia medica had much in common, although Siddha medicine placed greater emphasis on the use of alchemy for medical purposes. In his study of Indian materia medica on the Malay peninsula Colley found 83 plants being used locally. Traditional remedies used in Indian households were quite different from those used by the Chinese.[20] The result of this diversity was that Indian traditional medicine failed to achieve the prominence amongst Indian immigrants in the Eastern colonies that TCM did amongst the Chinese community.

Local practice interacted with Western medicine in many ways. Indian hospital assistants, dressers, and apothecaries administered western medicines in colonial hospitals. They were recruited in response to a growing demand by the colonial administration for English-speaking members of the indigenous population to work in subordinate roles. Most were recruited from India or Ceylon, since few in the indigenous populations of the other Eastern colonies spoke any English. One of the consequences of this interchange between western medical practitioners and Indian subordinate staff was a reduced prominence of traditional medicine amongst Indian immigrants, although it still flourished everywhere; it was familiar, affordable, and reassuring. But for outside observers it was largely an invisible presence only known to Indians themselves.[21]

The relative merits of practitioners from the different traditions were well-known in the different communities. In the Chinese community locally trained physicians were not considered to be on a par with those trained in China. Nevertheless, Dr. T. M. Ward, a British medical officer attached to the colonial garrison at Penang, noted that 'they enjoyed the confidence of the Chinese' and were regarded as superior to the European, though the European surgeon was thought 'better than his Chinese counterpart'.[22]

Apothecaries and Assistants in the Straits Settlements

Arrangements for the supply of medicines largely followed the pattern set in India (Chapter 8). The medical subordinates who supported the medical officers started as apprentices, initially becoming compounders or hospital dressers on completion of their apprenticeship. They could then be appointed sub-assistant surgeons, the highest rank they could attain. Titles changed in 1828, when sub-assistant surgeons were renamed 'apothecaries', with the lower rank termed 'assistant apothecaries'.[23] The title apothecary was thus a rank rather than a role. Whilst only Britons could be medical officers, medical subordinate ranks were open to both Britons and Indo-Britons (Eurasians).[24] When administered as part of India, the Medical Service in the Straits Settlements depended on the

Government of India for personnel to staff its hospitals. It was difficult to get replacements when vacancies arose as a result of death or ill-health; the Settlements were regarded as backward and service there as a 'hardship posting.'[25] Efforts were made to increase the establishment; in 1850 an application was made to Bengal for an assistant apothecary and a second dresser for Tan Tock Sing hospital. A reply was received nine months later; the services of Mr. Christopher Doyle, a senior hospital apprentice, were offered as assistant apothecary, but he would only be able to take up his appointment once he was 'relieved of the duties which he is at present performing with Her Majesty's 80th regiment at Darjeeling'.[26]

To address the recruitment problem the senior surgeon proposed that local boys be trained for the medical department. Apothecaries from India would, he pointed out

> be comparatively inefficient from ignorance of the language, habits and customs of the native patients under treatment, some being natives of different parts of China, speaking many different dialects, Cochin-Chinese, Siamese, Burmese, Malays, Bugis and Javanese, besides the natives of the Malabar and Coromandel coasts.[27]

He proposed to recruit apprentices from the Penang Free School, to train them, and to appoint them as assistant apothecaries once qualified. In March 1823 James Ash began training as a Compounder, and several further apprentices were appointed. The role of apothecaries and their assistants was spelled out in new regulations.[28] Both the apothecary and the apprentice were 'to live in the hospital compound. Both to be always present at the surgeon's visiting hours, and in no case both to be absent from the hospital at the same time'.[29] The duties of the assistant apothecary in Penang were described in 1830:

> He is placed in charge of the medicines at the Convict Hospital, from which the Native Pauper Hospital, Lunatic Asylum and Jails are also supplied. He has to attend to the Surgeon when he visits the hospitals, take down notes of any important case, register the prescriptions, see to the medicines compounded, superintend the Dressers, register the names, etc. of all the patients who come to hospital... He must be a person who is able to maintain strict discipline in the hospitals, which is sometimes no easy matter, particularly in the Convict Hospital.[30]

But by the late 1830s only two other local apprentices had been trained. Boys were unwilling to train for five years when the salary at the end was very poor. Those who did often left for private practice, and others threatened to resign if their salaries were not increased. In 1852 new rules for apprentices were published which placed greater emphasis on pharmaceutical duties.

> Apprentices of 3 years' standing will be expected to know how to read and write prescriptions in English, to determine the various medicines by sight, and to be able to tell their doses and qualities, as far as whether they belong to the class of purgatives, emetics, diuretics, etc... The passed apprentice (after 5 years) will be expected to have a thorough knowledge of materia medica, to know the components and proportions of all the official preparations, [and] to be able to tell the effects and the doses in which they are given.[31]

Although small numbers of young men continued to be trained the numbers were never sufficient; continued demands were made on the authorities in India. In 1860 the Governor reported to the Government of India that the apprenticeship scheme in the Straits Settlements had proved an utter failure, and that 'we shall in a great measure continue to be dependent for assistant apothecaries to fill any vacancies that may occur from the medical establishment either at Madras or Bengal'.[32]

In February 1849, a pupil at the Singapore Institution was selected as an apprentice, and after a three-month probationary period he was taken under the personal tuition of a surgeon 'for the purpose of becoming acquainted with the nature and properties of medicines'. The surgeon recommended that he receive his full allowance 'upon passing an examination in pharmacy at the end of the year'.[33] Two years later the salaries of apothecaries were raised with a view to improving recruitment. The senior surgeon reported that

> When they pass a satisfactory examination in pharmacy, show a general knowledge of medicine, the ability to read and write prescriptions, and acquaintance with the general routine of hospital duty, and can bleed and apply bandages in a skilful manner, I would raise their salary to 30 rupees a month.[34]

But by the 1850s there were only three locally trained staff in Singapore. A proposal was made that local boys be trained as apothecaries at Madras

Medical College, and this was subsequently approved by the Government.[35] By the 1860s difficulties in recruiting physicians and surgeons meant that apothecaries increasingly focussed on medical rather than pharmaceutical duties. The *Straits Times* lamented that 'Our hospitals are left too much to the control of the apothecaries. In scarcely any part of the civilised world will hospitals be found without resident physicians or surgeons or both'.[36] The dispensing of medicines was delegated to others. *The Daily Times* reported that 'The apothecary in charge has apprentices as assistants, whose duty we suppose is to aid in compounding and administering the medicines etc.'.[37]

Western Retail Pharmacies, 1825

Outside the government facilities a thriving retail trade in medicines developed. The earliest chemist's shop in Singapore (usually called 'dispensaries' or 'Medical Halls') was owned by Jose d'Almeida, a former surgeon on a Portuguese warship.[38] He had settled in Singapore in 1825 and opened a dispensary on the site of a godown (warehouse). His business began after Spanish and Portuguese vessels floundered during a monsoon and had to sell their cargo of medicines.[39] The early Singapore dispensaries were set up to cater for the needs of European troops and traders.[40] They were mainly located in areas where Europeans were concentrated. One was owned by J. I. Woodford, a local boy who had trained as an apothecary in Penang in the 1820s. The Penang Dispensary was one of the earliest western pharmacies in the Straits Settlements (Fig. 9.2). Woodford opened the Campong Glam Dispensary in Singapore and was listed as a 'chemist and druggist' in the 1864 *Singapore Directory*. Woodford advertised that 'medical prescriptions will meet with prompt and careful attention at all times. Medical chests can be filled up and supplied on short notice, and at moderate rates'.[41]

The retail pharmacy trade faced stiff competition from the medical profession, who continued to supply medicines and replenish medicine chests. Most doctors did their own dispensing or employed an unqualified assistant. Dr. J. Scott of the Straits Dispensary employed a William Neil as his dispenser. Dr. W. Maney of the Central Dispensary advertised that he was willing to visit the homes of parents and guardians to vaccinate children, and that medicine chests were replenished.[42] But

Fig. 9.2 Penang Dispensary, 1890 (Courtesy of National Archives of Malaysia)

some surgeons forsook medical duties totally in favour of pharmaceutical ones. In an 1845 newspaper advertisement, two surgeons, R. Little and T. J. Martin, declared that 'in consequence of the establishment of a government hospital, the undersigned have given up the Private Hospital attached to the above dispensary'. They offered their services to the owners and masters of vessels, stating that 'they, their passengers and crew can be attended at any hour of the day or night'. Furthermore, 'medicine chests are fitted up with the requisite medicines, with directions for a tropical climate, while particular medicines with their directions can be supplied [to] those visiting countries subject to peculiar endemics'.[43] In 1847 both names were included in the list of chemists and druggists.

Whilst the retail pharmacies stocked mainly European medicines, local medicines and those imported from countries including China, India, and Malaya (Bazaar medicines), were available from market stalls and other outlets. Both European and Bazaar medicines were available in the hospitals. Orders to medical officers stated, 'to seamen of private ships and other private European individuals, the medical officer in charge to furnish provisions and European and Bazaar medicines and other necessaries at the above rate'.[44] At first there was no control over the sale of medicines, or any regulation of pharmacy practitioners beyond the regulations of the IMS. Products were imported from Britain and elsewhere. Patent medicines were widely advertised with competition coming from Australian manufacturers. An advertisement for 'Woods' Great Peppermint Cure' listed all the 'Australian stores' where it could be obtained. The proprietors attempted to distinguish themselves from their British rivals: 'we wish to state that we do not depend on the questionable practice of publishing testimonials. We depend entirely on the virtue of our medicine and the good name that it will earn for itself'.[45]

By the end of the nineteenth century several British pharmacies had been established in Penang, Singapore, and the FMS. 'The Pharmacy' was opened in Singapore in 1902 and later run by G. Whyte Crawford. With no regulations preventing companies from owning pharmacies several firms such as Maynard and Co. opened several branches. They later merged with the Singapore Dispensary.[46] Their customers were largely middle-class Europeans holding the values of mid-Victorian Britain. But the European community increasingly drew apart from the Asian community. At the official level there were multi-racial dinners, balls, and celebrations, although people mainly relaxed amongst their own community. A growing gulf also developed between prosperous Asians and the general population.[47]

Pharmaceutical Associations

A new phase began after 1896 when the British government brought together four protected states in the Malay Peninsula—Perak, Pahang, Selangor, and Negeri Sembilan—to form the FMS, an arrangement that lasted until 1946. New legislation began to be passed that would affect pharmacy practitioners in both the FMS and the Straits Settlements. In the early twentieth century apothecaries continued to provide medical and surgical services in government hospitals, clinics, and mobile dispensaries, although health assistants and dressers prepared and supplied medicines.[48] Pharmacy remained firmly in the hands of the doctors.

Most early legislation in the Straits Settlements was based on British and Indian law.[49] For pharmacy the template for most legislation was usually the British Pharmacy and Poisons Act 1868, although colonial authorities had to deal with local issues as they arose. The most urgent need was to control the use of opium, and in 1894 an Opium Ordinance sought to control its availability and that of morphine. The PSGB appears to have played little role in advising about legislation in the region; it was more interested in its plants of medicinal value than the regulation of the profession.[50] Chandu (an opium preparation) was added to the Ordinance in 1900, and a Morphine Ordinance passed in 1904 restricted the prescribing of preparations containing them to 'qualified persons', i.e. chemists and druggists'. Further items were added in a 1905 Poisons Ordinance. Later Ordinances in 1910 and 1913 increased penalties for offenders and raised the reward for informers providing information leading to conviction.[51] A Deleterious Drugs Enactment was passed in 1928, and a Poisons List Confirmation Order only in 1935.

Pharmacists began working together in associations in response to the legislation rather than to shape it (Table 9.1). The first to be founded was the Penang Pharmaceutical Association in around 1902 by pharmacists in Penang and surrounding districts including Perak and Kelatan. It was followed three years later by the Straits Pharmaceutical Association, founded for pharmacists in Singapore and the southern part of the Federation of Malaya, particularly in Jahore.This was formed in response to passage of the 1904 Morphine Ordinance and the 1905 Poisons Ordinance. A month earlier the King Edward VII Medical School was given responsibility for preparing and examining candidates for the Pharmacy Certificate following its opening.[52] By 1910 trainee dressers had to pass an examination in practical pharmacy conducted at the medical school before being promoted. The declared aims of the Association were to 'protect and further the interests of pharmacy in general and the members of the Society in particular; and to encourage and to further the advancement of pharmacy and pharmaceutical education'.[53] Membership was open only to those holding PSGB qualifications. The ten pioneer members were all European, its first president being J. Mackenzie, who remained in office for several years.

The association's constitution was amended in 1937 to permit all registered pharmacists to become members. In 1946 plans were made for a unified Association, and on 31 May 1949 the name of the Straits Pharmaceutical Association was changed to the Malayan Pharmaceutical Association, with a Northern Division based in Penang and a Southern Division based in Singapore. In 1955 the total membership was around

Table 9.1 Professionalizing pharmacy in Eastern Division Colonies

Colony	*Year occupied by British*	*Year British occupation ended*	*Pharmaceutical society or association*	*Early pharmacy legislation*	*Legislation to control poisons and medicines*	*Pharmacy education*
Penang (Straits Settlements)	1786	Independence as Federation of Malaya 1957	Penang Pharmaceutical Association 1902	Registration of Pharmacists Ordinance 1903	Poisons Ordinance 1905	Local training
Malacca (Straits Settlements)	1795	Independence as Federation of Malaya 1957	Malayan Pharmaceutical Association 1949	Registration of Pharmacists Ordinance 1903	Poisons Ordinance 1905	Local training
Singapore (Straits Settlements)	1819	Independent Republic of Singapore 1965	Straits Pharmaceutical Association 1903	Registration of Pharmacists Ordinance 1903	Poisons Ordinance 1905	Government Medical School Singapore 1905
Seychelles	1794	Independent 1976	None	Pharmacy Act 1899	Pharmacy Act 1899	None
Ceylon	1795 (Madras Presidency)	Independent 1948	Pharmaceutical Society of Ceylon 1969	Ordinance as Madras	Ordinance as Madras	National Institute of Health Sciences 1959
Mauritius	1810	Independent 1968	Pharmaceutical Association of Mauritius 1979	Ordinance as Madras	Ordinance as Madras	None
Hong Kong	1841	Hand back to China 1997	Pharmaceutical Society of Hong Kong 1949	Pharmacy Ordinance 1908	Pharmacy and Poisons Ordinance 1916	Two-year course offered by Hong Kong University 1937
Labuan (North Borneo)	1846	Part of Malaysia 1963	None	None	None	None
Weihaiwei (North east China)	1898	Handed back to China 1930	None	None	None	None

30.[54] The first local president, Chong Ngiap Lye, held office from 1961–62. Following the separation of Malaysia and Singapore, a Malaysian Pharmaceutical Society was formed on 6 November 1967.[55]

A Dual Profession, 1927

In 1927 the principal medical officer, R. Dowden, made a clear distinction between the roles of those holding the PSGB's Minor qualification (chemists and druggists), and those holding the Major qualification (pharmaceutical chemists), of which there were none in the FMS at the time. A distinction was also made between the latter and analytical chemists. Dowden reported that pharmaceutical chemists were needed to deal with all matters relating to medicines: 'checking, preparation, import from England, training of dressers and dispensers, and the licensing of sales of medicines'.[56] In response to Dowden's request the British High Commissioner, Hugh Clifford, asked the Colonial Office to fund and find two pharmaceutical chemists from Britain. Their duties were to include taking charge of and inspecting all drug stores, drugs, and dressings in the government dispensaries, instructing the dressers in dispensing, advising on the purchase and storage of drugs, and preparing locally drugs then ordered from England. In the same year A. H. Millard and A. V. Hitch were appointed as the first pharmaceutical chemists to the FMS. Two further appointments were made later that year.

Single male pharmacists aged between 25 and 30 years were preferred. They were appointed on three-year contracts and had to achieve a minimum level of proficiency in Malay within two years. They were provided with furnished accommodation and first-class travel from and to England at the start and end of their contracts. On arrival they were posted across the states and acted as government inspectors. Victor Hitch had joined Boots as an apprentice in England in 1918, qualified as a pharmaceutical chemist in 1924, and worked in a British hospital pharmacy for four years.[57] In 1928 he was posted to Taiping Hospital as superintending pharmaceutical chemist for Perak and Pahang, reporting to the state Medical and Health Officer. His challenge was to implement a pharmacy service in Malaya along British lines with limited resources. His duties included preparing a six-monthly drug order for state hospitals and dispensaries and sending it to the Crown Agents in London, who supplied almost all medicines ordered for government institutions in the FMS.[58] Central medical stores in each state then distributed supplies to hospitals and clinics. The Crown Agents usually took a commission on the orders of 30 per cent of their value.[59] Hitch also controlled the import and supply of dangerous drugs and was required to teach basic dispensing to dressers. He also built a medical store and began the small-scale manufacture of quinine and other tablets to reduce the need for imports.[60]

Legal Recognition

It was only after the end of the Second World War that Britain introduced effective legislation concerning the use of medicines and the practice of pharmacy in the FMS.[61] In 1946, following the liberation of Malaya from the Japanese, all that was needed to open a pharmacy was a licence issued by the Chief Medical and Health Officer.[62] In 1946 the FMS was joined with Penang and Malacca and the Unfederated Malay States to form the Malayan Union, which two years later became the Federation of Malaya. This achieved independence in 1957 and was renamed Malaysia in 1963 when Sabah, Sarawak and Singapore joined, with Singapore becoming a republic in 1965.

The first law for the registration of pharmacists was the Registration of Pharmacists Ordinance passed in 1933, and the first Register of Pharmacists was opened on 22 May 1935; it listed only those holding a British qualification, although the first local pharmacists qualified in 1936 and 1937. The Ordinance was reviewed in the 1940s as part of wider medical reforms and the passage of separate medical and dental Bills. The proposed new Pharmacy Ordinance was subject to extensive debate in the Legislative Council because of a widespread belief that it was unnecessary. Some felt that existing Poisons and Deleterious Drugs Enactments—which established a link between pharmacy and the supply of drugs and poisons—provided sufficient authority for pharmacists to deal with the supply and dispensing of drugs. Others considered that pharmacy should be regulated in the same way as the medical and dental professions. The Director of Health and Medical Services considered that

> sooner or later the Colony must, for the proper protection of the public, concern itself with the better control of the dispensing of medicines and the sale of poisons. It is obviously desirable that such business should only

> be entrusted to persons who are specially qualified for it by a recognised course of study.[63]

A revised Registration of Pharmacists Ordinance was eventually passed in 1951. This established an independent Pharmacy Board and specified its membership of ten, at least three of whom were to be doctors, one of whom acted as chairman.[64] Doctors were still in overall control, and pharmacy was still far from autonomous. The Ordinance provided for the training and registration of pharmacists and established a Register of Pharmacists which recognized a range of qualifications. Those who worked in hospital dispensaries as dressers could take a 10-month top-up course in pharmacy at the College of Medicine in Singapore, and then serve as pharmacists in government hospitals. British qualifications were recognized along with those accepted by Britain under reciprocity agreements, including those obtained in South Africa, Australia, and New Zealand. The Ordinance also restricted use of the titles listed in the 1868 British Act and specified conditions for the 'keeping, retailing, dispensing and compounding of poisons and deleterious drugs by registered pharmacists and bodies corporate'.[65]

Here then was legislation that delivered most elements of the British model of pharmacy; the regulation of pharmacy was linked to the sale of poisons, with the Pharmacy Board having responsibility for both. The registration of pharmacists was undertaken by a Pharmacy Board rather than a Medical Board; and company pharmacies could own pharmacies. The authorities opted for a single level qualification rather the British dual register (which ended in 1954), and declined to follow the British model in having responsibility for maintaining the Register vested in a legally founded pharmaceutical society, instead opting for the South African and Victorian model of having an independent Pharmacy Board. The Board held its first meeting on 8 March 1952, and immediately established sub-committees to review the Poisons Ordinance, the Sale of Food and Drugs Ordinance, and the Dangerous Drugs Bill. A new Poisons Act was passed in 1952, five years before the Federation of Malaya was declared, although full implementation came only after independence.

Medical Halls and dispensaries in Hong Kong

When Britain acquired it in 1841 Hong Kong was a fishing port with around 20 villages and a few thousand residents under the jurisdiction of Canton (now Guangdong) province. The colonial Government was presented with great opportunities but also faced many challenges. With an open border, migrant workers in the mainland provided a convenient source of labour to fuel industrial and commercial development. They came in four main waves, the first occurring soon after the end of the Opium Wars. By 1864 the population had reached 125,504 including 4007 British and foreign residents. A second wave came between 1901 and 1911 following unrest on the mainland, with the population more than doubling from 206,162 in 1901 to 450,739—including 11,225 British and foreign residents—in 1911. A third wave came with the outbreak of the Anti-Japanese War in the 1930s in the mainland, and the population increased from 849,750 in 1931 to 1.82 million (including 24,125 British and foreign residents and around 750,000 refugees) in 1940.[66] The population dwindled to 600,000 during the Japanese occupation between 1941 and 1945 and bounced back to 1.8 million by the end of 1947. Another 400,000 arrived in 1949–1950 before the border was sealed after the civil war.[67]

Pharmacy in Hong Kong went through a process similar to that in the Straits Settlements.[68] Western pharmacies soon sprang up, although these were initially run by apothecaries who combined medical practice with the supply of medicines. Many were family concerns. Peter Young moved to Hong Kong from the Canton Dispensary in the late 1830s, set up a mat-shed apothecary's shop in 1841, and later became a partner in the Hong Kong Dispensary. This was followed by the Victoria Dispensary and the Medical Hall, which served the mainly expatriate community and western sailors.[69] James Young, a surgeon and brother of Peter Young, served as druggist and manager at the Hong Kong Dispensary. Its ownership passed to Dr. William Preston and subsequently to Dr. Thomas Boswell Watson.[70] In 1858 Watson brought his nephew Alexander Skirving Watson into the business, which became A. S. Watson & Co. It changed hands several times before being owned by John David Humphreys in 1879. It then passed to his son, Henry, at which point retail pharmacy became separate from medical practice. Henry Humphreys was sent to study in England, passed the PSGB's Major examination, and became a pharmaceutical chemist in 1889.[71] He returned to join his father at

A. S. Watson, and over 40 years the business grew to be a large chain of pharmacies.[72] The British model of company chemists owning large numbers of pharmacies thrived in Hong Kong; pharmacy ownership was not restricted to individual proprietors.

Retail pharmacy in Hong Kong was not the preserve of British citizens. Many Macanese moved to Hong Kong in the early 1840s as Hong Kong replaced Macao as the major re-export centre in southern China. Some Macanese served in the Colonial Government whilst others opened apothecary's businesses. Thomas Hunter established the Victoria Dispensary in Macao before opening a retail outlet in Hong Kong in 1846, later moving to larger premises in 1848.[73] Its British manager, George K. Barton, employed three Macanese assistants, one of whom, Joao Braga, replaced him as manager in 1849. Braga was assisted by another Macanese who had been an apothecary's assistant at the Hong Kong Dispensary before joining the Government Civil Hospital in 1856 as Apothecary.[74] Braga left the Victoria Dispensary in 1857 to re-open the Medical Hall Dispensary, ownership of which later changed to a German national, E. Niedhardt, who changed its name to 'German Dispensary'. Many other western pharmacies opened and closed in Hong Kong during the late nineteenth and early twentieth centuries, with consolidation later occurring under a handful of pharmacy groups. By 1940 around 38 British and locally trained chemists and druggists served the pharmaceutical needs of 20,000 expatriates and a sizeable proportion of the 1.8 million Chinese community.[75]

Pharmaceutical Practitioners

The first non-Chinese settlers in Hong Kong depended for medical care and pharmaceutical services largely on private practitioners. British, European, and Indian expatriates sought help from the Victoria Hospital operated by Drs. James Satchell and Richard Jones or other privately owned dispensaries. Visiting sailors could attend the privately operated Hong Kong Seamen Hospital, and poor labourers could seek treatment at the Hospital of the Medical Missionary operated by Dr. Benjamin Hobson. Morbidity and mortality were unusually high amongst early British colonial settlers, including military personnel, as a result of diarrhoea, dysentery, and malaria in Hong Kong. Four out of eight colonial surgeons died from various illnesses between 1847 and 1873.[76] Malaria was a particular problem during the hot and humid summer months of the early 1840s.

A Government Civil Hospital was opened to meet the needs of the expatriate community in 1849, and the colonial authorities appointed their first apothecary to the hospital in 1856. Alberto A. Bothelo received his training at the Hong Kong Dispensary and later worked as apothecary at the Victoria Dispensary. When he retired the Government opted to recruit a pharmaceutical chemist from Britain to strengthen the control of medicines in the colony. Pharmaceutical chemists had studied analytical chemistry and toxicology to a level not required of chemists and druggists. In 1879 Hugh McCallum was appointed as the first Apothecary and Analyst at the Government Civilian Hospital. He was from Banffshire in Scotland and had passed the PSGB's Major examination in Edinburgh. One of his duties was to check the quality and standard of the 'legal' opium sold in Hong Kong. In his first year he also conducted analyses of water and milk samples and searched for poisons in suspected poisoning cases, in addition to his duties as apothecary. The title 'apothecary' continued to be used in Hong Kong where it distinguished pharmaceutical chemists from chemists and druggists but involved no medical activity. McCallum was appointed as Sanitary Inspector in 1883.[77] The link between pharmacy and analysis continued well into the twentieth century. William Edward Crow replaced McCallum as Apothecary and Analyst in 1883, and was joined by another British pharmaceutical chemist, Frank Browne, as Assistant Apothecary. Browne had qualified in 1888 and replaced Crow in 1898: he became the first Government Chemist in charge of the Government Laboratory when the roles of Government Apothecary and Government Analyst were separated in 1913.[78]

Thereafter a succession of British pharmaceutical chemists was appointed as Apothecary and Analyst, with Ralph Edgar Cable serving from 1919 until 1940. Pharmacy in government institutions expanded, and in 1939 the Pharmacy Branch of the Medical Department employed 27 pharmacy practitioners with various qualifications. The Chief Pharmacist and three other pharmacists held the British Major qualification; 'charge dispensers' responsible for medicine supply in hospitals and dispensaries normally held the local chemist and druggist qualification; and 'probationer dispensers' working towards the qualifying examination were also employed, along with other staff.[79] Some government hospitals offered both western and TCM, as many Chinese retained a strong affinity to TCM. In 1938 attendance at TCM outpatient clinics at Chinese Hospitals remained high: nearly twice as many as those attending Government Hospitals providing only western medicine services. At Chinese Hospitals providing both TCM and Western medicine services, only a third of in-patients chose surgical operations requiring the use of anaesthetics, with the remainder opting for TCM therapy. Whilst appearing to reflect growing acceptance of western treatments by local patients, the fact that western doctors were funded by the Colonial Government but TCM doctors were not may help to explain the situation.[80]

Linking Pharmacy and Poisons Legislation, 1858

As in the Straits Settlements opium was widely traded and used in Hong Kong, and the colonial authorities took steps to regulate it. In 1858 an Ordinance for Licencing and Regulating the Sale of Prepared Opium was passed. It was enacted to ensure that 'preparing or selling prepared opium bona fide for medicinal purposes' was only carried out by 'a medical practitioner [or] chemist or druggist having a European or American diploma'.[81] It applied to both public hospitals and retail pharmacies but related only to the use of opium 'for medicinal purposes'. Whilst most colonial laws were based on British ones this Ordinance predated the first control of opium in Britain in the 1868 Act. The first Ordinance concerning chemists and druggists related to their professional status; a revision of the Jury List 1864 provided that 'chemists and druggists should be exempted from serving as Jurors'. The names of chemists and druggists licenced by the Medical Board were duly struck from the Jury List.[82] An Ordinance in 1886 allowed chemists and druggists to use stills: a licence could be issued 'to any Apothecary, Chemist, or Druggist…to keep and use on his premises, a still not more than eight gallons for the purpose of his trade only…'.[83] A 1903 Ordinance allowed them to sell certain poisons as part of a medicine: 'Chemists and Druggists…are exempted from the operations of Bye-laws when the article sold is an ingredient of any medicine dispensed by them'.[84]

There were frequently substantial delays between laws being passed in British and equivalent ones being passed in Hong Kong. The protection of titles and restrictions on sales of poisons provided in the 1868 British Act was finally enacted in a Hong Kong Pharmacy Ordinance in 1908. It provided that

> no person shall sell or keep an open shop for retailing, dispensing or compounding poisons' or use the title 'chemist and druggist or chemist or druggist, unless such person is a duly registered Pharmaceutical Chemist or Chemist and Druggist under...'the Pharmacy Act 1868'.[85]

Exceptions to the British qualification were made, as any person who had been practising as a chemist and druggist for some time and 'satisfied the Governor of his skill and competency' to conduct such a business could ask to be registered. The first Hong Kong Register of Chemists and Druggists was dominated by British pharmacists; of the 13 names, 12 were British and one was a local Chinese who had trained at the Government Hospital.[86] All 12 British chemists and druggists worked in retail pharmacies.

Changes in line with the 1908 British Poisons and Pharmacy Act came more rapidly. A Pharmacy and Poisons Ordinance was enacted in 1916 to revise the law relating to the registration of pharmacists and the sale and use of poisons.[87] It included a section formally acknowledging the right of companies to own pharmacies. Provided that the part of the business concerned with 'the keeping, retailing and dispensing of poisons'' was under the control of a superintendent who was a registered person, and that in every premises where such business was carried out it was conducted 'by a manager or assistant who is a registered person under the direction of the superintendent', such business was bona fide. The Ordinance also extended the list of scheduled poisons which could only be dispensed or sold by a registered person and for which written records were required.

Other Ordinances followed close on the heels of British legislation. In 1923 a Dangerous Drugs Ordinance based on the 1920 British Act was enacted to regulate the 'importation, exportation, manufacturing, sales and use of Dangerous Drugs in the Colony of Hong Kong'.[88] Following passage of the Pharmacy and Poisons Act 1933 in Britain, the 1916 Ordinance was repealed and replaced by a Pharmacy and Poisons Ordinance in 1937. Retail pharmacists in Hong Kong held the monopoly in the sale of opium substitutes. These initially took the form of morphine injections in the 1880s, but were replaced by heroin pills in the 1930s.[89] By 1939 the illegal manufacture and sale of these to 'coolies' (labourers), who could not afford the luxury of opium smoking as a 'leisure drug', had become a major problem. The sale of heroin pills was a major source of income as a result of a monopoly granted by the government to 'opium farmers' who were the opium wholesale licence holders.[90] Opium was finally classified as a dangerous drug soon after the Second World War, and the long history of the colony's notorious involvement with it came to an end.[91]

Extending Education

In Hong Kong, a dual system of pharmacy qualification effectively existed until at least the middle of the twentieth century. Holders of the British chemist and druggist qualification worked exclusively in retail pharmacies. Locally trained dispensers worked in government hospitals and clinics; the small number of British pharmacists holding the pharmaceutical chemist qualification were recruited directly by the government into senior roles. Training of dispensers was through apprenticeship in either a hospital or retail pharmacy. But dispensers were not holders of the chemist and druggist qualification, and had to work under the direction of a pharmacist. During 1913–1915 there was a high turnover of both British-trained chemists and druggists and dispensers in government institutions, and the need to train people locally as chemists and druggists became urgent. Regulations were made under the 1916 Ordinance for a local course and examination leading to registration as chemists and druggists.[92] But although registered locally, such people were not holders of the PSGB's Minor qualification, and their qualification did not allow them to work as chemists and druggists outside Hong Kong.

For locally trained chemists and druggists, a revised syllabus and prescribed training was detailed in a 1937 Ordinance.[93] The University of Hong Kong was invited by the Colonial Government to start a two-year full-time course. It was developed by a British pharmaceutical chemist, Ralph Edgar Cable.[94] The first-year intake of nine were external students not able to meet university entrance requirements. The first year's curriculum was the same as for medical students with classes in anatomy, biochemistry, and physiology. Pharmacy subjects were studied in the second year. Students' English language proficiency was often poor and five out of nine failed the first-year examination. The course was terminated at the request of the Government. But with increased demand for pharmacy services, it was re-instated at the Evening Technical Institute as a four-year part-time evening course in 1940. Each student was apprenticed to a community or hospital pharmacy where they worked during the day. Basic science was studied in the first year, with the second to fourth years focussed on pharmacy subjects. From 1950 examinations for the chemist and druggist qualification were held once or twice a year. Pharmacy classes ended, and the last examination was held in 1952.

They were replaced by a new two-year full-time Diploma in Pharmacy course at the University. It was organized by Thomas Mahon, a pharmaceutical chemist trained at the Victoria College of Pharmacy in Melbourne. He had been appointed as Chief Pharmacist of the Colonial government's Department of Medical and Health Service in January 1947 to be responsible for the training of local pharmacists.[95] The new course was similar to the old one but was more concentrated. Year one covered basic medical science and the second included five pharmacy subjects. After passing the examination students undertook a two-year apprenticeship at a hospital or community pharmacy. They then had to pass a pharmacy law examination, after which they could register as a chemist and druggist with the Pharmacy Board. The first intake of 12 students registered as chemists and druggists in 1956. But the number of jobs available was limited and many qualified people had to accept posts at the lower level of 'dispenser'. Of the first intake, four continued working for Government institutions as dispensers until 1962. Students in the second and third intakes changed to different courses and the course was closed again. Its failure to recruit was largely due to its being a Diploma rather than a Degree course, and it did not have reciprocal recognition with either Britain or other Commonwealth Countries.[96]

The Colonial Government continued to train dispensers for clinic and hospital pharmacy service, but relied on pharmacy graduates returning from Britain and elsewhere to maintain services in both the public and private sectors.[97] In 1959 the Pharmacy Board proposed a scheme to provide overseas scholarships for pharmacy students. One of the early recipients, Chi-Wai Pang, worked as a student dispenser for four years before being sponsored to study for a pharmacy degree in Australia in 1963. He qualified as a pharmaceutical chemist in Victoria in 1967, registered with the Hong Kong Pharmacy Board in 1968 and continued to work with the colonial government. He became Chief Pharmacist in charge of hospital services in 1985, and reliance on highly trained British individuals came to an end.[98]

With the effective transplantation of the British model of pharmacy in Hong Kong there was little necessity for pharmacists to work together to seek legislation to improve their standing or to seek autonomy from the doctors. For the most part relations with medical practitioners were amicable, but the matter came to a head in 1950 over the issue of doctor-dispensing. A Pharmaceutical Society of Hong Kong (PSHK) was formed in 1949 by a group of local retail pharmacists. Its first president was Cheng-Loon Hooi, who had graduated at Columbia University in New York in 1942 and registered in Hong Kong in 1947. He approached pharmaceutical societies in Britain and Australia for advice on initiating legislation to separate the roles of medical practitioner and pharmacist. But their experience was not considered relevant to Hong Kong, as dispensing in both countries was part of state health schemes. The PSHK failed to get assurances from the Government that their concerns would be addressed. A second attempt was made in 1976, but again the PSHK's proposals were dropped due to the small number of pharmacists involved. However, the failure to limit doctor-dispensing was widely believed to be due to strong lobbying by the Hong Kong Medical Association, whose members were anxious to retain their dispensing rights.[99]

Other Eastern Colonies, 1802

The pattern of pharmacy development in India and the Straits Settlements was largely repeated elsewhere in the eastern division. Ceylon was at first administered by Britain as part of the Madras Presidency, but it became a Crown Colony in 1802.[100] By the middle of the nineteenth century extensive tea plantations had been established, with planters importing large numbers of Tamil workers from south India to work the estates.[101] The diverse population grew rapidly. In 1881 the total population was 2.8 million, including 4,800 Europeans; by 1911 it had reached 4.1 million. In such mixed communities colonial rule brought with it a tightening of social and communal boundaries.[102] Clear distinctions were made on racial grounds; the British colonists favoured the semi-European Burgher people, descended from Portuguese, Dutch, and British men, along with some high-caste Sinhalese and the Tamils concentrated in the north. Strict differentiations were made between Sinhalese Buddhists and Hindu Tamils, and there was an intensification of caste-like divisions.

For the colonial authorities Ceylon was an outpost of India. Medical and pharmaceutical services were organized along similar lines, but few British pharmacists were set up in business there. The authorities were nevertheless anxious to pass legislation similar to that elsewhere in the Empire. In 1877 a draft Ordinance was proposed based on the British Pharmacy and Poisons Act of 1868. The colonial authorities considered that the list of poisons to be included might differ from that in Britain, and suggested adding chloral hydrate. But the British model linked restricting the sale of poisons to the regulation of pharmacists, and the lack of British chemists and druggists on the island was a problem. Under the Ordinance there would be no examinations for those wishing to have their names added to the list of authorized sellers of poisons. Instead 'respectable persons would be licensed for one year to sell poisons at the discretion of magistrates, penalties and imprisonment being imposed on any unlicensed person who sold poisons'.[103] In Ceylon the only privilege open to chemists and druggists was automatic inclusion on the list of authorized sellers of poisons.

Weihaiwei (now Weihai), was a small territory on the north east coast of China. It was under British rule between 1898 and 1930 when it was returned to China.[104] Britain implemented a form of indirect rule, but never became a colonial power on the East Asian mainland.[105] Although there is no evidence of British pharmacists in Weihaiwei they had a presence elsewhere in China, largely in the tracks of English medical missionaries. William Lockhart, a Liverpool-born doctor who worked for the London Missionary Society, went to work at the Canton Dispensary in 1839, and over the next 20 years opened western hospitals in Chusan, Hong Kong, Shanghai, and Peking.[106] Between 1910 and 1941 British pharmacists were recruited to develop pharmacy practice and education at the Peking Union Medical College Hospital (PUMCH).[107] Although Shanghai was never a formal British colony the 'British Concession' was established there in 1846, as it was one of five treaty ports created after the First Opium War. British chemists and druggists were amongst the first to set up dispensaries in the Bund area in 1856, including a branch of the Hong Kong Dispensary. Britain acquired the island of Labuan in 1846 to establish a naval station to protect her commercial interests in the region and to suppress piracy in the South China Sea. Basic medical services including a hospital were provided, but there are no records of any British pharmacists on the island.[108]

Mauritius was under British rule between 1810 and 1968. It had previously been a French possession, and French law and institutions were retained. The French language was used more widely than the English. The British administration rapidly introduced social and economic changes, and the development of medical and pharmaceutical services followed. Britain acquired the Seychelles in 1794. As in Mauritius, French customs and practices were allowed to continue. The British, like the French before them, saw the Seychelles as a useful place to exile troublesome political prisoners. Over the years, they became home to prisoners from Zanzibar, Egypt, Cyprus, and Palestine. The biggest grievance of the colonists was dependence on Mauritius, including for medical and pharmaceutical services.

Professionalization Delayed

A comparison of the development of pharmacy in the Straits Settlements and Hong Kong illustrates different challenges presented by size and homogeneity of population. In the nineteenth century, the Hong Kong government followed the laissez-faire policy of Great Britain closely, until changes were initiated towards the end of the century.[109] In Hong Kong it was relatively easy to transplant the British model of pharmacy into a largely Chinese population; western medicine competed only with TCM; in the Straits Settlements western medicine competed with a wide range of alternative medical traditions. Implementing the British model presented more difficulties, yet there were many similarities in the approaches taken.

In all the eastern colonies key aspects of the British model of pharmacy were implemented, including the linking of the control of poisons with the regulation of pharmacy. The British 1868 Pharmacy and Poisons Act was the template for most of the legislation passed. Companies were allowed to own pharmacies provided they employed pharmacists to run that part of the business concerned with the sale of poisons. And all colonies recognized a hierarchy of pharmaceutical qualifications, mainly consisting of the British chemist and druggist certificate and a lower-level dispenser qualification. All appointed British holders of the higher-level pharmaceutical chemist qualification for senior positions in pharmacy management, administration, and education. But it was in autonomy from the doctors that pharmacy in highly populated colonies differed from that in settler communities, Licences to practise were initially awarded by colonial governors rather than Medical Boards, and even when these were established there was often little point in seeking to create a separate Pharmacy Board when the numbers involved were so low. And with no legally established pharmaceutical societies the issue of whether registers should be kept by the society or a separate Pharmacy Board did not arise.

During the period of British rule complete autonomy from the medical profession was neither practical nor necessary. It was only the issue of doctor-dispensing in Hong Kong that brought the issue to a head. Pharmacists had little incentive to work together towards common goals, and professional organizations emerged only in the mid-twentieth century. Hierarchical structures clearly emerged, although the gap between dispensers and chemists and druggists slowly eroded as the educational levels of the former were steadily raised. On the face of it, the British legal framework for the control of poisons and the regulation of pharmacy was simply transplanted in the eastern colonies. But the practice did not always follow policy. In Malaysia, pharmacy services—unlike some other health services—developed little during the years of British rule, activities mainly being those concerned with the procurement, supply, and distribution of medicines.[110] Late in the British occupation of Hong Kong, major reforms had still to be made to bring the standard of western pharmacy up to that in other developed economies.[111] In practice, the full professionalization of pharmacy was delayed until after independence.

Notes

1. Seth, Suman. *Difference and Disease: Medicine, Race and the Eighteenth-Century British Empire* (Cambridge, 2018), pp. 190–1.
2. Magee, Gary B. and Thompson, Andrew S. *Empire and Globalisation: Networks of People, Goods and Capital in the British World, c.1850–1914* (Cambridge, 2010), p. 239.
3. Stockwell, A.J. 'Expansion in South East Asia', in Porter, Andrew (ed.) *The Nineteenth Century: The Oxford History of the British Empire* (Oxford, 1999), p. 372.
4. Ibid., p. 374.
5. Turnbull, C.M. *A History of Singapore 1819–1975* (Oxford, 1977), p. 33.
6. Lee, Y.K. 'The Early History of Pharmacy in Singapore', *Singapore Medical Journal*, 47 (2006), p. 436.
7. Stockwell, 'Expansion in South East Asia', p. 375.
8. Lee, Yong Kiat. *The Medical History of Early Singapore* (Tokyo, 1978), p. 215.
9. Lee, 'Early History of Pharmacy', p. 436.
10. Northrup, David. 'Migration: Africa, Asia, the Pacific', in Porter, *Nineteenth Century*, pp. 88–91.
11. *Straits Settlements and Federal Malay States Census* (Singapore, 1901).

12. Lee, 'Early History of Pharmacy', p. 437.
13. Tate, Muzaffar Desmond. *History of Medicine in Malaysia: The Foundation Years* (Kuala Lumpur, 2005), pp. 26–7.
14. Hose, Charles. *The Field-Book of a Jungle-Wallah* (London, 1929), p. 212.
15. Tate, *History of Medicine in Malaysia*, Note 37, p. 33.
16. Arasaratnan, Sinappah. *Indians in Malaysia and Singapore* (Oxford, 1970).
17. Tate, *History of Medicine in Malaysia*, p. 27.
18. Colley, Frederick C. 'Traditional Indian Medicine in Malaysia', *Journal of the Royal Asiatic Society Malaysian Branch*, 51 (1978), pp. 77–109.
19. Tate, *History of Medicine in Malaysia*, p. 34.
20. Colley, 'Traditional Indian Medicine', p. 78.
21. Tate, *History of Medicine in Malaysia*, p. 29.
22. Vaughan, J.D. *The Manners and Customs of the Chinese of the Straits Settlements* (Singapore, 1879), p. 22, cited in Tate, *History of Medicine in Malaysia*, p. 25. See also Purcell, V. *The Chinese in Malaya* (London, 1948).
23. Lee, *Medical History*, p. 8.
24. Ibid., p. 7.
25. Lee, 'Early History of Pharmacy', p. 437.
26. Lee, *Medical History*, p. 130.
27. Ibid., p. 10.
28. Ramasubban, R. 'Imperial Health in British India 1857–1900', in MacLeod, R. and Lewis, M. (eds) *Disease, Medicine and Empire: Perspectives on Western Medicine and the Experience of European Expansion* (London, 1988), p. 39.
29. Lee, *Medical History*, p. 72.
30. Lee, 'Early History of Pharmacy', p. 437.
31. Ibid.
32. Lee, *Medical History*, p. 13.
33. Lee, 'Early History of Pharmacy', p. 437.
34. Lee, *Medical History*, p. 12.
35. Ibid., p. 311.
36. *The Straits Times and Singapore Journal of Commerce*, 20 November (1869), p. 3.
37. Lee, *Medical History*, p. 70.

38. Paraidathathu, Thomas. *Pharmacy in Malaysia: History and Development* (Kuala Lumpur, 2020), p. 73.
39. Fong, T.S. *Pharmacy in Singapore: A Journey Through the Years* (Singapore, 2002), p. 21.
40. Bradley, Martin. *The Journey and Beyond: Community Pharmacy in Malaysia* (Kuala Lumpur, 2014), p. xx.
41. Lee, Y.K. 'Private Practitioners and Private Hospitals in Early Singapore, 1819–1872', *Singapore Medical Journal*, 46 (2005), pp. 489–503.
42. Lee, 'Early History of Pharmacy', p. 439.
43. *The Straits Times and Singapore Journal of Commerce*, 15 July (1845), p. 1.
44. Lee, *Medical History*, p. 24.
45. *Straits Times*, 15 July 1845, 1.
46. Bradley, *Journey and Beyond*, p. xx.
47. Turnbull, *History of Singapore*, p. 65.
48. Ibid., p. 9.
49. Ibid., p. 17.
50. They wrote to Henry Ridley, Director of Gardens and Forests for the Straits Settlements, informing him that he had been made a 'corresponding member'. The National Archives (TNA), London, RM 10/71, letter dated 12 May 1892.
51. Paraidathathu, *Pharmacy in Malaysia*, p. 34.
52. Lee, 'Early History of Pharmacy', p. 441.
53. Fong, *Pharmacy in Singapore*, p. 131.
54. *History of Malaysian Pharmaceutical Society* (Selangor, 2020). https://www.mps.org.my/index.cfm?&menuid=84. Accessed 20 April 2020.
55. Paraidathathu, *Pharmacy in Malaysia*, p. 40.
56. Ibid., p. 10.
57. Hitch, A.V. 'Apprenticeship, Qualification and a Taste of Hospital Pharmacy', *Pharmaceutical Journal*, 244 (1990), pp. 731–3.
58. Paraidathathu, *Pharmacy in Malaysia*, p. 48.
59. Ibid., p. 49.
60. Hitch, A.V. 'Working as a Pharmacist in Pre-War Taiping', *Pharmaceutical Journal*, 244 (1990), pp. 765–6.
61. Hitch, A.V. 'Post-war Malaya and Work with UNICEF', *Pharmaceutical Journal*, 246 (1991), pp. 20–1.
62. Paraidathathu, *Pharmacy in Malaysia*, p. 15.

63. Ibid., p. 29.
64. Ibid., p. 31.
65. Ibid., p. 30.
66. Chiu, Patrick. 'First Hundred Years of Western Pharmacy in Colonial Hong Kong', *Pharmaceutical Historian*, 46 (2016), p. 43.
67. Chiu, Patrick. 'Challenges and Opportunities for Western Pharmacy in Colonial Hong Kong, 1945–1984', *Pharmaceutical Historian*, 50 (2020), pp. 45–6.
68. Ibid., p. 48.
69. Waters, Dan. 'Hong Kong's Hongs with Long Histories and British Connections', *Journal of the Royal Asiatic Society Hong Kong Branch*, 30 (1961), pp. 219–56.
70. Chiu, 'First Hundred Years', Note 29, p. 46.32.
71. *Register of Pharmaceutical Chemists and Chemists and Druggists* (London, 1917), p. 17.
72. Chiu, Patrick. 'Henry Humphreys, 1867–1943: Visionary in Retail Pharmacy in Colonial Hong Kong', *Pharmaceutical Historian*, 48 (2018), pp. 77–81.
73. Chiu, 'First Hundred Years', p. 46.
74. *The Hong Kong Daily Press*, 24 August (1857), p. 2.
75. Chiu, 'First Hundred Years', p. 47.
76. *Biographical Dictionary of Medical Practitioners in Hong Kong: 1841–1941* (Hong Kong, 2020).
77. *Hong Kong Government Gazette* (HKGG) (1883), p. 203.
78. Atkinson, J.M., Wright, A. and Cartwright, H.A. *Twentieth Century Impressions of Hong Kong, Shanghai and Other Treaty Ports of China* (London, 1908), p. 265.
79. Chiu, 'First Hundred Years', p. 47.
80. *Medical and Sanitary Annual Report 1938* (Hong Kong, 1938), p. 38.
81. HKGG, 20 March (1858), No. 7.
82. HKGG, 20 March (1864), No. 3.
83. HKGG (1886), p. 216.
84. HKGG (1904), p. 243.
85. HKGG (1908), p. 447.
86. Ibid., p. 842.
87. HKGG (1916), p. 285.
88. HKGG Supplement (1923), p. 285.

89. *Royal Commission on Opium 1894–5* (London, 1895), pp. 205–11.
90. Costa, A.M. *A Century of International Drug Control, United Nations Office for Drug Control and Prevention* (Geneva, 2008), pp. 20–5.
91. Chiu, 'Challenges and Opportunities', p. 47.
92. Chui, 'First Hundred Years', p. 45.
93. HKGG Supplement (1937), p. 366.
94. Chiu, 'First Hundred Years', p. 47.
95. *Staff List, Hong Kong Government* (Hong Kong, 1967), p. 70.
96. Chiu, Patrick. 'Western Pharmacy Education and Practice in Colonial Hong Kong after the Second World War, 1945–1984', *Pharmaceutical Historian*, 51 (2021), pp. 18–24.
97. Chiu, 'Pharmacy Education', p. 19.
98. *Staff List, Hong Kong Government* (Hong Kong, 1990), p. 239.
99. Chiu, 'Pharmacy Education', p. 20.
100. Duffy, Michael. 'World-Wide War, 1793–1815', in Marshall, P.J. (ed.) *The Eighteenth Century: The Oxford History of the British Empire* (Oxford, 1998), p. 199.
101. Knighton, William. *The History of Ceylon: From the Earliest Period to the Present Time* (London, 2018).
102. Bayly, Susan. 'Colonial Cultures: Asia', in Porter, *Nineteenth Century*, p. 449.
103. *Pharmaceutical Journal and Transactions*, 8 (1877), p. 10.
104. Moore, Robin J. 'Imperial India, 1858–1914', in Porter, *Nineteenth Century*, p. 434.
105. Osterhammel, Jurgen. 'Britain and China, 1842–1914', in Porter *Nineteenth Century*, p. 149.
106. Hughes, A. *Biographical History of William Lockhart* (London, 2020). https://archiveshub.jisc.ac.uk/search/archives/0eaa4de9-e02a-3481-a32c-4cac0a324144. Accessed 25 April 2020.
107. Chiu, Patrick. 'British Pharmacists and Peking Union Medical College Hospital, 1910–1941', *Pharmaceutical Historian*, 47 (2017), p. 57.
108. Tate, *History of Medicine*, p. 38.
109. Harrison, Mark. *Diseases and the Modern World: 1500 to the Present Day* (Cambridge, 2004), p. 111.
110. Paraidathathu, *Pharmacy in Malaysia*, p. 16.
111. Chiu, 'Pharmacy Education', p. 23.

CHAPTER 10

The Australian Colonies: Adapting the British Model

This chapter considers the professionalization of pharmacy in six Australian colonies, where there were no previous European occupants, where most of those practising pharmacy had been trained in Britain, and where the indigenous population played no part. Many brought with them experience of the British model of pharmacy and the role of the PSGB. The processes by which pharmacists came together to form pharmaceutical societies, sought state support for the passage of Pharmacy and Poisons Acts, fended off attempts by the medical profession to control them, and succeeded in regulating and restricting pharmacy ownership, were intimately linked, and are best considered on a colony-by-colony basis. The colonies were settled over several decades, and pharmacy developed in different ways at different rates, with the foundation of pharmaceutical societies extending over 35 years.

In 1914 the Australasian Division of the British Empire consisted of twelve territories, eight of which constituted the Commonwealth of Australia: New South Wales, Queensland, South Australia, Tasmania, Victoria, Western Australia, Northern Territory, and Federal Capital Territory. The other four—New Zealand, Fiji, Papua, and Western Pacific—are considered in the next chapter. Most Australian colonies were separated by vast distances; around 2000 miles both west to east and north to south (Fig. 10.1). Even after intercolonial sea routes were devel-

S. Anderson, *Pharmacy and Professionalization in the British Empire, 1780–1970*, Medicine and Biomedical Sciences in Modern History,
https://doi.org/10.1007/978-3-030-78980-0_10

Fig. 10.1 Map of the Australian colonies of the British Empire, 1914

oped, colonial capitals were often closer to London than to each other; for most of the nineteenth century there was no overland communication between them. Federation of the colonies as the State of Australia took place in 1901. Northern Territory and Federal Capital Territory were created in 1911, having been separated from South Australia and transferred to federal control.

Australia's colonies had very different origins, and the first Europeans to arrive came from a wide variety of backgrounds, from transported convicts and religious dissenters to free settlers and those joining the gold rush. Their economic and political paths followed different routes, and the states evolved rather different cultures and norms. These were reflected in the way pharmacy developed, in the relationship between pharmacy and medicine, and in attitudes to education and training. For many years in Australia, there was no real distinction between pharmacy and medicine, with many doctors also practising pharmacy. As mainly pioneering settler colonies, pharmacy was in the hands of those with entrepreneurial spirit and flair—retail chemists who could adapt their range of merchandise and services to whatever opportunities arose. Intra- and intercolonial differences soon emerged.

Pharmacy and Transportation, 1788

When Captain James Cook reached the east coast of Australia in 1770 in *Endeavour* his botanist Joseph Banks confirmed that the landing spot at Botany Bay was 'sufficiently fertile to support a considerable number of Europeans'.[1] It presented the government with a possible solution to an emerging problem; British convicts had previously been transported to the American colonies, but with the outbreak of war in 1775 this was no longer possible.[2] New South Wales looked a good place to send them. The First Fleet sailed into Botany Bay in January 1788, although it quickly relocated to Sydney Cove where the soil and water supply were better. Convicts constituted a high proportion of the first colonists.[3] They were followed by many more; between 1788 and 1853 around 123,000 male and 25,00 female convicts were transported to New South Wales and Van Diemen's Land (Tasmania after 1855).[4]

In the early years of the colonies most of the drugs used were imported from England. The Society of Apothecaries in London had a regular contract to supply 'the convict establishment' in Australia from 1823. This entailed supplying some £500–£600 worth of drugs and medicines annually until the end of that decade, with a considerable proportion being sent on to Van Diemen's Land in 1830. The Australian emigration trade continued to be lucrative for the Society over the next decade; the cost of medicines per ship with 300 convicts was £14 7s 6d. In 1839 its United Stock was owed £4772 for medicines supplied to eight ships taking 2400 emigrants to the colonies. Settlers to Australian colonies intending to set up pharmacy businesses usually took with them their own medicine chests, often supplied by Apothecaries Hall. A well-stocked medicine chest could boost the valuation of their property, entitling them to a land grant. Colonial surgeons such as Dr. John Ferguson in Western Australia usually requested medicines from the Hall, as these were considered most likely to arrive in good condition; an order could take 15 months to be delivered.[5]

In 1815 the total number of convicts, guards, and settlers in Sydney and Hobart, Tasmania, was around 15,000 people.[6] Over 88 per cent of convicts arrived after the end of the Napoleonic Wars; half were transported for seven years, a quarter for life.[7] Most came from England, but about a quarter were from Ireland and about a quarter of them were women. Around one in twenty were from Scotland. Transportation, along with distance, cost, and other uncertainties, severely impeded free migration; in the 1820s convicts outnumbered free settlers by about three to one. Transportation to Australia stopped at different times; in Queensland it ended in 1840, but in Western Australia it stopped only in 1868.[8] Between 1850 and 1867 Western Australia received 9668 male convicts, much to the annoyance of the eastern Australian colonies where transportation had already ended.[9] There was never transportation to Victoria. By 1861 Australia was home to over a million colonists, whilst the Aboriginal population had dwindled to around a quarter of a million; by 1911 it was home to 4.5 million colonists.[10]

The list of medicines considered necessary for the health of convicts was kept to the absolute minimum. A list of drugs required for the use of the sick at the Moreton Bay Settlement in Brisbane when it was founded in 1824 was drawn up by the military commander, Lieutenant Henry Miller: only 17 preparations were considered necessary. The Convict Pharmacopoeia included two pills, four tinctures, and three ointments or balms.[11] Medical treatments still largely involved extensive use of bleeding, sweating, vomiting, and purging. An order was sent by ship to London; the replacement drugs arrived eleven months later. In Britain, pharmacopoeias in use at the time contained several hundred items, a substantial proportion of which was imported into the Australian colonies; for most colonists, some 200 pharmaceutical preparations were available.[12]

New South Wales: The Legacy of Transportation

Convicts and guards were at first in the care of naval doctors. But the convict ships often included people who had some limited medical or pharmaceutical experience. Amongst those who arrived at Sydney in 1815 was John Tawell, who had been a travelling salesman for a patent medicine company in England and had been transported for forging banknotes.[13] He presented himself as an apothecary from London although he had never been apprenticed and had no qualifications. He persuaded the authorities to let him prepare the medicines in the Sydney Hospital, and after five years he was granted a pardon by the Governor. In 1820 a newly established Medical Board granted Tawell a certificate to practise as an apothecary and to compound and dispense medicines; he became the first person in the British Empire to be licenced by a government-constituted body.[14] He opened the first chemist shop in Sydney, later retiring to England where he was hung for poisoning his mistress.[15]

The first private doctors in Sydney were convicts who had received their pardons. One, William Redfern worked at the Sydney Hospital but also had a private practice, supplying his private patients with medicines from the hospital stores.[16] John Neilson practised as a surgeon and druggist in 1833, running a shop with his brother William; and brothers Charles and Frederick McKellar combined the roles of physician, surgeon, and pharmacist. In Sydney in the 1840s there were at least six doctors who kept 'dispensing shops' where they could be consulted, whilst many others owned pharmacies where they employed druggists to undertake the dispensing.[17] Doctors usually took on apprentices to do the dispensing for them.[18] Pharmacy was viewed as part of medicine, and events in Britain were followed with interest. By 1844 there were several chemists and druggists in Sydney, and Ambrose Foss proposed that they form a pharmaceutical society, as a branch of the newly founded one in British.[19] There was no great threat from the doctors at the time, and the proposal received little interest.

In the new colony non-conformist churches served as important places where chemists and druggists could gather to discuss matters of mutual interest. Ambrose Foss was active in Sydney's Pitt Street Congregational Church, where fellow-worshippers included department store owner David Jones and newspaper proprietor John Fairfax, the owner of the *Sydney Herald*.[20] In addition to religious networks—both catholic and protestant—Freemasonry played an important role in the early development of pharmacy in New South Wales and Victoria, and by the time of federation it played a part in elections to the Council of the Pharmaceutical Society of New South Wales (PSNSW) and, later to the employer-pharmacists Pharmacy Guild, which had grown out of an earlier Master Pharmacists' Association of New South Wales. Social networks too played an important role in encouraging a collective identity, if not action. Later intercolonial conferences included dinners and sports activities as well as addresses, pharmacy balls, and chemists' golf days, which were often supported by local wholesaler and drug companies. They were very popular with pharmacists across Australian colonies.[21]

Threats from the Doctors

The situation changed abruptly at the start of the 1850s. The Australian Gold Rush of 1851 prompted a substantial increase in settlers, including several chemists and druggists (Figs. 10.2 and 10.3). In 1850 the first of several attempts were made by the medical profession to secure control over pharmacy. A Dr. Dickson attempted to secure 'a Bill for preventing the sale of virulent poisons, unless under particular circumstances'.[22] It was a Poisons Bill rather than a Pharmacy Bill, but the pharmacists saw it as an attack on their trade. The doctors considered that when pharmacists recommended a remedy for a cough or a headache, they were practising medicine, and that therefore they needed to be registered by a Medical Board consisting entirely of doctors.

Dickson's proposals were attacked in the press and by the Legislative Council. They were seen as an unwarranted restriction on trade, as well as an absurd attempt to guard against every possible abuse of poison, which was, claimed the *Sydney Morning Star*, only likely to benefit those opposed to the chemists.[23] It was also argued that 'if legislation upon the subject had been deemed advisable, such a course would have long since been adapted in England', reflecting a somewhat deferential attitude to 'the mother country'. With outside support, the pharmacists' efforts were enough to fight off the threat, at least for the time being. The Sydney society was the first in the Empire outside Britain, and its success demonstrated the value of pharmacists working together to fend off attempts at control by the medical profession.[24]

In the 1870s another attempt was made by the doctors to secure control over the pharmacists; this time Joseph Docker, a former surgeon

Fig. 10.2 Bray's Hill End Dispensary, 1872 (Courtesy of Holtermann Collection, Mitchell Library)

in the East India Company, submitted a Poisons Bill before Parliament, which would have placed control of the sale of poisons under the Medical Board.[25] To avoid this Parliament needed to be persuaded to place the sale of poisons under a separate Pharmacy or Poisons Board. This time the pharmacists were mobilized by William Townley Pinhey, an unqualified Englishman who had bought a pharmacy in Sydney in the early 1850s. He became one of Ambrose Foss's pupils and learnt the benefits of pharmacists working together, a view strengthened with the arrival of newly qualified pharmacists from Britain where the voice of the PSGB was beginning to be heard beyond pharmacists themselves. With the help of a newly reconstituted PSNSW, Pinhey successfully lobbied Parliament, and a Sale and Use of Poisons Act was passed in 1876.[26]

This Act created the first body with the title 'Pharmacy Board' in the British Empire, although it was actually a Poisons Board, as it had no responsibility for the registration of pharmacists; it did, however, keep a

Fig. 10.3 John Souter, chemist and druggist, Home Rule, New South Wales, 1872 (Courtesy of Holtermann Collection, Mitchell Library)

list of authorized sellers of poisons. It was neither independent of the Medical Board nor separate from the pharmaceutical society. Its members were the PSNSW's Council, but its chair was the Medical Board's president, although he rarely attended. Between 1876 and 1930 the posts of secretary of the Board and of the PSNSW were held by the same person. Although prosecutions were instituted for unauthorized sale of poisons, anyone could call themselves a chemist. The number of shops increased rapidly; in Sydney there were 51 in 1876, but 164 in 1895.[27] In 1893 across the colony there were 545 names on the list of authorized sellers of poisons with another 200 trading as chemists but not listed. Hospitals such as Sydney Women's Hospital trained their own 'nurse-dispensers', a role that continued well into the twentieth century.[28]

As the 1876 Act did not give it legal recognition or restrict the use of titles, the PSNSW prepared a Pharmacy Bill. But it failed to secure its incorporation in 1886, and later failed to persuade Parliament to pass the Bill. It faced considerable opposition from some doctors and many unregistered chemists, and a reshaped Pharmacy Bill was presented to Parliament in 1896. A petition noted that

> this colony is about the only British possession which has no Pharmacy Bill, and where under the present Poisons Act, anyone without the slightest knowledge can carry on a business as a chemist and druggist.[29]

But the Bill was blocked by the legislative Council over disputes over who should be registered. The PSNSW wanted it restricted to those whose names already appeared on the Board's list; politicians and the press favoured accepting everyone then trading as chemists. A compromise was reached, and the Pharmacy Act was passed in 1897, with all those who had traded for two years included in the Register, regardless of qualification or training. It was a decision that had lasting implications for the colony's society, and for pharmaceutical reciprocity within Australia.[30]

Victoria: A Separate Pharmacy Board, 1857

When the Port Phillip Settlement was founded in 1835 the early settlers included a number of medical and pharmaceutical practitioners, with some of the first pharmacy shops being opened by doctors.[31] George Wakefield imported and sold drugs and proprietary medicines and gave over-the-counter advice alongside his medical practice.[32] Some later employed chemists and druggists to run their shops, and sometimes doctors and pharmacists worked together in the same premises.[33] In October 1851 the *Pharmaceutical Journal* drew the attention of English chemists to the finding of Australian gold through an encouraging account of recent discoveries.[34] In the rush that followed many doctors and pharmacists headed to Victoria, where the doctors had taken steps towards professionalization in the 1840s.[35]

The doctors moved to control pharmacy in 1856; they took the familiar line, that in giving advice and suggesting medicines pharmacists were practising medicine and needed to operate under the authority of a Medical Board. A Bill was presented to the Victorian Parliament in 1856 for the registration of medical practitioners, which also proposed that vendors of drugs would have to obtain a licence from the Board. No one could

> Prescribe, advise to take or administer any medicine, or do or perform any surgical act or operation, not having obtained a certificate from the Board of examiners as a qualified medical practitioner.[36]

The druggists quickly recognized that their ability to supply medicines of their own making and under their own authority, and to perform minor acts of surgery, were under threat. The Bill failed, but the doctors tried again the next year, this time on the grounds that chemists could not be trusted to regulate the sale of poisons without statutory control. In early 1857 Dr. Tierney introduced a Bill 'to regulate the safe keeping and sale of arsenic and other poisons, and their administration'. It was based on the 1851 British Arsenic Act and would have forced chemists to label as poison and keep sales records for many substances that they frequently supplied, including opium, laudanum, and Dover's Powder.[37]

Chemists considered the proposals absurd and were incensed at not having been consulted. One of them, John Hood, had become a member of the Legislative Council in 1856 in order to see a Pharmacy Bill through Parliament, following the example of Jacob Bell in Britain (Chapter 2). On 16 February 1857, a group of Melbourne chemists met in the home of Samuel Croad, and elected a committee to draw up a petition against the Bill and rules for a pharmaceutical society.[38] The petition was handed to Hood, the committee lobbied other members of the Council, and Tierney subsequently withdrew his Bill. William Archer noted that rejection was achieved following 'the representations of the druggists, ably enforced by several of the more intelligent members of [the legislative] council'.[39] The doctors had now suffered two defeats in their attempts to keep the pharmacists under their control.

Freedom from Medical Control

By the end of 1857, the newly formed Pharmaceutical Society of Victoria (PSV) had over 90 members. Its aims were to be 'the promotion of education and scientific improvement, and the protection of their interests'.[40] It was noted in London that 'the laws and constitution of the Society resemble those of the PSGB'.[41] But the doctors were not giving up. In 1858 Dr. Embling submitted a Bill to register medical practitioners and tidy up the issuing of death certificates. But the registration clause was seen as again threatening the freedom of chemists to prescribe remedies. The PSV lobbied against it and the Bill was defeated. Thereafter, the PSV registered as a body corporate under the 1864 Companies Statute, giving it the same protection as if it were incorporated by an Act of Parliament. The *Chemist and Druggist* noted that 'under the new constitution, with its wider scope, there is no reason why the Society should not, in a few years...occupy a position as high and important as that of the PSGB'.[42]

It took 19 years for the PSV to secure a Pharmacy Act, as its effort had been 'rendered fruitless by changes in Government'.[43] Almost all the pharmacists in Melbourne had been trained in England, and by 1876 three of them had been elected to Parliament.[44] The Bill proposed a legally established Board of Pharmacy independent of the PSV. This was a significant departure from the British model of pharmacy, but in practice separation between Board and PSV was far from complete; there was substantial cross membership between the Board and the PSV Council. But its chair was a pharmacist rather than a doctor, so with the 1876 Act pharmacy in Victoria was finally free from medical control. A Pharmaceutical Register was established in 1877.[45] Training through apprenticeship followed by examination was the usual route, although classes began at the Victoria College of Pharmacy in 1882.[46] The decision to have a separate Pharmacy Board was highly significant, as all other Australian colonies except Western Australia followed the same path.

The Pharmacy Act was passed in Victoria four years before the 1880 legal case in Britain that led to the establishment of chain store pharmacies (Chapter 2). The same clause in the 1868 British Act which referred to a 'person' and did not prevent a company operating a pharmacy appeared in the 1876 Victoria Act. The Secretary of the Pharmacy Board, Harry Shillinglaw, arranged for the passage of an amending Act in 1885 to correct its deficiencies. The amendment defined 'person' in such a way that clearly excluded corporations from practising pharmacy, thus preventing the emergence of pharmacy chains in Victoria.[47] Poisons were controlled through a separate Poisons Act passed in the same year.

Queensland: Distinguishing the Competent from the Incompetent

The British colony of Queensland developed from the Moreton Bay penal colony. In 1841 it had a population of 200, only 67 of whom were free, but in 1844 the first chemist's shop was opened by William Kent in Brisbane.[48] As the colony developed it adopted earlier New South Wales legislation, and a Medical Board was established in 1860 to licence medical and pharmaceutical practitioners.[49] The first name was entered in the Register of Chemists and Druggists in Queensland on 3 October 1861.[50] By the 1870s Brisbane chemists found themselves up against severe competition from several directions. Wholesalers began also trading as retailers, general stores sold the same range of merchandise, and increasing numbers of individuals who were not certified by the Medical Board described themselves as 'chemists'. The Board showed little interest in pharmacy and saw certification as a means of keeping the chemists under control. Chemists doubted both their ability and their interest in distinguishing between the competent and the incompetent. In 1874 one of the chemists, R. T. Bellamy, asked the Medical Board for its help in arranging classes in chemistry for chemists' assistants from the government analyst. The Board promised to take up the matter, but nothing was done.[51]

As elsewhere dissatisfaction led to collective action. In October 1880, a meeting of chemists convened by Moses Ward was held in Brisbane, and rules for a society were drawn up.[52] These were approved at a meeting in November, and the Queensland Pharmaceutical Society (QPS) was founded. By 1881 it had a membership of fifty. Its main purpose was to wrest the control of pharmacy from the Medical Board. Its Council obtained help and advice from Shillinglaw in Victoria and immediately attempted to secure legislation in the form of separate Pharmacy and Poisons Bills. The Pharmacy Bill included proposals for a Pharmacy Board on Victorian lines consisting only of pharmacists, for protected titles reserved to the qualified, and for regulations regarding the supervision of training, examinations, and registration. The Poisons Bill dealt with control over the sale and supply of poisons.[53]

Moses Ward put the case to the Queensland Parliament; the Bills passed easily through the Assembly in 1881 but were blocked in the Legislative Council. The Medical Board was strongly opposed, claiming that chemists and druggists were incapable of accepting responsibility, and needed to be supervised by the doctors. The doctors, led by Dr. Bancroft, questioned whether there were enough capable chemists in Queensland to constitute a Pharmacy Board. The idea that chemists were capable of examining each other drew ridicule from the doctors; the chemists countered that many candidates, having done some study, would know a great deal more than the examiners.

The Bill became the subject of two enquiries by committees of the Legislative Council, where the civil war between the doctors and the pharmacists received public scrutiny. The Medical Board used scorn and prevarication to defend its power base. There were frequent promises to 'look into things'. But the pharmacists were not entirely without medical support; Dr. Kevin O'Doherty supported the measure, asking 'what was the use of a doctor if his chemist was unfit to compound the drugs prescribed?' And there was support from some politicians, including A. J. Thynne in the Council and Samuel Griffin, leader of the opposition in the Legislative Assembly. The Bill was presented as a necessary means of raising the standards of Queensland chemists. It finally won approval in a much-altered state, more as a consequence of politics that strength of argument. After four years of wrangling the Pharmacy Act was passed in 1884, and a Pharmacy Board established consisting of four chemists and three doctors.[54]

One of its first acts was to announce that it would recognize certificates issued by Pharmacy Boards in Victoria, New South Wales, and New Zealand, in expectation of reciprocity; British certificates were accepted automatically. But substantial differences existed; Queensland had not passed a Poisons Law, and unregistered persons could be left in charge of a pharmacy provided it was owned by someone who was. The Board eventually established oral, written, and practical examinations in chemistry, botany, and materia medica in 1886, but candidates were generally ill-prepared, and only three of the first twenty passed. As a result, the QPS offered formal classes and established a college of pharmacy; an English pharmaceutical chemist, R. C. Cowley, was appointed as director of the Queensland College of Pharmacy in 1888.[55]

South Australia: Pharmacy and Religious Dissenters

South Australia was different from other Australian colonies in being founded largely by non-conformist settlers, and many of the early chemists were themselves dissenters.[56] Its 1834 constitution placed authority in the hands of the South Australian Company. By 1839 the colony had its first chemist's shop in Adelaide, run by William Bickford who claimed to be a pharmacist. When he died in 1850 the business was run by his widow who sent her son to England to qualify. Francis Hardy Faulding from Yorkshire, the surgeon on a migrant ship, opened a pharmacy in Adelaide in 1844.[57] Within three years he operated a warehouse, and became South Australia's first pharmaceutical wholesaler.

South Australia established a Medical Board in 1844 to examine the qualifications of chemists and druggists as well as doctors, but of the 109 individuals who claimed to be medical practitioners, only 50 were judged of sufficient competence for inclusion on a register.

Following the Victoria Gold Rush many people moved on to South Australia. But the migration included many unqualified people, and there was no legislation to control their activities. The need to bring order to the selling of poisons and recognition of the qualifications of doctors, dentists and pharmacists saw legislators besieged with demands to pass appropriate laws. South Australia became the first of the Australian colonies to pass a Poisons Act in 1862, although their sale was not restricted to chemists and the Act did little more than impose labelling requirements.[58] In 1858 correspondence between pharmacists in Adelaide and Melbourne led to encouragement by the PSV to establish a branch in Adelaide, where, by the 1870s, the number of pharmacies had reached twenty, along with several run by doctors and three wholesalers, but no action was taken.

From 1884 there were increasing calls to establish a pharmaceutical society in Adelaide. In August 1885 W. J. Main called a meeting to discuss prices, education, and the need for a Pharmacy Act. An intercolonial pharmaceutical conference had been proposed, and Adelaide pharmacists needed to select delegates. They also needed to clarify the relationship between doctors and pharmacists. A subscriber to the *Chemist and Druggist of Australasia* in 1887 reported that 'here in Adelaide it is the custom for medical men to have the name and address of their chemist printed on their prescription'.[59] The editor asked whether it was the custom to pay 'the medical men a bonus on their prescription?' He was told that 'the commission allowed is one third the price of the medicine'. It was reported that the practice, called 'channelling', was 'condemned by nearly every leading chemist in Melbourne'.[60]

The aims of the South Australian Pharmaceutical Society (SAPS) were similar to those of the other Australian societies: 'to unite the chemists and druggists into one recognized and independent body, to protect their general interests and to advance the science and art of pharmacy'.[61] By August 1886 it had 105 members and associates and its leaders discussed submitting a Pharmacy Bill, although they had no political connections. Support within the membership was far from universal; dissenting voices considered legal restrictions to be a form of state aid and preferred to rely on the voluntary principle. But the Council persisted, and the South

Australian Pharmacy Act was passed in 1891. It followed the Victorian model of having a separate, independent Pharmacy Board for the registration of pharmacists, but 'following the British tradition, the Council declared that the Society was the body responsible for the education of pharmaceutical chemists'.[62]

Tasmania: The North-South Divide

The colony of Van Diemen's Land began in 1803 when separate settlements were established in Launceston and Hobart. There was no overland connection between them until 1818, and a deep north–south rivalry developed which was to have a negative impact on the development of pharmacy well into the twentieth century. At first many recalcitrant convicts were sent there; in 1821 they made up over half the white population. By 1823 law and order were improving and free settlers started arriving. Between 1820 and 1843 some 37 medical practitioners arrived, most taking up government appointments or opening private practices. More so than in other colonies, pharmacy practitioners in Tasmania operated under the tight control of the medical profession. Most worked as assistants to surgeons, although some opened on their own account: Michael Bates, a Yorkshire chemist, opened a shop in Launceston in 1825; John Wilkinson opened one in Hobart in 1829.[63]

In 1837 the Van Diemen's Land Legislative Council passed an Act to regulate the practice of medicine, although it made no claim for control over chemists. But under the 1842 Medical Act the Court of Medical Examiners was charged with maintaining a register of chemists; those who satisfied the Court's requirements were 'licensed to dispense medicines'. The Act enabled doctors to take action against those who were impinging on their private practice and vastly reducing their profits. The Court was the means by which the doctors retained control over the chemists. Regulations stipulated that those applying for a licence to practise as a dispenser of medicines must be examined orally by the Court of Medical Examiners and prove that they had been in practice for three years. They then received certificates indicating that they were legally qualified dispensers of medicines.[64]

The Medical Act also served as a Poisons Law. It restricted the sale of scheduled poisons to medical and pharmaceutical practitioners or to licenced storekeepers. It also prescribed labelling requirements, but the Medical Board had no inspectors, and the law was rarely enforced. Poisons controls were strengthened slightly in 1867 and again in 1886, but had little impact.[65] Chemists felt that they were denied any protection by the law, and following the intercolonial conference (below), they became aware of how far they lagged behind other states with regard to pharmacy education and legislation. Credentials awarded by the Court of Medical Examiners were not recognized outside the colony. By 1885 there were 34 registered chemists and druggists in Tasmania, and in 1887 the doctors consolidated their position by forming a Tasmanian branch of the British Medical Association.[66]

Faced with a number of common problems the chemists came together. But in Tasmania there were two main towns, and chemists in Hobart and Launceston battled for dominance. When vacancies arose on the Court of Medical Examiners, Hobart chemists sought to have them filled by chemists. In 1888 members of the Launceston Chemists' Association sent a letter of support to their colleagues in Hobart, suggesting that the nominees should be Hobart chemists. Four of the latter met four days later, and H. T. Gould proposed that 'an association be formed of all registered chemists in Tasmania, to be called the "Pharmaceutical Society of Tasmania" (PST), with councils in both Hobart and Launceston'.[67] This was carried, and an approach made to the Court, who resisted any suggestion that pharmacists should be included amongst their number. The society did not survive this initial failure.

Nothing changed until 1891, when issues around education, examination, and recognition became more prominent, and the Medical Board increasingly favoured a Pharmacy Act in line with other colonies. In 1890 it indicated that it would not hold further examinations until a legally recognized pharmaceutical society was formed to do so. Chemists' leaders from the north and south met again and formally inaugurated the society in 1891. It was based in Hobart but had northern and southern councils with very different views; when Hobart chemists set up classes in 1894, those in Launceston were annoyed, and membership of the society there dropped to seven. But the Court of Medical Examiners finally welcomed pharmacy examiners to its membership, having been persuaded by Shillinglaw and the head of the Victoria College of Pharmacy By 1894 membership of the society had increased to 38, and the Hobart Chemists' Association was revived to 'deal with trade matters outside the scope of the Society'.[68]

After 1897 only Tasmania lacked a Pharmacy Act. It was 1908, seven years after federation, when one was finally passed, marking the final autonomy of the pharmacists from the doctors.[69] The 1908 Pharmacy Act gave Tasmanian chemists a monopoly in the supply of medicines, but in 1930 an attempt was made to overturn some aspects of it. It was claimed to be the only one in the world that confined the sale of all medicines to qualified pharmacists.[70] The chemists launched a vigorous defence, with regular meetings being held with members of the Hobart government to put their case. When the Bill was finally brought to Parliament no seconder could be found and it failed. The chemists retained their monopoly.[71]

Western Australia: Following the British Example

Like South Australia, Western Australia was initially settled directly from England; it was vast and remote from the other Australian colonies. Some convicts had been transported there by 1829, but in 1834 there were still less than 1600 inhabitants, mainly in Freemantle.[72] The first chemist to arrive in Perth in 1833 was George Shenton, who had been apprenticed in Portsea, England.[73] It was another ten years before a second pharmacy was opened by Dr. John Shipton, a surgeon who previously practised in Fremantle. Doctor-owned pharmacies continued into the late 1880s; thereafter they became wholly owned and managed by pharmacists. There was also a dispenser at the hospital. But by 1887 the *Chemist and Druggist of Australasia* was noting that

> in pharmaceutical matters Western Australia is much behind her neighbouring colonies, there being no Pharmacy Act or Pharmaceutical Society in existence. Chemists and druggists have no special protection by law, there are only four establishments in Perth, and not one of these is attended by the owner.[74]

Some regulation did exist. A Medical Ordinance in 1860 established a Medical Board to register medical practitioners. But the Legislative Council retained a clause that originally appeared in the British 1815 Apothecaries Act (Chapter 2): 'Nothing in the Ordinance shall extend to or be construed to extend to prejudice or in any way affect the lawful occupation, trade or business of chemists and druggists and dentists'. The same clause had appeared in the British Medical Act of 1858 and in similar Ordinances in Victoria and New Zealand. But in the Western Australian legislation some words were omitted from the original text;

'so far as the same extend to selling, compounding or dispensing medicines'. In the early twentieth century the issue was tested in the courts. The doctors sought exclusive rights in offering treatment, claiming that under the Ordinance the omission of this phrase meant that chemists had no right to give advice or to prescribe remedies. They lost their case.[75]

A Sale of Poisons Act was passed in 1879 as it was desirable 'for the safety of the public that the retailing, dispensing or compounding of poisons should be subject to regulation and restraint'. But the law did not apply to medical practitioners or wholesalers, and it made the selling or keeping of open shop unlawful unless the person concerned had been licenced by a magistrate. The Act simply restricted the sale of poisons to those known to legal officers; it gave no monopoly or other privileges to chemists and druggists. This situation prompted pharmacists to come together, and a Pharmaceutical Society of Western Australia (PSWA) was formed in 1892. One of their first actions was to seek the passage of a Pharmacy Act. They succeeded within two years; but in doing so they rejected the Victorian model of a Pharmacy Board in favour of the British model, in which the society itself was the registration body.[76]

In commenting on the Act, the *Chemist and Druggist* in London saw the decision to reject a Pharmacy Board in favour of the British model as a positive one. It noted that

> The distinction between the Pharmacy Board and the Pharmaceutical Society which prevails in the other Australasian Pharmacy Acts is abandoned in this one, and consequently the Pharmaceutical Society of Western Australia becomes the administrative body.[77]

The advantage of the British model, it noted, was the greater freedom that combining the functions provided in deciding who should be allowed to practise. 'In all the other colonies, without exception' it continued

> persons holding the certificates of the British Society are entitled to registration, and the Pharmacy Board or Pharmaceutical Society has no control over the matter. The Western Australian Society has secured greater power. People will be able to register who hold certificates or diplomas of competence from any society or college or Board of Pharmacy recognised by the regulations. It is open to the Society to refuse to recognize British certificates, while at the same time, it will have power to recognise certificates from Great Britain and Ireland, any of the colonies, or from Europe.[78]

The Act became effective on 1 March 1895. Pharmacists in Western Australia wanted to be regarded on equal terms with those in eastern colonies and sought advice from other colonies and Britain. Why they rejected the Victoria model has been a matter of some conjecture. Haines suggests that it may have been an attempt to save the newly founded society; that it was a means of rationalizing costs; that it might have been a show of independence from the other Australian states; or that it was simply a wish to follow the British model more closely.[79]

Inter-Colonial Rivalry and Cooperation

Pharmacy in the Australian colonies had a common British heritage, although that heritage changed as pharmacy in Britain evolved. The colonies had different economic and political experiences, and different identities emerged. Until the late nineteenth century most maintained closer links with Britain than they did with each other. Differences became consolidated as laws were passed; no colony was prepared to pass pharmacy laws identical with those of another.[80] The examination and registration of pharmacists differed, so that qualifications recognized in one colony were not recognized in others. By the 1880s, with rising prosperity across the country, federation of the Australasian colonies was being discussed. Pharmacists began to consider the implications for pharmacy. In 1883 the Victorian Society proposed a conference to explore ways in which Australian pharmacy might be made more uniform.[81]

The omens were not good. A degree of reciprocity was established when the PSNSW began to accept the certificates of any state that would recognize its registration.[82] But friction between Victoria and New South Wales came to a head in 1884 when the Victoria Pharmacy Board declined to accept the certificates issued in Sydney. New South Wales struck back by refusing to accept Victorian qualifications; it proposed an alternative conference in Sydney, but in October 1886 the first intercolonial conference went ahead in Melbourne.[83] It was attended by representatives from New Zealand and all the colonies except Western Australia. There were no representatives from Britain, and with no clear lead being given on education there was uncertainty about who should determine what needed to be known by aspiring pharmacists.[84] The conference failed to reach agreement about common examination standards, but it did succeed in drawing attention to the bitter distrust between New South Wales and Victoria that continued for many years.[85] And it highlighted the fact that pharmacy development was at different stages in the various colonies, and that Victoria was ahead of the others. But there were some positive outcomes: it helped develop a pride in pharmacy and a concern for standards in education; it encouraged those colonies that did not have them to form pharmaceutical societies; and it prompted societies that had not done so to press for pharmacy and poisons laws.

Attitudes to education and training were at the heart of the distrust between pharmacists in Victoria and New South Wales; the first pharmacy courses in Australia were arranged in Victoria. In 1884, 53 candidates presented themselves for examination. An observer in Melbourne regretted 'that the other colonies view with such utter indifference any attempt at uniformity in pharmaceutical education'.[86] In the same year a pharmacist in Sydney noted

> it is a matter of great regret that [the Technical College] is not used to a greater degree by chemists and druggists. In the chemistry class the trade is not represented, and the botany class, which was started especially for the benefit of chemists' assistants and apprentices, is in a similar condition. The reason is very plain. As long as a knowledge of these subjects is not demanded in the Society's examinations, so long will the study of them be neglected.[87]

Many of the British pharmacists emigrating to Australia had taken the Minor examination after attending a crammer school. Some of their pupils went to London to do the same: the Westminster College of Chemistry and Pharmacy had students from Adelaide, Melbourne, Sydney, Brisbane, and Tasmania.[88]

Titles and journals also added to distrust between colonies. Late in 1884 the PSV renamed itself the Pharmaceutical Society of Australasia, much to the annoyance of other colonies, and in 1886 it issued its own publication, the *Australasian Journal of Pharmacy*.[89] During the 1890s attempts were made to arrange further intercolonial pharmaceutical conferences.[90] But nothing came of them and little changed, as existing arrangements were laid down in colonial laws, and any change would require action by the various colonial parliaments. One observer concluded: 'I do not see what further can be done unless the Victorian Board of Pharmacy can be persuaded to accept the registrations of all the other colonies'.[91] The best hope was federation, with the six colonies being brought together as the State of Australia in 1901. The first Australian Pharmaceutical Conference was held in Adelaide in 1907, and thereafter they were held biennially in different states. Intercolonial rivalry continued late into the twentieth century; a federal Pharmaceutical Society of Australia (PSA) was founded only in 1977.[92]

Education Standards: Reciprocal Recognition of Qualifications

The time interval between passage of the various Pharmacy Acts laid the seeds for growing animosity between pharmacy practitioners in the various Australian colonies. The barriers were slowly lowered, and each admitted to the register some individuals who would not be able to obtain registration in states with earlier Acts. A chemist registered in one colony might find that his qualifications were not recognized in any other. Whilst all the colonies accepted British qualified pharmacists to their registers,[93] large numbers had no diploma at all, having been registered simply on the basis of having been in business before the passing of the relevant Act.[94] This led to arguments about the relative merits of each state's qualifications, and to legal entanglements when individuals appealed against decisions made against them.

In 1881 the Victorian Board applied unsuccessfully to Britain for recognition of its qualifications; in London reciprocity with colonial qualifications was not a pressing issue. It was claimed that even if chemists wanted to move between colonies, intercolonial reciprocity of diplomas would affect very few of them. One of the first acts of the Queensland Pharmacy Board in 1885 was to state that it would honour certificates issued by Pharmacy Boards in Victoria, New South Wales, and New Zealand, in the expectation of recognition of its own. English certificates were accepted automatically. But attempts to enforce the new law highlighted that Queensland had no poisons legislation, and that unregistered people could be left in charge of a pharmacy provided it was owned by a registered chemist.[95]

In the 1890s both the Queensland and Western Australia Pharmaceutical Societies also discussed whether they should seek reciprocal recognition with the PSGB. But any such request would draw attention to the lack of reciprocity between Australian colonies. In 1897 the QPS secretary wrote to his opposite number in Western Australia saying that he was

> directed to inform you that this Society desires to support your efforts to promote reciprocity with Great Britain, and we are of opinion that reciprocity should be established throughout Australia before we approach the Mother Country...Will the 'Old Country'...admit Australians when Australians will not admit each other to their respective provinces?[96]

The answer was no. A request made to the PSGB for reciprocal recognition in 1897 was unsuccessful (Chapter 12). Proposals for a pragmatic reciprocity were considered within Australia in 1899, but the PSV again rejected overtures from New South Wales for a conference on reciprocity.[97] Pharmacy leaders in New South Wales concluded that they had little choice but to raise education standards. In 1902 they wrote to Victoria saying they were now willing to register all persons on current registers in all states, but indicated that in future new applicants would need to fulfil a number of conditions. It was the breakthrough that was needed; other states indicated their support for the proposals, and the PSV finally agreed. Reciprocity arrangements were quickly agreed by New South Wales, Victoria, and South Australia, but in Queensland, Western Australia, and Tasmania state laws needed to be amended first. Mutual recognition of state qualifications in Australia and reciprocity of registration was finally achieved just prior to the First World War.[98]

Intercolonial recognition was a prerequisite for reciprocity with Britain, and the two states with colleges of pharmacy—Victoria and Queensland—reached agreement with the PSGB in 1913, following passage of a Byelaw which allowed colonial pharmacists to join the British register.[99] But it was not the end of distrust between pharmacy leaders in the Australian states. In Victoria pharmacists jealously guarded their separate standards until the 1950s; and the PSGB continued to have great influence over education in the Australian states by threatening to withdraw its reciprocity agreements.[100]

Professionalization and the British Model

Professionalization in the Australian colonies generally followed a familiar pattern. Although early pharmacy practitioners were mainly of British origin, local pharmacy leaders sometimes found it necessary to adapt the British model to local circumstances. Early chemists were mostly sole proprietors often isolated from each other, but they soon found it necessary to work together to fight moves by doctors to take control of pharmacy. Pharmaceutical societies were founded largely independently of each other over a period of 35 years, between 1857 and 1892 (Table 10.1) and sometimes built on existing loose trade associations that had been founded in an attempt to regulate prices and to restrict competition from other groups. Each society adopted an organization modelled based on that of the PSGB, although the desire simply to imitate Britain was not always sufficient to sustain their growth, as in Sydney in 1844.[101]

The societies played the central role in establishing the autonomy of pharmacy and its independence from medicine. They helped to change the relationship between chemists and medical practitioners from 'one of servant and master to something resembling mutual respect'.[102] In his study of the early history of the PSV between 1857 and 1918, Boyce noted that its success was due in large measure to its ability to successfully develop communication channels and the creation of strong institutional constructs such as journals and conferences. He found that setting itself a wide range of objectives and having a broad agenda worked much to the organization's advantage: 'the PSV succeeded in shaping the public conceptions of the profession in order to influence an emerging regulatory regime'.[103]

A small but influential group of pharmacists sought support for the passage of legislation through local parliaments, but their success in getting it varied. They were seeking not only autonomy from the doctors but also protection against other traders who sold patent medicines; against the dispensaries of Friendly Societies which dispensed medicines at lower cost; and—in the case of New South Wales—against company chemists. Whilst legislation was based on the British Pharmacy Act 1868, priority was usually given to the control of poisons.[104] Between 1876 and 1879 each colony introduced controls over the sale of poisons, and there was often a long gap between the passage of poisons laws and Pharmacy Acts to regulate the pharmacy profession. In Tasmania the Poisons Act predated the Pharmacy Act by 41 years. Although advice was sometimes sought from other Australian colonies, most had stronger links with Britain than they did with each other. But even when passed the laws were often weak; the first legislation in New South Wales and Queensland failed to limit the practice of pharmacy to registered chemists, leading to admission to the register of unqualified people.[105]

The establishment of a Pharmacy Board separate from a pharmaceutical society was a major departure from the British model. Although initiated by the PSV it was replicated by all others except Western Australia

Table 10.1 Professionalizing pharmacy in the Australian Colonies

Colony	*Year founded*	*State of Australia*	*Pharmaceutical society or association*	*Early pharmacy legislation*	*Early poisons legislation*	*Pharmacy education*
Victoria	Part of NSW 1839. Colony 1851–1901	1901	Society founded 1857	Pharmacy Act 1876	1876	Victoria Pharmacy College 1882
New South Wales	1788	1901	Society founded 1876	Pharmacy Act 1897	1876 (Sale and Use of Poisons Act)	Materia medica at Sydney school 1899
Queensland	Part of NSW 1824 Colony 1859–1901	1901	Society founded 1880	Pharmacy Act 1884	1888	Queensland College of Pharmacy 1888
South Australia	Part of NSW 1788; Province 1836. With Northern Territories 1863	1901	Society founded 1885	Pharmacy Act 1891	1862	Courses available in Adelaide
Tasmania (originally Van Dieman's Land)	1855	1901	Launceston Chemists' Association 1888. Tasmania Society 1891. Hobart Chemists' Association 1894	Pharmacy Act 1908	1867	Courses available in Hobart and Launceston

(continued)

Table 10.1 (continued)

Colony	*Year founded*	*State of Australia*	*Pharmaceutical society or association*	*Early pharmacy legislation*	*Early poisons legislation*	*Pharmacy education*
Western Australia	Settlement 1791–1829. Swan River 1829–32. Colony 1832–1901	1901	Society founded 1892	Pharmacy Act 1894	1879	Apprentice-ship only before 1901
Northern Territory of Australia	Part of South Australia 1863–1911 when joined Federation	1911	As South Australia	As South Australia	As South Australia	As South Australia
Federal Capital Territory	Part of NS Wales Created 1911. Canberra founded 1913	Australian Capital Territory 1938	As New South Wales	As New South Wales	As New South Wales	As New South Wales

along with British colonies elsewhere. Lessons were learnt from the British experience regarding company chemists only in Victoria, which sought amending legislation to prevent companies opening pharmacies. No such law existed in the other colonies. Chain pharmacies were operating in New South Wales by the 1870s, and by 1922 Boots had registered an Australian subsidiary in Sydney and later one in Queensland. But in 1936 Boots succeeded in registering a company in Victoria despite extensive attempts to prevent it. They were to find equally fierce opposition in New Zealand.

Notes

1. The National Archives (TNA), Kew. HO 7/1, 10 May (1785), no page numbers.
2. Ekirch, A. Roger. *Bound for America: The Transportation of British Convicts to the Colonies, 1718–1775* (Oxford, 1987).
3. Hughes, Robert. *The Fatal Shore: A History of the Transportation of Convicts to Australia, 1787–1868* (London, 1987).
4. Robson, L. *The Convict Settlers of Australia* (Melbourne, 1981), p. 9.
5. Hunting, Penelope. *A History of the Society of Apothecaries* (London, 1998), p. 182.
6. Denoon, Donald, with Wyndham, Marivic. 'Australia and the Western Pacific', in Porter, Andrew (ed.) *The Nineteenth Century, The Oxford History of the British Empire* (Oxford, 1999), p. 548.
7. Harper, Marjory. 'British Migration and the Peopling of the Empire', in Porter, Andrew (ed.) *The Nineteenth Century, The Oxford History of the British Empire* (Oxford, 1999), pp. 78–9.
8. Denoon, 'Australia and Western Pacific', p. 549.
9. Harper, 3,120,039,242 'British Migration', p. 78.
10. Denoon, 'Australia and Western Pacific', p. 549.
11. Pearn, John, Petrie, Andrew F. and Petrie, Gwynneth M. 'An Early Colonial Pharmacopoeia: A Drug List and Its Materia Medica for an Australian Convict Settlement', *The Medical Journal of Australia*, 149 (1988), pp. 630–4.
12. TNA, CO/201/153. File no. 106435; 106957; pp. 253–4.
13. Birks, G.F. 'An Early Australian Chemist and Druggist', *The Australasian Journal of Pharmacy*, 36 (1921), 550–3.
14. Haines, Gregory. *Pharmacy in Australia: The National Experience* (Deakin, 1988), p. 23.
15. Miller, Geoff. 'Famous Murderer Caught by the Wire', *Pharmaceutical Journal*, 269 (2002), pp. 905–7.
16. Nichol, William. 'The Medical Profession in New South Wales, 1788–1850', *Australian Economic History Review*, 24 (1984), 120–31.
17. Burnby, J.G.L. 'The Early Days of Pharmacy in Australia', *The Pharmaceutical Journal*, 241 (1988), pp. 56–60.

18. Lewis, Milton and MacLeod, Roy. 'Medical Politics and the Professionalization of Medicine in New South Wales, 1850–1901', *Journal of Australian Studies*, 12 (1988), pp. 69–82.
19. Ibid., p. 58.
20. Haines, Gregory. *The Grains and Threepenn'orths of Pharmacy: Pharmacy in New South Wales 1788–1976* (Kilmore, 1976).
21. Ibid.
22. Haines, *National Experience*, p. 86.
23. Ibid., p. 87.
24. Haines, *Pharmacy in New South Wales*, p. 21.
25. Haines, Gregory. *A History of the Pharmacy Board of New South Wales* (Sydney, 1997), p. 32.
26. Burnby, 'Early Days', p. 58.
27. Haines, *National Experience*, p. 138.
28. Haines, *Pharmacy Board*, p. 43.
29. Haines, *National Experience*, p. 139.
30. Ibid.
31. Burnby, 'Early Days', p. 59.
32. Dyason, Diana. 'The Medical Profession in Colonial Victoria, 1834–1901', in Roy MacLeod and Milton Lewis (eds) *Disease, Medicine and Empire: Perspectives on Western Medicine and the Experience of European Expansion* (London: Routledge, 1988), p. 198.
33. Sissons, A.T.S. *Australasian Journal of Pharmacy*, 49 (1968), p. 401.
34. Anderson, Stuart. 'Australia and the Gold Rush Pharmacists', *The Pharmaceutical Journal*, 291 (2013), p. 656.
35. Dyason, 'Medical Profession', p. 205.
36. Haines, *National Experience*, p. 82.
37. The Bill made no provision for restricting the sale of poisons to chemists, but in 1857 nor did any law anywhere in the Empire. Noted by F.C. Kent and quoted in Haines, *National Experience*, p. 82.
38. Haines, *National Experience*, p. 83.
39. Archer, William. Cited in Haines, *National Experience*, p. 82.
40. Feehan, H.V. *Bond and Link: Pharmacy Organizations and Education in Victoria, Australia, 1857–1977* (Melbourne, 1978).
41. *Pharmaceutical Journal and Transactions*, 18 (1858), p. 83.
42. *Chemist and Druggist*, 26 (1884), p. 158.

43. *Pharmaceutical Journal and Transactions*, 3 (1872), p. 960.
44. Haines, *National Experience*, p. 84.
45. Wells, David. 'All Wells That Ends Wells', *Pharmacy History Australia*, 18 (2002), p. 8.
46. Titcher, Warren. 'The Titcher Story: A Pharmaceutical Dynasty', *Pharmacy History Australia*, 15 (2001), p. 3.
47. Lloyd, Alistair. 'Pharmacy in the Australian Colonies: The British Influence, Part 1', *Pharmaceutical Historian*, 18 (1988), p. 8.
48. Haines, *National Experience*, p. 38.
49. Ibid., p. 40.
50. West, Radley. 'Moses Ward: His Influence on Pharmacy in Queensland', *Pharmacy History Australia*, 20 (2003), p. 9.
51. Haines, *National Experience*, pp. 89–90.
52. West, 'Moses Ward', p. 9.
53. Haines, *National Experience*, p. 90.
54. Ibid., p. 92.
55. Ibid., p. 93.
56. Ibid., p. 50.
57. Ibid., p. 54.
58. Ibid., p. 58.
59. *Chemist and Druggist of Australasia* (1887).
60. Haines, *National Experience*, p. 94.
61. Ibid., p. 95.
62. Attiwill, Keith. 'Historical Glimpses of Pharmacy in Australia', *Australian Journal of Pharmacy*, 38 (1957), p. 529.
63. Haines, *National Experience*, p. 29.
64. Ibid.
65. Ibid., p. 127.
66. Ibid., p. 30.
67. Ibid., p. 127.
68. Ibid., p. 129.
69. Miller, Geoff. 'Tasmania: Green Pastures for the Browne Family of Pharmacists', *Pharmacy History Australia*, 16 (2002), p. 5.
70. *Australian Journal of Pharmacy*, 30 November 1960, 1184.
71. Miller, 'Tasmania', p. 4.
72. Haines, *National Experience*, p. 44.
73. McWhinney, Alan. *A History of Pharmacy in Western Australia* (Perth, 1975).
74. Haines, *National Experience*, p. 48.

75. Ibid., p. 49.
76. Ibid., p. 134.
77. *Chemist and Druggist*, 45 (1894), p. 821.
78. Ibid.
79. Haines, *National Experience*, p. 134.
80. Ibid., p. 35.
81. Ibid., p. 94.
82. Matthews, Leslie G. *History of Pharmacy in Britain* (London, 1962), p. 144.
83. *Australasian Journal of Pharmacy*, 1 (1886), pp. 397–416.
84. Haines, *National Experience*, p. 100.
85. Ibid., pp. 95–96.
86. *Chemist and Druggist*, 26 (1884), p. 61.
87. Ibid.
88. Wills, G.S.V. *A Jubilee Souvenir: The Work of GSV Wills and the Westminster College of Chemistry and Pharmacy* (Stratford-upon-Avon, 1899), p. 198.
89. Lloyd, 'British Influence, Part 1', pp. 7–8.
90. Haines, *National Experience*, p. 140.
91. Ibid., p. 149.
92. Ibid., p. 426.
93. Lloyd, Alistair. 'Pharmacy in the Australian Colonies: The British Influence, Part 2', *Pharmaceutical Historian*, 19 (1989), pp. 6–8.
94. Ibid., p. 151.
95. Haines, *National Experience*, p. 92.
96. Ibid.
97. Ibid., pp. 153–6.
98. Ibid., p. 79.
99. Lloyd, 'British Influence, Part 2', p. 6.
100. Haines, *National Experience*, p. 157.
101. Ibid., pp. 142–5.
102. Haines, *National Experience*, p. 77.
103. Boyce, Gordon. 'A Professional Association as Network and Communicating Node: The Pharmaceutical Society of Australia, 1857–1918', *Australian Economic History Review*, 39 (1999), p. 258.
104. *Pharmaceutical Journal and Transactions*, 46 (1886–1887), p. 1073.
105. Haines, *National Experience*, p. 78.

CHAPTER 11

New Zealand and the Western Pacific: Ownership and Company Chemists

In 1914 the Australasian Division of the British Empire included not only the Australian colonies but also New Zealand, the Fiji Islands, Papua, and some Western Pacific islands (Fig. 11.1). Britain had a presence in the region from the mid-nineteenth century, with the first settler ships arriving in New Zealand from Britain in 1840. Early settlers there included chemists and druggists from Britain and Ireland. A few found their way to the smaller islands, although many of the pharmacists who settled there moved on from Australia or New Zealand. By the time the first ones arrived in New Zealand first steps in the professionalization of pharmacy and its separation from medicine had already been taken in Britain; the PSGB was founded just a year after the first settlers left to start new lives. This chapter considers the development of pharmacy in a settler colony in the post-apothecary era.

Pioneer chemists and druggists took with them the British model of pharmacy, along with the stock and fittings needed to start a business on the other side of the world. Unlike Australia, New Zealand was a single colony, with several settlements established at different places around the same time. It had a single Parliament with a full legislative programme, although it was more concerned with the control of poisons and the openness of trade than with the regulation of pharmacy. In drafting their Pharmacy Bill in 1880 New Zealand pharmacists used much of the same

S. Anderson, *Pharmacy and Professionalization in the British Empire, 1780–1970*, Medicine and Biomedical Sciences in Modern History,
https://doi.org/10.1007/978-3-030-78980-0_11

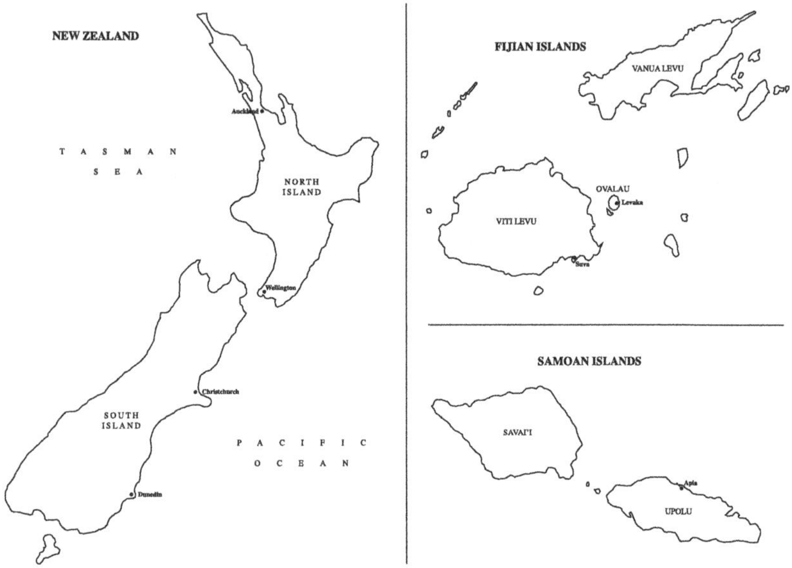

Fig. 11.1 Map of New Zealand, Fiji and Samoa, 1914

wording that appeared in the British Pharmacy Act of 1868. As such it did not require that each pharmacy must be owned by an individual proprietor; indeed, prior to 1954 there were no restrictions on pharmacy ownership. But events in the 1930s demonstrated the consequences of failing to do so in legislation. The New Zealand Parliament was reluctant to limit the emergence of company chemists and was unwilling to agree amending legislation to restrict them. Attempts to limit the expansion of company chemists led to the closest parliamentary scrutiny of pharmacy anywhere in the Empire.

Pharmacy and Migrant Ships

The extent of migration from Britain meant that, for pharmacists in migrant ports, supplying the migrant ships with medicines was a lucrative business. But in the early 1850s it became highly contentious; the issue was the quality of the medicines being supplied. Until then medicines and other stores required on board these vessels were supplied under contract with local chemists.[1]

The people responsible for ensuring that relevant regulations were complied with were the Colonial Land and Emigration Commissioners; they claimed to have received numerous complaints about the bad quality of the medicines supplied, although they neither inspected the medicines themselves nor arranged for others to do so. It was claimed that inferior and sometimes worthless medicines were often provided. In response the Commissioners issued an Order in 1852 that in future all ships chartered by them must obtain their supplies of medicines only from Apothecaries Hall in London.

This action generated an angry response from chemists and druggists at some seaports. They were, according to Jacob Bell, 'excluded from a branch of trade which some of them had been in the habit of carrying on in a creditable manner and with satisfaction to those who had employed them'.[2] Many felt that the new regulation was a stigma that fell on the whole body of chemists and druggists. Those in London met with the Land and Emigration Commissioners to present their case. Shortly after, Liverpool chemists and druggists petitioned the Commissioners about 'the hardship and impolicy of your Order' and pleaded with them to revoke it. 'We are aware that this Order has arisen out of a well-founded complaint of the medicines previously supplied', they wrote

> but we feel sure that your object can be attained without giving a monopoly to an establishment which has not exclusive claims to public confidence... In Liverpool, whence emigration takes place on so large a scale, there can be no difficulty in making the necessary arrangements.[3]

The London chemists suggested that inspectors be appointed who were capable of judging the quality of medicines supplied. But the Commissioners rejected this suggestion and ignored the pleas of the Liverpool chemists. The question was referred to two eminent professors for their opinions. They agreed with the Commissioners, concluding that 'the best safeguard for the interests of the class of emigrants' was to obtain the medicines from an accredited establishment such as Apothecaries' Hall.[4] The battle was lost, and individual chemists and druggists were no longer able to tender for the contracts, although other business opportunities were available. Most trading, whaling and other non-migrant vessels were equipped with medicine chests that contained a wide range of items, and some migrant families took domestic medicine chests with them.[5]

Pioneer Pharmacists in New Zealand

New Zealand was regularly visited by explorers, traders, and missionaries from the late eighteenth century.[6] Around 1828 Methodist missionaries established a mission station to prepare and dispense remedies for the indigenous population; Archdeacon Brown had an extensive medicine chest at Tauranga in 1838.[7] Plans were made for British settlement, and in 1839 ships carried the first settlers into Port Nicholson, Wellington, on the North Island. Dr. John Dorset, the principal surgeon, brought sufficient pharmaceutical supplies to meet the expected initial needs of the new colony. The signing of the Treaty of Waitangi in 1840 by Māori chiefs and British representatives brought New Zealand into the British Empire, prompting a migration rush. By mid-1840 over 1000 people had arrived; by 1841 there were 2500 Europeans living around the harbour, and by 1843 there were 4000. British settlements sprang up across the two islands. But disputes over the Treaty and settler demands to acquire land led to a series of New Zealand Wars from 1843. The imposition of European economic and legal systems led to most of the land passing to European ownership, and most of the indigenous population became impoverished. The colony established its own Parliament in the 1850s, becoming a self-governing Dominion in 1907.

Amongst the new settlers were New Zealand's first chemists or druggists (Fig. 11.2). Charles Decimus Barraud from Southampton opened a pharmacy in 1840 in Wellington, where he operated for 38 years. Archibald Anderson travelled south from Wellington with a stock of medical supplies and opened a pharmacy in Dunedin. He later sold the business to another pharmacist, John Sutton, who had arrived in 1850. Two years later Sutton moved to larger premises which he called 'Medical Hall'. In 1862 he sold it to T. M. Wilkinson, who was followed into the business by his son and grandson. The first pharmacist in Auckland was William McKenzie, a Scottish immigrant who was in business by 1845.[8] An Irish pharmacist, Graves Aickin, opened a pharmacy in Auckland, becoming a strong advocate of the need to raise the standard of pharmacy education.[9] The first pharmacist to open a business in Canterbury was Arthur Bayfield, who arrived with the first group of pilgrims in December 1850 with 'a young wife, one son and sufficient stock to open a pharmacy'.[10] Like many others he became an active member of the community, as volunteer fireman, vestryman of the Anglican Church, member of the Colonists' Society, and sub-postmaster to Lyttelton.

Fig. 11.2 Tyerman Chemist and Druggist Store, Wanganui, North Island (Courtesy of Alexander Turnbull Library, National Library of New Zealand)

The flow of immigrants from Britain continued for several years. They included James Butler Swann, who gave up a settled life in England to start a new life north of Christchurch with his family in 1862.[11] According to his granddaughter's biographer, the business thrived.[12] The family later established pharmacies in Fiji and Samoa. New settlers needed to have considerable financial capital behind them, and had to take with them sufficient equipment and supplies to set up shop once they got there. This included 'bottles, corks, boxes, wrapping and note paper, labels, account forms, scales, glass measures, pill machines, plaster irons, marble slabs for preparing ointment and shop fittings'.[13] If they lost an item in transit it could take up to six months to obtain replacements from Britain. Their first premises were usually roughly erected sheds, later replaced by timber structures. They offered a wide range of pharmacy and non-pharmacy products and services, often doubling as grocers. Some advertised 'paints, oils, white lead, pickles, stationery, spices and curry powder, as well as tonics, blood purifiers, mixtures and packaged pills'.[14] There were many more chemists and druggists than doctors, dentists, opticians, and veterinary surgeons, and they often had to acquire new skills including extracting teeth, sight testing, and providing treatments for animals.

Collective Action: Founding Pharmaceutical Organizations

During the 1860s and 1870s pharmacists in the new colony were preoccupied with commercial survival, but slowly the services offered became more professional. With increasing security some turned their attention to issues of professional control and educational standards. Although some control was introduced over the sale of poisons there was initially no regulation of pharmacy. The first steps were taken by pharmacists in Wellington. On 22 August 1878 twenty-seven of them—nearly all those in Canterbury—met, and J. C. Brook proposed 'that the chemists of Canterbury form themselves into a Pharmaceutical Society in conjunction with Wellington and other New Zealand kindred Societies'.[15] The proposal was carried unanimously, and discussions held with pharmacists elsewhere. The first combined meeting of pharmacists in the colony was held on 28 January 1879, convened by Charles Decimus Barraud. The meeting agreed that a Pharmaceutical Society of New Zealand (PSNZ) should be formed, with headquarters at Wellington, and local committees at Auckland, Christchurch, and Dunedin.[16] Barraud became its first president. To avoid parochialism and jealousy the venues for the annual meetings would be rotated and decided by lot.[17]

In the absence of legally constituted branches, regional associations were formed (Table 11.1). By 1888 there were four; a Midlands Pharmaceutical Association centred on Christchurch, an Otago Pharmaceutical Association based in Dunedin, a Central Pharmaceutical Association at Wellington; and Auckland had an active Chemists and Druggists Association . During the 1890s the associations grew stronger, taking up a wide range of political, professional, and domestic problems, but inevitably differences began to emerge. In 1896 a proposal was made by the Auckland association to link the associations into a federation, but it was then realized that they admitted registered members only, whilst the other associations admitted unregistered persons and wholesalers. Graves Aickin drew attention to shortcomings in legislation relating to education: 'we

Table 11.1 Pharmacy in New Zealand and Western Pacific Colonies

Colony/Branch	*Year British colony*	*Pharmaceutical society or association*	*Early pharmacy legislation*	*Early poisons legislation*	*Pharmacy education*	*Pharmaceutical journals*
New Zealand	1840	Pharmaceutical Society of New Zealand 1879	1880	1887; Sale of Food & Drugs 1936; Poisons	Classes in chemistry and botany 1910	*Pharmaceutical Journal of New Zealand* 1929
Auckland (Branch)	-	Auckland Chemists and Druggists Association 1888	-	-	Auckland correspondence course (Pryor)	-
Christchurch (Branch)	-	Midlands Pharmaceutical Association 1888	-	-	Classes in chemistry and botany	-
Dunedin (Branch)	-	Otago Pharmaceutical Association 1888	-	-	1962; course at University of Otago	-
Wellington (Branch)	-	Central Pharmaceutical Association 1888	-	-	1960; course at Petone Institute of Technology	*Sharland's Trade Journal*. Official for Pharmacy Board 1899
Fiji Islands	1874	None	Ordinance 1880	Ordinance 1880	-	-
Papua	1888	None	As Australia	As Australia	-	-
Samoa	1885	None	Ordinance	Ordinance	-	-

have not succeeded in raising the educational standard to the proficiency required to claim reciprocity with Australian states or Great Britain'.[18]

As voluntary organizations with no legal standing, the associations struggled to survive. By 1896 the Wellington and Christchurch associations had folded, but interest revived in Auckland in 1913 when a Northern Pharmaceutical Association was established. This worked well until the early 1930s when it was absorbed into the newly formed Chemists' Service Guild of New Zealand. This had its origin in the Jenkin Case in Britain (Chapter 2), which decided that the PSGB did not have the authority to determine the conditions of service of pharmacists, and resulted in the creation of a Retail Pharmacists Union (RPU). In New Zealand it was recognized that its Pharmacy Board would no longer be able to do so either. In a letter to the *Australasian Journal of Pharmacy* in 1929 William Davis from Australia made a persuasive case for the formation of a similar body to the RPU; it was a call to pharmacists to work together to defend their commercial interests.[19] An Australian Chemists' Guild was established in Sydney in 1930, and a Chemists' Service Guild of New Zealand was launched in 1933. It was to play a leading role when company chemists became an issue.

Pharmacy Legislation

For the New Zealand Parliament regulating pharmacy was not a high priority, but in order to reduce harm from poisons a Sale of Poisons Act was passed in 1866, although this was repealed and replaced by a new Sale of Poisons Act only five years later, in 1871. This stated that 'persons keeping open shop for the retailing, dispensing or compounding of poisons shall be registered'.[20] But this was no Register of Pharmaceutical Chemists; instead, the Governor appointed a registrar for each province, and the register was simply a list of vendors of poisons. Anyone could apply for a licence, although doctors were excluded; Sect. "Pharmacists in Papua New Guinea" stated that 'persons registered under the Medical Practitioners Registration Act of 1869 shall not be registered under this Act'.[21] Chemists and druggists were licenced to sell poisons but did not have a monopoly in doing so. The Register of Vendors of Poisons included many people with few or no qualifications.

One of the first actions of the newly formed PSNZ was to seek protection in law through a Pharmacy Act. Shortly after its inaugural meeting a Pharmacy Bill was drafted and introduced into Parliament. This was more a Pharmacy and Poisons Bill than a Pharmacy Bill in that it linked the regulation of pharmacy with the sale of poisons, and would have given the pharmacists a monopoly in their supply. The MP who moved the second reading in 1880 told the House that it was similar to legislation in Britain and Victoria and used the same phrases:

> it is expedient to prevent ignorant and incompetent persons from assuming or pretending to be pharmaceutical chemists, and that a register should be kept by some legally authorized officer of such persons.[22]

In drafting their Bill New Zealand pharmacists could draw on experience in both Britain and Australia relating to pharmacy organization. Should the registration of pharmacists be the responsibility of the PSNZ itself, as in Britain? Or should there be a separate Pharmacy Board, as in Victoria? The first required that the PSNZ itself be legally recognized: the second only that the Board be legally established.

New Zealand followed Victoria in opting for an independently appointed Pharmacy Board. There was some opposition to the Bill in Parliament although this related to monopoly in the sale of poisons rather than to regulating pharmacy. At the Select Committee stage it was revised such that pharmaceutical chemists would not have the sole right to deal in poisons. With that change it became the Pharmacy Act 1880, and on 1 January 1881 pharmacy in New Zealand achieved official professional status.[23] The Act legally established a Board of Pharmacy whose members were to be appointed by the Governor in Council, at least for the first three years. This meant that nominally Board members were appointed by the government independently of the PSNZ, but in practice those appointed were the members of the Council of the PSNZ at the time, with Charles Decimus Barraud, the president of the PSNZ, being its first chairman. The Board consisted entirely of pharmacists. The Act also made provision for the inclusion of homoeopathic chemists on the Register of Pharmaceutical Chemists as well as chemists and druggists already in business.

Until the 1880 Pharmacy Act, pharmacy in New Zealand remained remarkably unregulated. A Medical Board appointed under a Medical Act maintained a register of those authorized to practise medicine.[24] But it had no interest in taking control of pharmacy or of maintaining a register of qualified practitioners. Nor were any attempts made to claim that pharmacists were practising medicine by prescribing over-the-counter medicines, as they had in both Australia and Great Britain.[25] Yet the lack of regulation was a problem. One pharmacist recalled that there was

> nothing to prevent anyone, however uneducated or untrained, from dispensing or dealing with medicines, the nature and uses of which they may be entirely ignorant, a procedure...eventually becoming fraught with public danger through gross abuse.[26]

The PSNZ was in no doubt that further legislation was needed to improve training, raise educational standards, introduce compulsory examinations, maintain a register of qualified practitioners, and achieve reciprocity in the recognition of qualifications with other colonies.

Registration of the Unqualified

The 1880 Act fell far short of the aspirations of New Zealand pharmacists in establishing a monopoly in the supply of poisons. The Pharmacy Board were aware of shortcomings in the Act, and in 1883 sought an amendment 'to improve its efficiency'. But the Government declined to introduce further legislation, since to do so would create a 'protected' interest. The Board were nevertheless keen to bring under their control individuals whose names appeared on the Register of Vendors of Poisons but were not pharmaceutical chemists. They made repeated attempts to amend the 1880 Act to strengthen its rules and regulations regarding the sale of poisons between 1884 and 1895, but all failed to achieve government support.[27] The Board did however make efforts to improve pharmacy training, and in 1889 a four-year apprenticeship was introduced.

The PSNZ introduced other attributes of an aspiring profession soon after its foundation. A pharmacy library had been opened in 1880 at Christchurch, where the PSNZ's second meeting was held in 1882. They agreed that publishing their own journal would not be economic, and instead a link was established with the British *Chemist and Druggist*. Graves Aitkin of Auckland was appointed as the New Zealand correspondent. Other periodicals soon appeared, including from 1886 a house journal called *Sharland's Trade Journal*. This had informed editorial content and restrained advertising, and in 1899 the Pharmacy Board adopted it as its official journal. New Zealand pharmacists also received the *Australasian Journal of Pharmacy*, but complained about its lack of interest in New Zealand issues. A *Pharmaceutical Journal of New Zealand* was launched in 1929, and in 1952 it was granted recognition as the official journal of the PSNZ.[28]

In 1896, the government finally gave its endorsement to changes that allowed for the registration of certain individuals who had been in business before 1 January 1880, and in 1897 further provision was made for the registration of unqualified people who had dispensing experience with medical practitioners. At the end of 1898, the Government unexpectedly introduced a new Bill which allowed for the registration of people with no qualifications and little experience. It was rapidly approved and became the Pharmacy Act 1898. This gave the PSNZ legal recognition as a body corporate, and redefined the relationship between the PSNZ and the Pharmacy Board: 'the affairs of the Society shall be managed and controlled by a Board to be elected under the provisions of this Act and to be called the Pharmacy Board of New Zealand'.[29]

A new Pharmacy Act in 1908 made membership of the PSNZ by all registered pharmacists obligatory, twenty-five years before membership of the PSGB was made obligatory in Britain. Further legislation followed; a Pharmacy Amendment Act was enacted in 1912, following passage of a Prohibition of Secret Commissions Act in 1910, to prevent pharmacists entering into unethical arrangements in return for gifts or considerations. But it also provided for the registration of enrolled managers employed by laymen proprietors, corporations, companies, and Friendly Societies.[30] A Pharmacy Act in 1939 was passed 'to make better provision for the registration and control of pharmaceutical chemists' and specified that the Pharmacy Board would consist of eleven directly elected pharmacists—nine pharmacy owners and two employee pharmacists.

Company Chemists in New Zealand, 1923

The 1880 Pharmacy Act was passed in the very year that the case of the Pharmaceutical Society of Great Britain v. The London and Provincial Supply Association was decided by the House of Lords in London (Chapter 2). In drafting their Act New Zealand pharmacists could not have guessed the consequences of the PSGB losing its case. Much of the wording in their Act followed that in the 1868 British Act in referring to a 'person'; the London case ruled that companies as well as persons could own pharmacies. In the light of the British ruling New Zealand pharmacists attempted to follow the example of Shillinglaw in Victoria by seeking the passage of an amending Act to correct its deficiencies. But they were unsuccessful; the government was more concerned about the registration of those practising pharmacy without qualifications. The 1898 Pharmacy Act had nothing to say about the ownership of pharmacies; there were no restrictions on the ownership of pharmacies, and the door was left open for company chemists.[31]

For most New Zealand pharmacists, ownership was not a major concern until 1923, when the Boots Company in England expressed an interest in entering the New Zealand retail pharmacy market. In the early years of the twentieth century Boots had expanded rapidly in England, with the number of branches increasing from 251 in 1901 to 560 in 1914.[32] In 1923 they registered the name Boots the Chemists (New Zealand) as a company with a capital of £100, claiming that their purpose was to protect the name, and that they had no intention of starting a retail business in New Zealand. Until then Boots had a thriving export market for their products in New Zealand, but their actions were not well received, and their goods were boycotted.[33] Pharmacists raised their concerns with the Pharmacy Board and called for action.

In July 1923, the Pharmacy Board considered possible amendments to the 1898 Pharmacy Act in order to limit the growth of company chemists, and a draft Bill went to the Minister in October 1924. But the pharmacists were told that it could not be considered that session. It was the first of many attempts to constrain the power of corporate pharmacy in New Zealand. Over the next 12 years discussions were held with health department officials and deputations were sent to Ministers, but all efforts failed 'because of political dilatoriness or apathy'. The Bill was amended, but health officers continued to object to those clauses aimed at curbing the activities of company chemists.[34]

In October 1935, a high-level Boots delegation arrived to investigate possible trading conditions. They concluded that prospects were favourable, and recommended that business be commenced. Suitable sites were acquired in Auckland and Wellington. To fight the new threat a committee consisting of representatives from retail pharmacy, wholesale druggists, and Friendly Societies was quickly assembled by New Zealand pharmacists. The concerns of younger pharmacists were expressed by Walter K. Hounsell of the Auckland and Otago Chemists' Assistants Union:

> Are we entitled to expect a reasonable opportunity to earn a livelihood in the profession to which we have devoted so much study, time and money, or are we to be sacrificed for the benefit of the shareholders of a combine whose only interest in the welfare of the Dominion will be the amount of profit they can take out of New Zealand?[35]

The answer would be decided by Parliament. Their petition was presented to Parliament on 23 October 1935, the principal speaker being Mr. H. J. S. Rickard, representing the Pharmacy Board of New Zealand. They sought support for an Amendment Bill which would restrict the ownership of pharmacies and prevent the intrusion of company pharmacy. They presented eight key arguments in support of restriction: they emphasized that New Zealand pharmacists were not seeking a monopoly; they noted that the ratio of pharmacists to population in Great Britain was one in

4000 whilst in New Zealand it was 1 in 2500; that the profits of a large overseas company would go not to New Zealand but to the company's overseas shareholders; that for those about to qualify, company pharmacy would destroy their chances of successfully operating their own businesses; and they added that New Zealand pharmacists had established a College of Pharmacy without government support.[36]

Prime Minister Involved

Both the British and New Zealand governments were drawn into the issue. The British Government applied pressure to allow Boots into New Zealand. On 16 October 1935 the Secretary of State for Dominion Affairs sent a telegram to the New Zealand government:

> We have learnt with concern of this development and... we earnestly hope that this legislation [restricting company pharmacy] will not be proceeded with, having regard to the discriminatory effect that it would have upon an important United Kingdom interest.[37]

The Governor General (the senior British representative in New Zealand), dutifully sent a telegram to the Prime Minister, G. W. Forbes, expressing his regret at the possible legislation.[38] One week later the Prime Minister told New Zealand pharmacists that the government 'did not want to cause ill-feeling in the old country by putting hurried legislation through without a thorough investigation'.[39] A report by those present at the meeting noted that the Prime Minister 'referred to negotiations with the Imperial Government for the purchase of New Zealand produce, and he would not do anything which might upset such arrangements'.[40]

But not all British pharmacy organizations supported the Boots position. Wholesalers and manufacturers recognized that they would face reduced sales. The wholesale Drug Trade Association of Great Britain briefed Mr. A. H. Johnstone, KC, to present its observations and to support the position of the Pharmaceutical Society of New Zealand. British manufacturers were the main source of supply for medicines in New Zealand, and they would be seriously affected by a large expansion of Boots retail outlets. Pharmacy leaders too understood that New Zealand pharmacists were seeking a situation that they themselves had failed to secure in Britain. In 1935 the *Pharmaceutical Journal* drew readers' attention to the fact that the number of pharmacists in New

Zealand had reached saturation point, and that British wholesale suppliers and manufacturers were likely to be seriously affected by any control of company chemists in New Zealand.[41]

Parliamentary Committee of Enquiry into Pharmacy, 1936

In Parliament the arguments, whilst persuasive, were summarily dismissed. Boots appeared to have won. But one member of Parliament, Downie Stewart, suggested that Boots be prevented from extending any further into New Zealand, beyond its first two branches, until the issue of company pharmacy had been thoroughly investigated by a Parliamentary Committee. The suggestion had general support, with some members concerned that company pharmacy might not only destroy a legitimate business but also the profession of pharmacy. The Parliamentary Committee of Enquiry into Pharmacy began work on 7 May 1936, but not before Lord Trent (Jesse Boot's son) met with the Pharmacy Board in person at his own request on 10 February 1936. He claimed that the New Zealand pharmacists had forced the issue, and sought an amicable agreement.[42]

The Parliamentary Committee considered evidence from the PSNZ, from wholesale druggists' and retail chemists' associations, and from Boots. It sat for five days over two weeks and heard from Counsel representing each of the parties. Boot's petition claimed that 'it would be unjust, contrary to the principle of allowing modern chain store distribution to develop, and contrary to the interests of the purchasing public, to prevent the company from opening'.[43] Counsel for the PSNZ claimed that 'that statement flies directly in the face of the experience of every civilized country wherein the chain store business has developed'. It was not necessary for New Zealand pharmacists to show that chain store pharmacy was an evil in both Britain and America. They were, however, concerned that chain stores might be

> introduced into the virgin field of New Zealand, whose total population was no larger than the city of Glasgow. Unlike many congested and densely populated cities in Britain, the population of New Zealand was small and widely scattered.[44]

The crux of the issue was crystallized by D. S. Dodds, a pharmacist in Christchurch with extensive experience in chain stores overseas. 'The operations of chain company pharmacy were governed by profit. Current personal attention given by New Zealand pharmacists to customers would become secondary'.[45] He strongly believed that individual control and ownership were preferable to chain store operations. R. A. Cometti, representing the Wholesale Druggists' Association, noted that 85 per cent of the wholesalers' stock came from Great Britain and Australia, and wholesalers could be seriously affected by any large organization with a strong overseas base.[46]

In submitting themselves to review by Parliamentary Committee New Zealand pharmacists found their practices and activities exposed to intense scrutiny. The committee missed nothing; its recommendations were extremely wide ranging. There were many criticisms, including over-supply of pharmacists leading to unemployment and low wages, and the pricing of prescriptions using an 'average flat rate' which was anti-competitive. On the other hand, they clearly recognized that if Boots were allowed unrestricted entry into New Zealand, both wholesale and retail pharmacy services would be seriously disrupted without providing the coverage of current arrangements. The key recommendation was that pharmacists in New Zealand should be given six months to 'put their house in order'.[47] They unanimously accepted the recommendations and set about doing so.

The Porirua Case, 1962

Boots kept their two pharmacies, in Auckland and Wellington, and by 1954 they had opened another four.[48] In 1955 a Boots team from Nottingham, including Michael Peretz, visited New Zealand and Fiji to explore opportunities to open more. They were not welcome. New Zealand pharmacists made their opposition clear in an open letter published in the *Pharmaceutical Journal of New Zealand* later that year.[49] But despite the obstacles placed in their way, by 1962 Boots owned and operated eight pharmacies in New Zealand.[50] Peretz and a colleague returned in September 1962, and Boots made an application to the Pharmacy Board under the New Zealand Pharmacy Act to open a pharmacy in the new town of Porirua north of Wellington on the North Island.[51]

The Pharmacy Board refused the application, believing that any further expansion by Boots would have a damaging effect on retail pharmacy practice in New Zealand. Boots decided to test the matter in the Courts.[52] They lost the early rounds but appealed first to the Supreme Court. By 1966 the case of Boots the Chemists (New Zealand) v. The Chemists' Service Guild of New Zealand had reached the New Zealand Court of Appeal. The case was again lost, and Boots made a final appeal to the Privy Council in Britain. In 1969 the Privy Council judged in their favour, and the Porirua branch opened.[53] In New Zealand pharmacists sought to limit any further expansion. A new Pharmacy Act was passed in 1970 which allowed the continuation of existing company pharmacies whilst preventing the establishment of new ones. The Act specified that, wherever possible, pharmacies should be 'carried out by independent pharmacists owning and conducting their own businesses'. The intended expansion could not take place, and Boots eventually withdrew entirely from New Zealand.

Inter-Colonial Relations and Reciprocity

The decision of the PSGB to refuse reciprocity in the recognition of pharmaceutical qualifications with Victoria in 1881 had focussed attention on educational standards throughout the Australasian colonies. New Zealand pharmacists were in close contact with their colleagues in Australia from the beginning. Their representatives attended the first intercolonial pharmaceutical conference in Melbourne in 1886, and the *Australasian Journal of Pharmacy* reached pharmacists across the region. The core issue addressed at the conference had been the reciprocal recognition of qualifications. For New Zealand pharmacists, reciprocity with both the Australian colonies and with Britain remained an important issue for much of the twentieth century.[54]

An application for reciprocity from New Zealand to Victoria was refused in 1896. A suggestion was then made for a uniform examination between the two, but this was not taken up. Reciprocity was discussed in 1905 in *Sharland's Trade Journal* under the title 'how to become a pharmacist in the Colonies'. It demonstrated that there were still considerable differences in pharmacy educational standards across the Empire. In 1911, with the PSGB moving towards recognizing colonial pharmaceutical qualifications (Chapter 12), Edward Smith in Auckland declared that reciprocity between Australia and New Zealand was an important first step in securing reciprocity with Great Britain.[55] He noted that 'for every three New Zealanders who would go to Australia, one Australian would come to New Zealand'.[56] Oversupply would not be a problem. By then most New Zealand pharmacists supported reciprocity with the Australian states, an arrangement that was eventually implemented after the First World War. As educational standards were raised across the Empire more and more intercolonial reciprocity agreements were reached. By the late 1940s pharmacists registered in New Zealand were eligible for registration in Australia, Great Britain, and the Republic of Ireland, but not in South Africa or Canada.[57]

At least some of the reciprocity problems resulted from the wording of pharmacy legislation. On one occasion the New Zealand Pharmacy Board declined an application for registration by John Bronte, a pharmaceutical chemist trained and registered in Ireland. They noted that the 1880 Pharmacy Act referred only to 'Great Britain', not to 'Great Britain and Ireland'.[58] This anomaly was later remedied by passage of a Pharmacy Amendment Act in 1887.

In 1953 the PSGB sought to amend reciprocal agreements then in place to stem the registration in Britain of pharmacists deemed to have completed less-demanding courses elsewhere. The New Zealand diploma course was judged no longer acceptable. In 1967 the PSGB informed the New Zealand Pharmacy Board that in future reciprocal arrangements would be limited to those who held the BPharm degree from the University of Otago and to those registered before 31 December 1967. The deadline was extended to enable existing students to register. In response to concerns in 1969 the Council of the PSGB despatched its president, registrar, and the chair of its education committee on a mission to visit schools of pharmacy in Australia and New Zealand and to report back. It resulted in the restoration of reciprocal recognition; to be registered in Britain, a person needed to have completed one year's employment as a registered pharmacist in New Zealand or Australia, to have completed four weeks of employment in a British pharmacy under the supervision of a pharmacist on the British register, and to have studied the law relating to pharmacy, medicines, and poisons in the United Kingdom.[59]

Professionalization and the British Model of Pharmacy

As a settler colony and later as a Dominion, New Zealand was able to establish its own Parliament and make its own laws, and it did so largely in the image of Britain. This applied as much to the regulation of pharmacy as it did to other areas; in many respects the professionalization of pharmacy followed the British model. New Zealand pharmacists had been remarkably successful in founding a national pharmaceutical society before going on to draft a Pharmacy Bill to give the society a legal foundation. They were helped by two key factors: New Zealand was a single colony, not a collection of disparate colonies founded at different times, as was Australia; and in Britain, earlier Medical and Pharmacy Acts had already established medicine and pharmacy as separate and distinct professions. New Zealand pharmacists did not face repeated attempts by the doctors to control them through a Medical Board, as had their colleagues in Australia.

But there were differences. Control of poisons was an incidental effect of the Pharmacy Acts, rather than being subject to a linked Pharmacy and Poisons Act. It was not until 1936 that New Zealand passed a separate Poisons Act and checks were made on the implementation of regulations.[60] New Zealand did not operate a twin register of chemists and druggists and pharmaceutical chemists with a higher qualification. Rather, the government's priority was to ensure that all those licenced to sell poisons—however qualified—be included in the register of chemists and druggists. New Zealand did however largely follow the British model regarding the separation or combining of regulatory and representative functions: the regulatory authority was the PSNZ's Council which was described as the Pharmacy Board. The 1970 Pharmacy Act formally replaced the Board by the Council.

It was with regard to company chemists that New Zealand pharmacists may have regretted following the British model a little too closely. In using the wording of the British Pharmacy and Poisons Act 1868, those involved in drafting the New Zealand 1898 Pharmacy Act did not make explicit that a pharmacist must only own one pharmacy, thus excluding the ownership of chains of chemists' shops by companies. Yet by 1898 the impact of company pharmacy in Britain on the practice of pharmacy was already very clear. Although—like their colleagues in Victoria—New Zealand pharmacists sought amending legislation over several years, their efforts were unsuccessful: Parliament was indifferent to their concerns. At the heart of the arguments was the key tension between pharmacy as business and pharmacy as profession. During the Boots enquiry, the pharmacists emphasized their professional interests, whilst Boots framed the issue around the protection of pharmacists' business interests.[61] In having their entire activities subjected to the exacting scrutiny of a parliamentary committee, pharmacists in New Zealand paid a heavy price for using British legislation as the basis of their own.

New Zealand pharmacists were fortunate in that most of the issues surrounding professionalization had already been resolved elsewhere before they sought their own Pharmacy Act in 1880. Boundaries were well established between pharmacy and medicine, and they were able to secure the support of the state in obtaining legislation. They were successful in securing a monopoly in the sale of poisons, but at the cost of an uncomfortable compromise in the shape of the qualifications of some of those that Parliament insisted must be included on the register.

Pharmacy in the Fijian Islands

The first Europeans in Fiji were probably the crew of a shipwreck in 1800. Thereafter a steady stream of ships arrived from Sydney, and in 1835 two missionaries from London reached the islands. They were soon followed by others who set up a base around Levuka on the east coast of Ovalau, a small island off one of the larger islands, Viti Levu. Difficulties occurred between settlers and the local chief, Cakabau, making it a challenging place.[62] But disruption to supplies of American cotton during the Civil War 1861–1865 resulted in high cotton prices, and encouraged its production in other parts of the world. The promise of cheap land, abundant labour, and large profits drew many Europeans to the Fijian islands, where they would be less constrained by government laws and taxation. It was the start of a 'Fiji cotton boom'.[63]

Enthusiastic but exaggerated reports appeared in the Australian and New Zealand press. An editorial in the *Herald of Melbourne* in 1868 declared:

> To the small capitalists, the men prepared to encounter danger and privation, Fiji presents a brilliant future. Hundreds of young men in Europe and in this community are now eagerly looking for some such opening as Victoria and neighbouring colonies presented twenty years ago.[64]

Whilst benefits were played up—it was claimed that planters could expect returns within two year—difficulties were dismissed. In an 1869 pamphlet

'Ceres' asserted that 'labourers could easily be imported from the adjacent islands to the Fiji group under three-year contracts'.[65] Difficulties in obtaining land from Fijian landowners were also played down.

One of those persuaded to make the move was James Butler Swann; his family arrived from New Zealand in 1867.[66] He became a cotton planter, but the venture failed, and in 1870 he bought the chemist's business of Dr. Riley in Levuka. He remained there for the next twenty-four years, building it up until 1894 when he fell ill and retired.[67] His three sons were all apprenticed to him, qualified with the PSGB's Minor certificate in London, and ran pharmacies in Fiji or Samoa. His eldest son, Arthur, returned initially to his father's shop in Levuka, but opened a pharmacy in Suva on Fiji's main island when the government moved the capital there in 1882 (Fig. 11.1). He built up the business over the next 43 years, until his own death in 1926. His father's business was taken over by the youngest son, William. By then several other pharmacies had opened in Fiji. Thomas Parker had originally emigrated from Ireland to Melbourne before establishing a chemist's business in Queensland before moving again to Fiji. He worked as a chemist in Levuka in the mid-1870s.[68]

The 'Mild Despotism'

When Swann arrived in Fiji pharmacy was firmly in the hands of the doctors. In the 1860s an Ordinance had been passed which made it compulsory to obtain a permit from the medical officer before any poison could be purchased. By the 1870s attempts were made to bring arrangement into line with those in Britain. In 1881 a new Ordinance was passed 'to regulate the registration of medical practitioners and apothecaries, and the sale of poisonous medicines'. The *Chemist and Druggist* noted that it adopted 'Colonial and English practice, simply making it necessary for the chemist to keep a record of all poisonous articles supplied otherwise than on the prescription of a medical man'.[69] It reserved its criticism for the regulation of pharmacy. 'But though this Ordinance remedies one defect of the discarded enactment, it leaves others not less prominent still in force'. It noted that

> as the Government of Fiji is a mild despotism...it would seem that very great watchfulness is required lest the great power possessed by the authorities in Crown colonies should be abused.[70]

Their concerns related to restrictions placed on pharmacy practice, the size of penalties, and lack of recognition of qualifications from other colonies. 'Only pharmacists who can produce the diploma of the PSGB can perform all acts and functions connected with the practice of pharmacy', it continued, 'and the certificate of all other examining bodies, such as that of the Pharmacy Board of Victoria, are quite ignored'. Pharmacists who moved to Fiji had mainly trained in Australia. The penalties attached to the illegal practice of medicine were severe. 'If a pharmacist looks at the tongue of a man, woman, or child, suffering from the most ordinary ailment, and presumes to recommend any simple remedy, he is liable to a month's imprisonment or a 10 pound fine'.[71]

The matter was taken up by the *Fiji Times*. It asked:

> While the apothecary is required...to produce the diploma certifying that he has satisfied the PSGB as to his being fully competent...; that while Colonial certificates are absolutely refused...; and that while the holders of such qualifications...are required to pass an examination before the chief medical officer; does it not seem an anomaly that... the Ordinance should distinctly prohibit the exercise of those functions permitted by the Imperial and Colonial Acts, and restrict the duties of the successful applicant to matters which any shop-boy could perform after a year's probation behind the counter? Yet this is the position in which the pharmaceutists were placed under the old, and in which they are continued under the new Ordinance.[72]

Holders of Colonial pharmaceutical certificates were required to pass an examination in Fiji before being allowed to perform even limited duties. The same did not apply to other professions:

> The doctor's colonial diploma is accepted; the Chief Justice is content with the lawyer's certificate of admission to the Colonial Bar; the Marine Board is satisfied with the mariner's Colonial qualification. Why then is the English certificate of the pharmaceutist accepted while the Colonial diploma is regarded as so much wastepaper? If a man was competent in Australia or New Zealand, surely he is competent in Fiji; then wherefore this invidious distinction between the English and Colonial certificates?[73]

The authorities had no answer, and the issue remained unresolved until the early twentieth century. But British pharmacy thrived nevertheless, and in 1944 Boots opened its first shop in Fiji.[74]

Pharmacy in the Western Pacific Islands

In 1877 islands in the western Pacific under British control were brought together for administrative purposes into a single colonial entity, the British Western Pacific Territories (BWPT). They included the British Solomon Islands and the New Hebrides (now Vanuatu) in Melanesia and the Pitcairn Islands and Tonga in Polynesia, along with the Gilbert and Ellice Islands (now Kiribati and Tuvalu).[75] Samoa was under the joint protection of the British, Germans, and Americans between 1889 and 1900, and came under New Zealand jurisdiction after the First World War (Fig. 11.1). British administrators and medical officers in the region brought with them the familiar rules and procedures of Empire. Medical Ordinances were passed, licences to practise were issued, and regulations were laid down to control the sale of poisons.

There were few British pharmacists in the Western Pacific, but one such was William John Swann, the second son of James Butler Swann. He was only three when the family migrated to New Zealand, and only eight when they moved to Fiji. In 1876 his father sent him to England to study pharmacy, probably at one of the crammer colleges in London.[76] On his return to Fiji he initially worked in the family shop, but soon became restless. In the mid-1880s he was offered the post of apothecary on an American Navy vessel that weaved its way through the South Seas islands. He developed great proficiency in dealing with the indigenous peoples of the Pacific.[77]

By 1885 Apia in western Samoa had been proclaimed an international settlement by Britain. After a few years at sea William opened a chemist's shop there in 1889, but within a few weeks Samoa had been struck by a hurricane.[78] Despite unrest throughout the island his business flourished. He married the daughter of a village chief and they had three daughters, one of whom, Aggie Grey, became famous as the founder of a hotel in 1933. As well as being a pharmacist, William acted as a doctor and as a dentist to the Samoans, reportedly extracting teeth with considerable skill. Germany annexed Western Samoa between 1899 and 1914, but again Swann's business prospered as the arrival of new traders resulted in a thriving economy.[79]

The German occupation illustrated the uncertain nature of British colonialism in the western Pacific and the frequently difficult relations between European powers. Swann soon found himself with stiff competition. A German chemist, G. Sabiel, was soon in business nearby and proposed to run Swann out of Apia. He approached the Customs Department for special privileges as a German national. But R. P. Berking, the German-born Collector of Customs, would not agree to Sabiel's proposals, ruling that Swann already had his business established in Samoa 'where everyone has the same rights'.[80] With the outbreak of war in 1914 the Germans departed. New Zealand occupation favoured British traders, and Swann secured lucrative contracts to supply military dispensaries.[81] William Swann died in 1936 at the Government Hospital in Apia at the age of 77.[82]

Those pharmacists who ventured beyond Australia and New Zealand thus found themselves subject to a variety of regimes, from the 'mild despotisms' of Fiji to the control of colonial Ordinances as in Samoa. But most ran their pharmacies along British lines, using British formularies and selling British patent medicines, along with the traditional range of products available from most British pharmacies. But with small numbers of pharmacists on each island there was little need to establish pharmaceutical societies, to create Pharmacy Boards, or to arrange pharmacy education. And if this was true in the western Pacific it was even more so in Papua.

Pharmacists in Papua New Guinea

Papua New Guinea occupies the eastern half of the large island of New Guinea, about 150 kilometres to the north of Queensland, Australia (Fig. 10.1). They are separated by the Torres Strait in which lie a large number of small islands. In 1883 Queensland attempted to annex the southern half of eastern New Guinea without the approval of the British government. But when Germany established settlements in north New Guinea, a British protectorate was proclaimed in 1884 over its southern coast and adjacent islands. British New Guinea was annexed in 1888 and placed under Australian authority. Under a 1905 Act, Papua became a British possession under Australian control, remaining so until 1975. It included islands in the Torres Strait amongst which were Prince of Wales Island, Thursday Island, Moa Island, and Badu Island (Mulgrave Island).

There was little economic activity for the indigenous population on the smaller islands beyond hunting and fishing.[83] Nevertheless the expatriate community often grew to the point where it was large enough to sustain a small drugstore. On Thursday Island one was operated from at least the early 1890s by John Turnbull, who had been elected a member of the Pharmaceutical Society of Queensland in 1892 (Fig. 11.3). The store was located on the ground floor of the building and had a large display window. The upper floor was probably his home. It was an Australian pharmacy offering the island's residents a range of goods and services similar to that which was available to those in less isolated locations.[84]

On Badu Island pearl hunting developed during the 1870s, and by the 1880s many of the islanders had become wage earners. Head-hunting ceased following the arrival of missionaries and the adoption of Christianity.[85] One of the missionaries sent to Papua New Guinea by the London Missionary Society in 1888 was Frederick William Walker

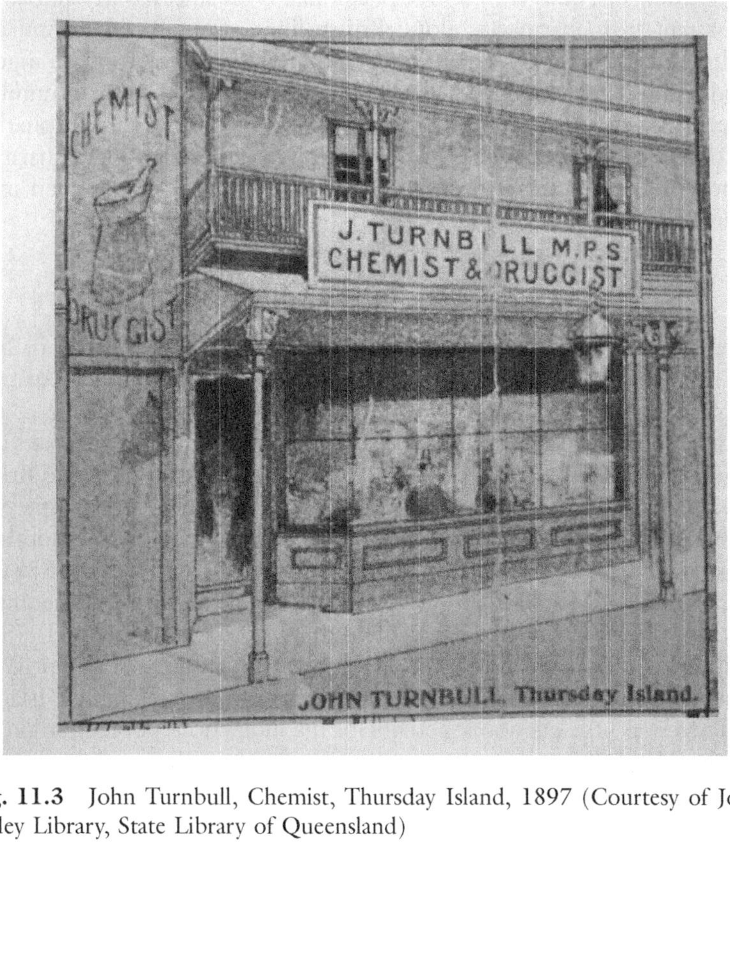

Fig. 11.3 John Turnbull, Chemist, Thursday Island, 1897 (Courtesy of John Oxley Library, State Library of Queensland)

(1860–1926) from Hull, whose younger brother, Charles Herbert Walker (1864–1925), had recently qualified as a chemist and druggist, having taken the Minor examination in Edinburgh in 1887. Charles first moved to London, and in 1889 he was manager of a pharmacy in Finsbury Park.[86] But the following year he set off for Australia, and in December 1890 he was with Pattison & Co. in Sydney. He stayed there for a few years, but in 1893 he travelled northwards to Kwato Island, Papua New Guinea, to join his brother.

The missionaries soon discovered that

> their efforts to civilise and uplift the natives were largely neutralised by the unscrupulousness of the white traders...They were utterly dishonest in all their dealings with the natives, defrauding them at every turn, both in buying and in selling.[87]

Charles Walker and his brother decided to start trading with them on a non-profit making basis, founding a company. Charles remained there for about five years before returning to England to run his own chemist's business in London in 1898. But in 1906 he sold up and re-joined his brother in Papua. He settled on Badu Island where he spent the rest of his life, and "devoted his medical and business knowledge to the benefit of the people". There was no doctor on the island, and 'his training as a pharmacist proved most beneficial to sick folk far and near, and the natives welcomed him gladly'.[88]

After twenty years in the region his health failed, and he died in the hospital on Thursday Island in 1925. The indigenous people of Badu requested that his body be buried in their cemetery, and that a 'Day of Memory' be arranged. They decided 'without any outside pressure or suggestions' to erect a suitable memorial. It consisted of a lighthouse with a roof of white quartz brought from another island and decorated with coral. They also arranged a great feast with invitations sent to all islands. Boats 'passed from island to island and brought a great crowd of people to honour the memory of the White Man whom they had learnt to love'.[89] On the day itself a great crowd processed from the mission house to the monument, and the unveiling ceremony was conducted by the Protector of Aborigines from the Queensland Government. There can have been few pharmacists in the British Empire who gained such high esteem and gratitude from the indigenous people as Charles Herbert Walker.

Notes

1. Bell, Jacob and Redwood, Theophilus. *Historical Sketch of the Progress of Pharmacy in Great Britain* (London, 1880), p. 223.
2. *Pharmaceutical Journal and Transactions*, 12 (1852), pp. 417–20.
3. Ibid., p. 419.
4. Bell and Redwood, *Historical Sketch*, p. 224.
5. Coombes, Reg. *Pharmacy in New Zealand: Aspects and Reminiscences* (Wellington, 1981), p. 14.
6. Byrnes, Giselle (ed.) *The New Oxford History of New Zealand* (Oxford, 2009).
7. Drummond, Alison and L.R. *At Home in New Zealand* (Auckland, 1967).
8. Frost, M. *Behind the Dispensing Screen: Early New Zealand Pharmacists* (Hamilton, 2004).
9. Coombes, *Pharmacy in New Zealand*, p. 18.
10. Ibid.
11. Anderson, Stuart. 'Pharmacy and Migration: James Butler Swann (1834–1901) and His Three Sons in New Zealand, Fiji and Samoa 1862–1936', *Pharmaceutical Historian*, 50 (2020), pp. 1–15.
12. Eustis, Nelson. *Aggie Grey of Samoa* (Adelaide, 1979), p. 13.
13. Coombes, *Pharmacy in New Zealand*, p. 19.
14. Ibid.
15. Matthews, L.G. *History of Pharmacy in Britain* (London, 1962), p. 145.
16. Coombes, *Pharmacy in New Zealand*, p. 20.
17. Ibid., p. 21.
18. Ibid., p. 28.
19. Ibid., p. 201.
20. *New Zealand Gazette*, 31 (1871).
21. *Pharmaceutical Journal and Transactions*, 2 (1871), pp. 757–9.
22. Ibid.
23. Coombes, *Pharmacy in New Zealand*, p. 16.
24. Colebourne, C. 'Health and Illness, 1840s–1990s', in G. Byrnes (ed.) *New Oxford History of New Zealand* (Oxford, 2009).
25. Belgrave, M. 'Medicine and the Rise of the Health Professions in New Zealand, 1860–1939', in L. Bryder (ed.) *A Healthy Country: Essays on the Social History of Medicine in New Zealand* (Wellington, 1991).

26. Farquharson, C.H. *Historical Outline of New Zealand Pharmacy* (1929), p. xx.
27. Coombes, *Pharmacy in New Zealand*, p. 24.
28. Ibid., p. 110.
29. Ibid.
30. Ibid., pp. 25–9.
31. Ibid., p. 150.
32. Chapman, Stanley. *Jesse Boot of Boots the Chemists: A Study in Business History* (London, 1974), p. 86.
33. Ibid., p. 201.
34. Coombes, *Pharmacy in New Zealand*, p. 150.
35. Ibid., p. 151.
36. Ibid., p. 146.
37. British Secretary of State for Dominions Affairs, Telegram to the Governor General of New Zealand, 16/10/35, National Archives, Wellington, Department of Health, Series 1, File 182/16.
38. Lord Bledisloe, Governor General of New Zealand, Telegram to the Prime Minister, 18/10/35, National Archives, Wellington, Department of Health, Series 1, File 182/16.
39. Pharmacy Board and Friendly Societies, Deputation to Forbes, Prime Minister, 23/10/35, National Archives, Wellington, Department of Health, Series 1, File 182/16.
40. Anon. Unofficial Report of a Delegation to the Prime Minister, the Rt. Hon. G.W. Forbes, 18/10/35, National Archives, Wellington, Department of Health, Series 1, File 182/16.
41. *Pharmaceutical Journal*, 134 (1935), p. 148.
42. Coombes, *Pharmacy in New Zealand*, p. 147.
43. Ibid.
44. *Chemist and Druggist*, 125 (1936), p. 60.
45. Coombes, *Pharmacy in New Zealand*, p. 147.
46. Ibid.
47. Ibid., p. 158.
48. Norris, Pauline. 'Changes in New Zealand Pharmacy, 1935–1945', *Pharmacy in History*, 37 (1995), p. 172.
49. *Pharmaceutical Journal of New Zealand*, 27 (Oct 1955).
50. Walgreens Boots Alliance Archive, WBA/BT/13/32/1.
51. Ibid.

52. By 1962 Michael Peretz was on Boots' Executive Management Committee. Interview with Stuart Anderson 29 June 1995, British Library Sound Archive, C816/07/01–03, 2B121.
53. The Privy Council proceedings and judgement are available at the Walgreens Boots Alliance Archive in Nottingham, WBA/BT/13/32/1. Other papers relating to the case are held at The National Archives, London, File PCAP 6/1818/4, but were closed for 100 year until 1 January 2068.
54. Coombes, *Pharmacy in New Zealand*, p. 139.
55. Ibid.
56. Ibid.
57. Ibid.
58. Farquharson, *Historical Outline*, p. 18.
59. Ibid., p. 140.
60. Ibid., p. 104.
61. Norris, 'Changes in New Zealand Pharmacy', p. 167.
62. Chambers, J.H. *A Traveller's History of New Zealand and the South Pacific Islands* (London, 2003), p. 112.
63. Stokes, E. 'The Fiji Cotton Boom in the Eighteen-Sixties', *New Zealand Journal of History*, 2 (1968), pp. 165–77.
64. Ibid.
65. 'Ceres'. *The Fiji Islands as a Field for Emigration* (Melbourne, 1869).
66. Eustis, *Aggie Grey*, p. 14.
67. *Chemist and Druggist*, 59 (1901), p. 405.
68. *Fijian Almanac & Directory* (1873), p. 28 and (1874), p. 32.
69. *Chemist and Druggist*, 23 (1881), p. 523.
70. Ibid.
71. *Australian Supplement to the Chemist and Druggist*. Cited in ibid., p. 445.
72. *Fiji Times*, 13 July (1881).
73. Ibid.
74. Greenwood, Anna and Ingram, Hilary. '"The People's Chemists": The Walgreens Boots Alliance Archive', *Social History of Medicine*, 31 (2018), p. 886.
75. Campbell, I.C. *A History of the Pacific Islands* (Queensland, 1990).
76. Anderson, 'Pharmacy and Migration', pp. 9–12.
77. Eustis, *Aggie Grey*, p. 15.

78. Chambers, *Traveller's History*, p. 117.
79. Eustis, *Aggie Grey*, pp. 40–2.
80. Ibid., p. 29.
81. Ibid., p. 54.
82. *Auckland Star*, 138 (1936), p. 12.
83. Haddon, A.C. *Reports of the Cambridge Anthropological Expedition to the Torres Straits*, volume 6 (Cambridge, 1908).
84. Randall, Brian. *John Turnbull, Chemist, Thursday Island* (Brisbane, 2017), p. 1.
85. Beckett, Jeremy. *Torres Strait Islanders: Custom and Colonialism* (Cambridge, 1987).
86. Hersom, Andrew. 'A Pharmacist in Papua New Guinea: Charles Herbert Walker (1864–1925) and His Missionary Brother', *Pharmaceutical Historian*, 51 (2021), pp. 93–5.
87. Masters, P.F. 'Charles Walker (1864–1925)', undated note, Archives of the Hull Pharmaceutical Association, Hull History Centre, pp. 1–2.
88. Ibid., p. 1.
89. Hersom, 'Pharmacist in Papua New Guinea', p. 95.

CHAPTER 12

Conclusion: Colonial Pharmacy, Professionalization, and Reciprocity

This final chapter has three main aims: it reflects on the experience of professionalization of pharmacy in the British Empire described in previous chapters, against the five markers of professionalization derived from Bennell's analysis of pharmacy in Ghana (Chapter 1)[1]; it explores the extent to which a 'British model' of pharmacy developed in the colonies, based on the five markers of pharmacy Britishness described in Chapter 2; and it concludes with a description of how pharmacists in British colonies across three continents came together for collective action in their search for the reciprocal recognition of their qualifications with the British authorities.

For pharmacists in the British Empire, the transition from an occupation—usually referred to by themselves as the 'drug trade'—to a profession seeking a market monopoly was a process that progressed at different rates and with varying degrees of success in the different colonies. At the heart of the issue was the relationship between pharmacy and medicine. In many colonies the making and supply of medicines was initially a legitimate and necessary part of the doctor's role. If they delegated this responsibility, they expected that those who carried out these functions would do so under their supervision. Occupational boundaries in the colonies were highly fluid: many of those who opened shop for the

S. Anderson, *Pharmacy and Professionalization in the British Empire, 1780–1970*, Medicine and Biomedical Sciences in Modern History,
https://doi.org/10.1007/978-3-030-78980-0_12

sale of medicines were doctors; chemists and druggists who did so usually had to also sell a wide range of other merchandise to make a living.

Great differences existed between settler colonies and those with large indigenous populations. Yet even within the settler communities there were deep divisions. These were most prominent in relation to the reciprocal recognition of qualifications. Despite the apparently common values and norms shared by pharmacists trained in Britain, and the influence both direct and indirect of the PSGB, the criteria for registration as a pharmacist in one colony often differed substantially from those in another. Differences encompassed the length of an apprenticeship, the subjects to be studied, and the examinations to be taken. The result was usually refusal by one colony to accept the qualifications awarded by another. Collective action was needed to resolve the issue.

Colonial Pharmacy and the Metropole

The relationship between Britain and the colonies operated at many levels, but the PSGB played a central role in shaping pharmacy across the Empire. It communicated directly with pharmacists scattered across the globe; it was involved in the formation of pharmaceutical societies and in the drafting of legislation; and it gave advice directly to colonial authorities. Writing in 1962, the pharmaceutical historian Leslie Matthews suggested that

> though it is difficult to trace the direct influence which the Pharmaceutical Society has exercised upon the formation and the growth of the independent societies overseas – for it has no authority over or oversight of any of them – its indirect influence has been strong…Many Societies in the Commonwealth have adopted a constitution and bye-laws similar to those of the Society in Great Britain.[2]

There were nevertheless key differences in the way colonial pharmacists came together, whether as loose voluntary associations focussed mainly on commercial issues, or as pharmaceutical societies anxious to achieve professional autonomy by means of Pharmacy Acts.

The PSGB never proposed establishing overseas branches, as had the British Medical Association (BMA).[3] In 1885 the BMA prided itself on its status as 'an Imperial institution, extending its ramifications to all parts of the Empire and linking together the members of the profession, wherever found, in one body'.[4] In Britain, local pharmacy associations were established soon after the foundation of the PSGB. One of their main functions became the setting up of branch schools of pharmacy.[5] But they were local voluntary associations rather than formal branches of the Society, except in Scotland. The 1852 Pharmacy Act had granted Scotland a separate Board of Examiners with candidates examined at Edinburgh.[6]

The appointment of examiners became a major obstacle when the issue of reciprocity in the recognition of colonial pharmacy qualifications was raised. It was only in 1921 that the PSGB Council decided that the formal establishment of branches might provide a benefit to members and a much-needed stimulus to membership. In January 1922 it was agreed that each member would be assigned to a named branch, and by the end of the year 114 had been established.[7] Every part of the United Kingdom was covered; some replaced local pharmaceutical associations, whilst others operated alongside them. But none were established in any of the colonies.[8]

The influence of the PSGB was often mediated through the involvement of the Colonial Office, colonial administrators, and occasionally ministers and prime ministers. Its advice was usually filtered through official government or colonial channels. Initial meetings between pharmacy representatives and state officials frequently resulted in advice to proceed along the line taken in Britain. In Newfoundland in 1910 St. John's pharmacists who met the Prime Minister, Sir Edward Morris, to discuss the development of pharmacy were advised to form a pharmaceutical society and secure its legal recognition by means of a Pharmacy Act.[9] This replicated the process of professionalization followed in Britain. A Newfoundland Pharmaceutical Society was duly founded and given legal recognition by means of a Pharmacy Act (Chapter 3). But Morris was not the first Prime Minister to get involved with pharmacy; the Queensland Premier, Sir Hugh Nelson, was recruited to help secure the reciprocal recognition of pharmacy qualifications during a visit to Britain (below). And the Prime Minister of New Zealand, G. W. Forbes, became embroiled in debates surrounding the expansion of company chemists (Chapter 11).

Networks and Pharmacy

Associations between early modern colonial science and trade often provided the basis for the networks that connected the British metropole to the colonies.[10] Networks played as much a part in the recruitment and training of pharmacy practitioners as they did in the international drug trade. European colonial projects had much in common; networks were important for the trade in medicines between Portuguese colonies in Brazil and Angola and the imperial capital of Lisbon. Indeed, Ben Breen suggests that much of the success of the British Empire in the early modern period depended on it usurping Portuguese drug networks.[11]

Much of the early drugs trade was based on networks of religious affiliation. Sebastian Kroupa has shown the importance of Jesuit missionary networks in the development of colonial pharmacy. The Bohemian Jesuit pharmacist Georg Joseph Kamel was sent to the Philippines in 1688, and soon entered into communication with English and Dutch correspondents, and exchanged knowledge and materials with figures in the East Indies and Europe.[12] Whilst for Spain and Portugal religious networks were usually Catholic, for Britain they were Protestant and nonconformist. In pharmacy they included Quaker networks in the West Indies, dissenter networks in South Australia, and Church of England connections in New Zealand. Religious affiliation provided a basis for the trust essential for successful long-distance business transactions. Other networks also played their part, including missionary connections in the western Pacific and Freemasonry in South Africa.

Science networks played an important part in the development of pharmacy across the Empire. They transcended religious affiliations: the Jesuit Georg Kamel communication with the parson-naturalist John Ray and the English apothecary James Petiver.[13] Such networks reinforced the scientific basis of pharmacy—that it had a strong claim as a specialist area of knowledge and was much more than 'the drug trade'. In 1844 Jonathan Pereira, the PSGB's professor of materia medica, had suggested that the PSGB could take a leading role in science (Chapter 1).[14] He proposed the formation of a scientific committee for the promotion of pharmaceutical knowledge, for the elucidation of the natural history and origin of substances used in medicine. He declared that:

> By the establishment of a committee on pharmacology in the mother country, an opportunity would be obtained of bringing into notice the

various medicinal substances produced in the different portions of this great empire.[15]

The first meeting of the committee took place in January 1845, with several sub-committees appointed to carry out its objectives. It became a major hub for the exchange of information about plant and other medicines across the Empire.

In 1852 George F. Schaft, a pharmaceutical chemist from Bristol, suggested that scientific meetings be held for members of the PSGB to address 'the lack of scientific knowledge' amongst some members, and the 'desire to became better acquainted with the new discoveries'.[16] The first meeting of the British Pharmaceutical Conference was eventually held in 1863, and it quickly became a network for those interested in scientific developments in pharmacy. Pharmacists across the Empire soon joined it, both to keep themselves informed about scientific developments and to emphasize their scientific credentials. For pharmacy, connectivity was greatly enhanced by weekly journals including the *Pharmaceutical Journal* and the *Chemist and Druggist*, which became key vehicles for the exchange of information and opinions between pharmacists across the Empire.

Pharmaceutical Associations

Previous chapters have illustrated how the markers of professionalization based on Bennell's analysis were intimately linked.[17] Establishing pharmaceutical societies, escaping from the control of the doctors, and achieving the passage of a Pharmacy Act, were closely associated objectives. Pharmaceutical societies were usually voluntary bodies made up of people drawn together for collective action, whether to protect the trade or to avoid control by the doctors. They differed in how quickly they achieved legal recognition. In Britain this came within two years of foundation, but in some colonies, societies remained voluntary until at least the 1930s. Those with legal foundations could seek legislation that gave them powers to decide qualification criteria and to maintain registers, as in Britain. Where pharmaceutical societies remained voluntary, separate legally recognized Pharmacy Boards were needed to maintain registers of qualified practitioners. These were not always entirely independent; in practice there was often cross membership between the Pharmacy Board and the Council of the society.

As in Britain, the passing of a Pharmacy Act giving pharmaceutical bodies powers over who to admit to registers took pharmacy out of control by the doctors. But there were also strong incentives for pharmaceutical societies to seek legal recognition through a Pharmacy Act and to secure responsibility for regulating the profession themselves. Income from examination and registration fees was essential to survival for many of the pharmaceutical associations. The gap between association formation and passage of the Act was crucial: too short and the society might quickly become redundant and face rapid extinction, having achieved its main purpose (as in Queensland and Tasmania); too long and the society might find it impossible to sustain itself (as in New South Wales in the 1840s).[18]

Pharmaceutical organizations across the Empire invariably depended on the efforts and enthusiasm of a small group of committed individuals. Many complained about the apathy of the majority, with meetings often being postponed because of a lack of quorum. Many pharmacists simply declined to join the societies, weakening their influence and authority.[19] It was a problem as much in the metropole as the periphery; the Chemists and Druggists Trade Association had to convene a special meeting in 1887 in order to prevent its collapse.[20] In 1898, the president of the New South Wales Pharmaceutical Society deplored the fact that so little interest was taken in the Society.[21] And in South Africa it was noted that it was only when the pharmacists' interests were directly threatened that members showed any interest and pulled together.[22]

Autonomy from Medicine

In Britain, the separation of pharmacy from medicine took many years to become absolute; both were practised by a wide variety of practitioners. There was a plethora of licencing bodies for the medical profession; until the mid-nineteenth century there were nineteen in the United Kingdom.[23] There were no separate ones for pharmacy—it was an integral part of medicine. Colonial practice closely followed that in Britain; pharmaceutical qualifications were initially validated by Medical Boards. Following the separation of pharmacy from medicine in Britain, the process occurred at different times in different ways in the various colonies, and in some places not at all. Pharmacy in settler communities was often different from that in other colonies, and some followed the

British model more closely than others, adapting it where necessary to their own needs.

Whilst separation usually occurred as a result of a Pharmacy Act—which often followed attempts by doctors to exert their control over pharmacy—the pharmacists' case was frequently helped by shortcomings in Medical Acts, or by the inability of doctors to judge pharmaceutical competence. Medical Acts gave powers to doctors to maintain registers of chemists and druggists, but in practice they did little to raise standards of practice. In Victoria the Medical Registration Amendment Act 1862 gave the Medical Board powers to weed out unqualified practitioners, but a drafting error prevented it from deleting from the Register the name of anyone guilty of a misdemeanour, gross misconduct, or presenting false credentials.[24] But Australia's medical experience was not readily comparable with that of other British colonies in the nineteenth century. Australian doctors considered themselves to be part of the British medical profession; they shared a similar professional education and aspirations, and believed that all qualified practitioners should have an equal right to practise.[25] The locus of the profession was the colony rather than the Empire. In Australia pharmacy practitioners also shared a common training, and faced similar challenges.

Whilst the aim of the pharmacists was to escape the clutches of the doctors, issues at the boundaries between the professions continued to cause tensions between them. On the one hand was dispensing by doctor; on the other was counter-prescribing by pharmacists. These were the more obvious expressions of Larkin's concept of 'occupational imperialism',[26] the tendency of medical occupations to trespass on the areas of responsibility, or to poach the skills, of neighbouring disciplines. Redefining boundaries often proved difficult; doctors who dispensed their own medicines were continuing a practice stretching back centuries; pharmacists who undertook counter-prescribing were following in the footsteps of their predecessors, the apothecaries.

The separation of pharmacy from medicine in Britain and the colonies is perhaps ironic, given that this was the opposite of the PSGB founder's intention. Jacob Bell hoped to develop pharmacy as a discrete section of medicine, but—as Holloway suggests—passage of the 1852 Pharmacy Act in Britain made the exclusion of pharmacists from the medical profession inevitable. Before it, chemists and druggists could still be regarded as an integral part of it, along with physicians, surgeons, and apothecaries. Without it, the 1859 Medical Register might have been published in four sections rather than three. As Holloway notes, 'there is nothing inevitable about the present division between the practice of medicine and that of pharmacy'.[27] It was socially and politically constructed by three Acts of Parliament; the 1852 Pharmacy Act, the 1858 Medical Act, and the 1868 Pharmacy Act. These determined the relationship between pharmacy and medicine not only in Britain but throughout the Empire.

The Britishness of Colonial Pharmacy

The extent to which pharmacy in the Empire might be considered 'British' has been considered in this book in relation to what made it different from pharmacy in the rest of Europe, based on the different paths followed in Britain and Europe concerning the distinction between pharmacy and medicine. This kind of 'Britishness' was created in Britain itself rather than in the colonies. The concept of Britishness has now been extensively studied by imperial historians and others. Some have contested the idea that it was something created at home and diffused to the colonies.[28] They suggest that 'Britishness' was born in the colonies as British settlers and expatriates set about recreating the familiar world they left behind.[29] Britishness took shape at the periphery and was reinforced at the centre.[30] A developing sense of global Britishness aided rather than thwarted economic and social integration in the fifty years before the First World War.[31] Yet the Britishness of pharmacy is a reflection of the transplantation of the British model in the colonies.

The five markers of the British model of pharmacy (Chapter 2) indicate that in many parts of the Empire the model was closely followed, although all colonies adapted it to local circumstances. British pharmacists found themselves in a wide variety of social, economic, and political conditions during the mid- to late-nineteenth century, and needed to respond appropriately. Some aspects of the model were the result of errors or omissions; lessons were learnt in some British colonies but not others. In most colonies, pharmacists achieved autonomy from the doctors by means of a Pharmacy Act. Some made the pharmaceutical society the registration body, but few followed the British model in having a two-tier register of pharmaceutical chemists and chemists and druggists. Most did however follow Britain in linking the regulation of pharmacy with the control of poisons, and company chemists eventually emerged in most British colonies.

Pharmacy and Poisons

In Britain linking pharmacy with poisons legislation was a pragmatic response to the need for a Pharmacy Bill in the 1860s (Chapter 2). It was not an inevitable link, yet the Pharmacy and Poisons Act of 1868 became the foundation of pharmacy and poisons legislation in colonies throughout the Empire. These Acts usually confirmed the pharmaceutical society as the registering body, established a number of restricted titles, and gave pharmacists a monopoly in the supply of certain poisons used as medicines. Where differences from the 1868 Act were considered necessary, positive action was necessary to make the changes. Some colonies did things differently, as when Victoria opted for a Pharmacy Board separate from the pharmaceutical society as the registering authority. Others followed the wording of the British Act to the letter, and found themselves repeating errors later exposed in that legislation, most notably with regard to company chemists.

Pharmacy legislation was the subject of extensive discussion between the metropole and the periphery. As Leslie Matthews noted

> the legislation of the United Kingdom has been followed in large measure in many of the countries of the Commonwealth, from the time when they were styled the 'Colonies,' and there has been frequent consultation on points of mutual interest.[32]

Before 1850 many of the early Ordinances mirrored the British Apothecaries' Act of 1815, in specifying that it did not apply to chemists and druggists: after 1870 the 1868 Act formed the basis of legislation throughout the Empire. Yet the Act was to cause endless difficulties for pharmacists everywhere: its nature confirmed the link between pharmacy and poisons, and its wording allowed for the ownership of pharmacies by companies.

In Britain the working of the Pharmacy and Poisons Acts—and the linking of the regulation of pharmacy with the control of poisons—came under parliamentary scrutiny in 1926, and the Departmental Committee on the Poisons and Pharmacy Acts reported in 1930. It concluded that linking the regulation of pharmacy with the control of poisons was not in the public interest.

> It is to be observed that the scheme of the Poisons and Pharmacy Acts does not distinguish the law relating to the practice of pharmacy from that

> relating to poisons. Both are frequently dealt with in the same section. In both cases the Privy Council is the Central Authority; in both cases the Pharmaceutical Society of Great Britain exercises the control...The distinction between these two things...is obvious...Their severance, while it presents no difficulty, offers, in our opinion, great public advantages.[33]

Their main objective was 'to simplify the law by distinguishing between pharmacy and poisons'.[34] A Bill was presented to Parliament in 1931 where it was 'very substantially recast', partly because of objections by the medical profession that it had not been consulted.[35] But the link between the two remained explicit when a combined Pharmacy and Poisons Act was passed in 1933, with Part I headed 'Pharmacy' and Part II 'Poisons'. Medicines in Britain continued to be regulated under poisons legislation until passage of the Medicines Act in 1968.

Company Chemists

The decision of the House of Lords in 1880 that companies could legally own pharmacies had resulted from wording in the Pharmacy and Poisons Act 1868 that assumed that chemists' shops would always be owned by a 'person' (Chapter 2). The Act had only been in force for ten years by the time the case first went to court, and shifting patterns of retail trade were already apparent when the Bill was drafted. It is difficult not to conclude that there was a lack of foresight regarding the possible future interest of companies in the pharmacy business. It might however have been more difficult to foresee that losing the case would lead to the rapid growth of pharmacy chains, or that shortcomings in British legislation might have repercussions across the Empire.

For those in the colonies the message was clear. Specific legislation was needed to prevent the emergence of chain store pharmacies, by indicating that one pharmacy must be owned by one pharmacist; legislation needed to be carefully phrased. In South Africa, chemists' chains emerged in the absence of legislation to prevent them, and in New Zealand initial legislation failed to prevent the development of chains. New Zealand pharmacists were to find that securing amending legislation after the event was not only extremely difficult but also would lay them open to intense parliamentary scrutiny. In Victoria, on the other hand, amending legislation to prevent them was enacted before the issue arose, and the development of chain pharmacies was prevented, at least until legislation changed again.

Across the Empire opposition by retail chemists to chain stores was by no means universal. Arguments for and against chains raged throughout the colonies. Arguments in favour emphasized the fact that chains tended to drive prices down, thus making as many medicines as possible available to a grateful public. Those against argued that it would drive the small independent proprietor out of business, reducing the access of many people to a vital service. But as the chains grew so did concern in pharmaceutical circles about their power and influence, both commercially and politically. There was no going back, and chain store pharmacy remained a major source of conflict and distrust for many pharmacists across the world throughout the twentieth century, whilst providing employment to many others.

Education and the Dual Register

It was the 1868 Act that created the dual register in Britain, consisting of chemists and druggists and a smaller group of pharmaceutical chemists (Chapter 2). This was a pragmatic response by the PSGB to the need to increase membership and to bring the chemists and druggists on board. It soon became two levels of qualification by examination, the Minor and the Major. In many ways it reflected a class structure in British pharmacy. Those engaged in retail trade as chemists and druggists were a diverse group; at one end of the spectrum was the poor corner shopkeeper; at the other were high class establishments such as that of Jacob Bell in London's Oxford Street.[36] Whilst the lower-level chemist and druggist certificate was what was needed to open a shop, the higher-level pharmaceutical chemist qualification increasingly became a matter of status, with holders taking academic, industrial, and administrative positions. Colonies often found it difficult to deliver a level of training even close to that of the chemist and druggist certificate, and colonial authorities without large settler populations frequently turned to those who had obtained the pharmaceutical chemist qualification in Britain to lead and develop their pharmacy services.

For pharmacy, education was the key to professionalization. When the PSGB was founded in 1841 education was to be the grounds on which government approval and support would be sought.[37] Jacob Bell made it explicit:

> The Society was established for the purpose of...raising the character of those who practise pharmacy in Great Britain. It is proposed to attain this end, first, by uniting all the chemists and druggists into one body; secondly, by introducing a system of pharmaceutical education; thirdly, by claiming for the body thus organized and educated, such protection and privileges as the qualification of the members would entitle them to possess.[38]

The 'system of pharmaceutical education' developed in small steps over a period of many years. Those who went out to the colonies had a wide range of pharmaceutical qualifications. Some were apothecaries who had completed a lengthy apprenticeship, whilst others described themselves as 'chemists and druggists' but had no formal qualifications. It mattered little, since in many colonies there were initially few laws or regulations relating to pharmacy, and anyone could set up a pharmacy business. The situation differed enormously between colonies however, many later developing their own arrangements for pharmacy training and registration, usually based on the British model consisting of apprenticeship followed by examination. But there was little uniformity even in Britain. A multiplicity of courses was available that varied in coverage, teaching standard, and facilities.[39] It was only in 1918 that the principle of a compulsory course of study was adopted. In 1929 most pharmacists in Britain still followed an apprenticeship leading to the basic chemist and druggist qualification.[40]

Until the end of the nineteenth-century colonies accepted those with PSGB qualifications without question, but they slowly introduced their own arrangements for pharmacy education. In Australia these were largely based on the British model, although there were substantial differences between colonies.[41] The same was true of Canada; moves were made to regularize pharmacy training in Quebec in 1879. Ontario established educational standards and organized training facilities in 1882, but in Alberta a pharmacy course was established only in 1912.[42] In South Africa an informal school of pharmacy had been established in Cape Town in 1885,[43] whilst formal pharmacy classes were not inaugurated in Natal until 1908.[44] In India, although formalized pharmaceutical instruction began in 1860, only a handful of dispensers qualified.[45]

Seeking Reciprocity

As education and training standards rose throughout the Empire, the 'monopoly of knowledge' increased, and pharmacy established increasing distance between itself and medicine. In some colonies standards of pharmacy training matched and occasionally surpassed those in Britain. It was only a matter of time before one of them requested reciprocity in the recognition of pharmacy qualifications with the PSGB. The first request arrived in autumn 1881. The *Chemist and Druggist* reported that a letter had been received asking the PSGB's Council 'to consider the possibility of arranging reciprocity of certificates awarded by the Victoria Pharmacy Board and the PSGB'.[46] It was referred to the PSGB's Library, Museum, Labour, and House Committee who examined the qualification requirements in Victoria. They comprised

> a four-year apprenticeship, a preliminary examination identical with that of the PSGB, an examination at Melbourne University or other school or college recognised by the Board in elementary and practical chemistry, materia medica and botany, and a final examination before the Pharmacy Board in practical pharmacy, undertaken when the candidate has served his apprenticeship and attained the age of twenty-one years.[47]

This was an impressive training and was considerably more extensive than that in Britain at the time. The PSGB committee noted that the Registrar had no power to place names on the register except those appearing on certificates from Boards of Examiners whose names had been approved by the Privy Council. But there was nothing in the law to prevent the PSGB appointing examiners in the colonies and having their names approved by the Privy Council. It was a convenient excuse to take no action. The *Chemist and Druggist* noted: 'We rather regret the somewhat brusque reception which seems to have been given to the communication from the Pharmacy Board of Victoria... we are sorry that the question has not been at least submitted to the whole Council for discussion'. It thought it unlikely that many Australian pharmacists would want to come to England. 'Pharmacists are much more likely to be exported from Great Britain to the Colonies than imported from them'.[48]

The PSGB's blunt refusal postponed the need to address the issue for fifteen years. The amendment to the 1868 Act needed to allow the PSGB to make Byelaws that would allow colonial certificates to be recognised could have been included in the 1898 Pharmacy Acts Amendments Act (Chapter 2). But it was not, and the issue gained momentum throughout the Empire. There was general recognition that if the PSGB were to be persuaded to allow pharmacists with colonial qualifications to join the British register, then colonies in the same country would have to be recognizing each other's qualifications.[49] This was no easy task, as Australian pharmacists found in 1886 (Chapter 10). In the colonies there was growing awareness that collective action across colonies would be needed to persuade the PSGB to seek the changes necessary for it to recognize colonial qualifications.

The Queensland Initiative, 1896

The initiative was taken by the Pharmaceutical Society of Queensland. On 14 July 1896 they wrote to pharmaceutical authorities in the other five Australian colonies, in New Zealand, Cape Colony, and in six Canadian colonies, urging a combined appeal to the PSGB regarding reciprocity of British and Colonial certificates. They noted that

> although on inquiry it might be deemed inadvisable to grant reciprocity in every case where the British certificate is accepted, there are already some of the colonial licensing boards whose qualification is quite equal to that of the parent Society, and to them reciprocity should in justice be granted.[50]

The petition received active support from all colonies except New Zealand, who did not feel ready to participate in such an arrangement. The Queensland Initiative was reported in the *Pharmaceutical Journal*. It noted:

> Reciprocity in Certificates is at present a topic of considerable interest in the British Colonies, the height of their ambition being to obtain recognition of their certificates by Great Britain. Even were the fitness and desirability of such recognition admitted in this country, nothing could be done in the matter until special Parliamentary powers were obtained, inasmuch as no power of recognising the certificates of external examining bodies is conferred by the Pharmacy Acts.[51]

It also reported a call in the *Australasian Journal of Pharmacy* for a common intercolonial examination 'as a necessary preliminary'. It suggested that intercolonial reciprocity was of more pressing importance than reciprocity with the 'old country', and wondered why anyone might wish to move to Britain. 'To people living at this side of the world, it seems odd to think that colonials should ever entertain any desire to practise in such a worn-out spot'.[52]

The Queensland Initiative prompted renewed efforts to achieve inter-colonial reciprocity, and individual colonies undertook to make individual requests for reciprocity with the PSGB. Within a few months separate applications had been made by Quebec, Queensland, and Victoria. It also prompted the PSGB to make its own enquiries. In late 1896 it sent letters 'to pharmacy officials at the outposts of Empire on the matter of imperial reciprocity' to gauge the level of interest in such an initiative.[53] In the event of a positive response, it would consider what action might be taken to enable the free movement of pharmacists both between Britain and the colonies. Quebec submitted their request for 'an interchange of certificates' between themselves and the PSGB in January 1897. It was discussed by the Society's Council, where—the *Chemist and Druggist* reported—the president announced that 'the only reply could be that they had no power to receive certificates in lieu of their qualifying examination. They would require a new Act in order to grant what the letter asked'.[54]

The Premier's Visit

Following the failure of the Quebec appeal, Queensland submitted theirs. In their letter to the PSGB dated 4 June 1897, the Queensland Board emphasized the matter of justice whilst threatening to end the automatic recognition of the British qualification.

> The Council is chiefly actuated by a desire to obtain for the chemists of this colony justice in respect to the relations with chemists from home, who are admitted without examination, and further by a desire to avoid the alternative, which would be the repealing of the section recognising the certificates of the PSGB.[55]

The Queensland Premier, Sir Hugh Nelson, was asked to take up the matter on his visit to Britain to participate in Queen Victoria's Diamond Jubilee celebrations on 22 June 1897. The *Chemist and Druggist* noted that 'the opportunity is a magnificent one for the Society. It is a tribute to it that it should be regarded in distant parts of the empire as the "parent society"'. The PSGB needed 'to take the initiative in a scheme of pharmaceutical imperialism... But we may say frankly that the Council is almost destitute of enthusiasm about the matter. It has little power of initiative'.[56]

It was after the Queensland Premier's visit that the *Chemist and Druggist* was

> pleased to note a change in the tone of the President of the PSGB regarding reciprocity in respect to certificates between the Society and the various other pharmaceutical bodies of the British Empire...on Wednesday, when the Pharmacy Board of Victoria made application on the same subject, the President wisely said that the matter was most important, and consideration of it was referred to the Law and Parliamentary Committee.[57]

But when that committee reported back to the Council it recommended that no attempt be made at obtaining reciprocity powers between Britain and the colonies until a scheme of intercolonial reciprocity had been established.[58]

By July 1897, the Pharmacy Board of Victoria had submitted a second request to the PSGB, having previously had lengthy correspondence with the Pharmacy Board of Queensland, and agreed with them the need for a common examination process as a means of establishing intercolonial reciprocity. Their letter suggested that, as fifteen years had elapsed since the application was first made, and as 'the Imperial Government was so desirous of knitting closer the bonds that united the various parts of the Empire', it would be a fitting opportunity for an amendment to be made to the Imperial Pharmacy Laws so as to 'enable the parent Society to accept the graduates of colonial colleges whose education and examinations were such as to fulfil the required conditions'.[59]

But the PSGB's response was unchanged:

> Your letter ...has been carefully and sympathetically considered. The general conclusion...is that the time is not yet ripe for any general amendment in this direction...Byelaws have recently been approved by the Privy Council effecting a considerable improvement in the standard of general education required from those who...seek to enter pharmacy in Great Britain ...It is felt, however, that at present it would be almost impossible to ensure in the different parts of the Empire a practical uniformity in the scope and stringency of the technical examinations...My Council

> regrets that public opinion is not sufficiently advanced to enable us to obtain powers for a compulsory curriculum even in Great Britain.[60]

The Inquiry into Colonial Pharmacy Education, 1897

After these negative responses, the *Chemist and Druggist* questioned the PSGB's position.

> There is nothing in the Pharmacy Acts which specifically permits acceptance of certificates in Latin, English, and arithmetic, such as have been accepted for many years by the Board of Examiners in lieu of the Preliminary Examination. The Council and the Boards shelter themselves behind the technicality that no certificate is accepted which is not approved by the examiners, and as the Minor examination comes under the same section (viii) of the 1852 Pharmacy Act as the Preliminary examination, the same procedure ought also to obtain.[61]

The PSGB's intransigence had been exposed. The *Chemist and Druggist* instituted its own inquiry into the state of pharmacy education and training across the Empire—something that the PSGB had declined to do—with a view to considering the possibility of one certificate qualifying the holder in any part of the Empire. On 22 January 1897, the editor sent a handwritten letter to pharmacy bodies across the Empire.[62] He received replies from most colonial authorities and concluded that 'there is a desire on the part of the pharmaceutical authorities in the extreme parts of Greater Britain to come more closely into touch with the mother-country'.[63] Details of qualification requirements were published over several weeks, identifying where differences lay (Table 12.1). It concluded its series with a call for pharmacy to emulate arrangements applying to medicine. 'Precedent in regard to medical registration has shown that there is utility in recognition of Canadian, Colonial and Indian qualifications, and we ask no more for pharmacy than that the Pharmaceutical societies at home should follow lines similar to those adopted by the General Medical Council'.[64]

By 1898 a change in tone was apparent. By April reciprocity had been arranged between Victoria and the Cape. In his address to PSGB members in August the President declared

Table 12.1 Pharmacy education, examination, and registration in the British colonies, 1897

Colony	*Preliminary examination*	*Years of practical training*	*College education*	*Intermediate examination*	*Registration examination*
Ontario	Algebra required but not Latin	Four years	Two sessions at college	College examination	Qualifying examination
Quebec	French, geography and history required	Three years	Two sessions at college	College examination	Qualifying examination
Nova Scotia	Evidence of sound education	Three years	None	No	Examination
Manitoba	Completed school examinations	Four years	Two sessions at college	College examination	Qualifying examination
New Brunswick	Completed school examinations	Three years	None	No	Qualifying examination
British Columbia	Completed school examinations	Three years	None	No	Qualifying examination
Victoria	Exam as in Britain but before apprenticeship	Four years	Curriculum Compulsory	No	Similar to PSGB Minor examination
South Australia	Exam as in Britain but no restrictions	ditto	Lectures provided but not compulsory	No	Similar to PSGB Minor examination
Queensland	Exam as in Britain but no restrictions	Three years	Lectures provided but not compulsory	No	Similar to PSGB Minor examination
New South Wales	Exam as in Britain but no restrictions	Three years	Lectures provided but not compulsory	No	Similar to PSGB Minor examination
Western Australia	Exam as in Britain but no restrictions	Four years	None	No	Similar to PSGB Minor examination

(continued)

Table 12.1 (continued)

Colony	*Preliminary examination*	*Years of practical training*	*College education*	*Intermediate examination*	*Registration examination*
Tasmania	Exam as in GB but before apprenticeship	Four years	None	No	Similar to PSGB Minor examination
New Zealand	Exam as in Britain but no restrictions	Three years	None	No	Similar to PSGB Minor examination
Barbados	Good school training	Three years	Optional training in chemistry	No	Like the Minor (no botany)
Jamaica	Good school training	Three years	Optional training in chemistry	No	Like the Minor (no botany)
Trinidad	Good school training	Three years	Optional training in chemistry	No	Like the Minor (no botany)
Cape Colony	Preliminary examination	Three years	None	No	Similar to PSGB Minor examination
Natal	Preliminary examination	Three years	None	No	Similar to PSGB Minor examination

Source *Chemist and Druggist*, 51 (1897), pp. 122–3 and 239–40

> I look forward with hope and confidence to the time when there will be increased unity amongst all English-speaking pharmacists and pharmaceutical societies, and when there will be sufficient uniformity in the various qualifications to enable reciprocity to exist amongst them all.[65]

But the *Chemist and Druggist* continued its campaign. 'It is exceedingly desirable that all opportunities should be taken to secure colonial pharmaceutical reciprocity with the view to ultimate uniformity in pharmaceutical qualification for the British Empire'.[66]

In June, the Council of the Queensland Society renewed its bid directly through the Government.[67] By mid-1899 a draft of a new Poisons and Pharmacy Bill had been prepared which included powers to make Byelaws. Reciprocity between the British and the Irish societies was to be established by the Act, but this arrangement was 'opposed for several reasons' by the PSGB and was later omitted.[68] But the continued refusal of the PSGB to recognize colonial qualifications resulted in inevitable reprisals. At a meeting of the New South Wales Pharmacy Board in May 1899 a pharmacist holding the Major certificate of the PSGB applied for registration. With little or no discussion his application was refused, on the grounds that the Board would recognize no such certificates until reciprocity was established between themselves and Britain.[69]

Enabling Reciprocity

The mechanism by which reciprocity in the recognition of pharmaceutical qualifications would be enabled would be through a Section in the new Poisons and Pharmacy Bill. The PSGB would be 'empowered to make byelaws establishing a curriculum, dividing the examinations, and making provisions for the acceptance, for registration in lieu of examination, of colonial diplomas'.[70] This part of the Act underwent amendment as it went through the House of Commons, and was taken through the House of Lords by the Secretary of State for the Colonies. Hansard reported that it provided

> for the registration, upon payment of the prescribed fee, as pharmaceutical chemists or chemists and druggists under the Pharmacy Acts, 1852 and 1868, without examination, of any persons holding colonial diplomas or of qualified military dispensers who produce evidence...that they are persons of sufficient skill and knowledge to be so registered.

Following certain amendments, the Bill was agreed and became the Poisons and Pharmacy Act 1908.[71] But the PSGB was in no hurry to implement reciprocal recognition, not least because of its concerns about the qualifications being awarded in some colonies. Some were considered inadequate despite high failure rates. In 1908 the Colonial Pharmacy Board in Cape Colony expressed regret that 'so many candidates came up for examination without any apparent preparation'.[72] The PSGB's tardiness on the issue was raised in the House of Commons in 1911. The Home Secretary was asked whether byelaws had yet been made. Winston Churchill replied that before doing so the PSGB 'has to dispose of certain preliminary questions connected with the educational curriculum of pharmaceutical chemists and druggists, and that the Society is now actively engaged in those questions'.[73] They took another 4 years to pass the Byelaw, in 1912.[74]

In 1911 the PSGB finally declared that they wished to establish 'reciprocal relationships with similar societies in other parts of the Empire'.[75] The proposal was considered by pharmaceutical associations across the Empire. Whilst most were keen to do so, others acknowledged that they still had much to do before being ready. In Newfoundland—still outside the Canadian federation—druggists recognized that they could not meet the necessary educational criteria.[76] A letter from the pharmaceutical association's secretary to the Colonial Secretary in 1912 conceded that 'it is impossible for us to entertain any reciprocal relations whatever'.[77] Twelve months later he added that 'we in this Colony are much in the position of the members of the PSGB in the years immediately succeeding 1867, and conditions in Newfoundland have not been such as to make a high standard of pharmaceutical knowledge possible'.[78]

Over the next two years the Society considered the curriculum and procedures necessary to implement the Byelaw. A note spelling out the details for 'pharmaceutical registration in the British Empire' eventually appeared in the *Pharmaceutical Journal* in August 1913. Twelve months later an updated version covering the whole world appeared under the title 'pharmaceutical registration abroad'. The fact that it took so long to implement was largely due to the diversity of pharmacy education, examination, and registration arrangements across the Empire.[79] By 1914 some Australian states had agreed reciprocity either of the whole register and on an examination basis. But only the two states with colleges of pharmacy—Victoria and Queensland—reached agreement with the PSGB, and with the outbreak of the First World War further progress came to an end.

Towards a Universal Pharmaceutical Qualification

The PSGB continued to grapple with reciprocal recognition of qualifications for many years, and in 1929 the possibility of an imperial pharmaceutical qualification was explored after the League of Nations proposed that member states consider the feasibility of developing internationally recognized qualifications. Governments referred the matter to relevant bodies in their countries; for pharmacy in Britain it was referred to the PSGB. The Council restricted its discussions to a standard qualification for the Commonwealth.[80] The issue was considered at the British Pharmaceutical Conference, and a detailed analysis of pharmacy education across the empire was presented by Herbert Skinner, a former PSGB president. He concluded that Britain needed to put its own house in order first. 'The present moment is one for Great Britain to re-examine its own syllabus, its own system of training, set that in order, and present an example to the others who, in my opinion, would willingly follow'.[81] Further discussion was deferred to a later meeting, but the idea of an imperial pharmacy qualification seems to have been quietly dropped.[82]

In the 1950s the PSGB still exercised great influence over pharmacy education and standards in the Australian states by threatening to withdraw its reciprocity agreement. As the colonies achieved independence from the late 1950s the PSGB was all too aware of the state of pharmacy across the Empire. In 1965 a Commonwealth Foundation was established to facilitate the creation of Commonwealth professional associations. The PSGB convened a meeting in London attended by pharmacy representatives from 28 countries and a Commonwealth Pharmaceutical Association (CPA) was formally established in 1970, with a headquarters and secretary provided by the PSGB.[83] The CPA established a Council consisting of one delegate from each member country, and formulated policy guidelines covering education, the role of the pharmacist, ethics, legislation, and the dispensing of medicines, amongst others. In collaboration with the International Pharmaceutical Federation (FIP) it subsequently worked towards the establishment of a universal standard for pharmacy education.[84]

British Pharmacy's Imperial Legacy

This book has demonstrated that the British model of pharmacy was not simply transplanted into colonies across the Empire. Even in those colonies where this was the intention, adaptation to local circumstances were made. The professionalization of pharmacy did not follow the same pattern in each of the colonies, with diverse political and economic factors leading to delays in or even the failure of professionalization. In some colonies pharmacy failed to emerge as a profession distinct from medicine during the period of British rule, whilst in others doctors maintained a firm control over it. A variety of actors were involved in these processes, but in many the PSGB had a central role. Pharmacy was more British in some colonies than others.

Britain's pharmacy's legacy to the Empire can be contrasted with that of other empires. Studies of pharmacy in the colonies of other European powers generally note that pharmacy and medicine existed as separate professions from the beginning of their Empires, as reflected in French medicine in Vietnam,[85] German pharmacy in southeast Asian missions,[86] pharmacy in French colonies,[87] and pharmacy in Goa during the Portuguese period.[88] The role of colonial pharmacists in disseminating western medicines, and both their positive and negative impacts on the health of indigenous populations, has been explored in several colonial contexts. A study of increasing recognition of the benefits of biomedicine in Mexico in the 1930s noted that 'nurses, physician aids, and pharmacists who claimed to know medicine, but who in fact had only partial knowledge, damaged the science of biomedicine and contributed to the lack of faith in its benefits'.[89] The legislation of some European countries was more focussed on the indigenous population than others. Spanish colonial legislation aimed to improve healthcare for indigenous peoples, and elite physicians in Puerto Rico were required to ensure medical care was available in poor areas where there was none.[90] Colonial legislation relating to both medicine and pharmacy tended to be reactive and pragmatic.

British pharmacy in the colonies was more focussed on trade than profession. For the expatriate community, the presence of modern British retail pharmacies across the Empire offered comfort and reassurance, whilst presenting potent symbols of colonial cultural capital to the indigenous population. In the metropole, public awareness of imperial exploits was high and retail pharmacies were present in every neighbourhood. Dispensing analogies presented cartoonists with a ready source for imperial imagery. Federation in Canada and Australia and union in South Africa were closely followed in Britain, and cartoons with a pharmaceutical theme were readily understood (Fig. 12.1).

The 'Unfinished' Profession

This book has demonstrated that—despite the passage of British pharmacy legislation, the active role of the PSGB, and the efforts of British settler pharmacists—pharmacy in the British Empire was at very different stages, not only in 1914 but also at the time of independence. Yet recent studies of the professionalization of pharmacy question its completeness even in Britain. To echo John Darwin's analysis of Britain's imperial adventure as the 'Unfinished Empire' it is perhaps not inappropriate to describe pharmacy as the 'unfinished profession'. This book has illustrated its experience in very different forms of Empire. Pharmacy's journey, from being 'unorthodox' and on the 'fringe' of medicine to being an

Fig. 12.1 The Imperial Dispensary: a pharmacist making up a prescription for a kangaroo, representing Chamberlain's advocacy of the Commonwealth of Australia, 1900 (Courtesy of Wellcome Collection)

autonomous profession, has taken different routes in different colonies, with many providing examples of pharmacy as 'incomplete professionalization',[91] as not possessing all the characteristic traits of a profession.[92] Some colonies were less successful than others in maintaining occupation unity, in exercising control over the sale of poisons and in establishing an appropriate balance between profession and trade, despite the fact that doctors and lawyers have been shown to be no more 'professional' than pharmacists.[93]

Whilst the separation of pharmacy from medicine may be complete, obstacles remain concerning its claim for 'complete professionalization', particularly its continuing association with trade, and its heavy reliance on an item largely generated by another profession—the prescription. Changing the status quo would have substantial implications for pharmacy education and practice, with pharmacists returning to the Medical Register as the fourth branch of the medical profession, as Jacob Bell had intended. If the 'progress of pharmacy' in the Empire was delayed by Britain, its origins are to be found in the tortuous path of professionalization in Britain itself. Reflecting on the 1704 Rose Case in 1927, Thompson noted that

> owing to the curious evolution of the apothecary into the medical practitioner, and the chemist combining with the druggist and absorbing the trade side of the apothecary's business, the position of pharmacy in England remained so much behind that of other countries until after the middle of the nineteenth century.[94]

In the twenty-first century all professions are changing rapidly. Richard and Daniel Susskind suggest that in a society where the internet is ubiquitous, we will no longer need doctors, teachers, accountants, and the rest (including pharmacists) to work in the way they have previously. Occupations will need to look very different in the future; there will be a new generation of 'open-collar workers', and the various occupations that have worked so hard to achieve 'professional' status over the last two centuries will have to make way for a new breed of 'para-professionals'.[95] The professionalization of pharmacy clearly remains a work in progress.

Notes

1. Bennell, Paul. 'Professionalization: The Case of Pharmacy in Ghana', *Social Science and Medicine*, 16 (1982), p. 601.
2. Matthews, L.G. *History of Pharmacy in Britain* (London, 1962), p. 141.
3. Johnson, Terence J. and Caygill, Marjorie. 'The British Medical Association and Its Overseas Branches: A Short History', *The Journal of Imperial and Commonwealth History*, 1 (1973), pp. 303–29.
4. *British Medical Journal*, II (1885), p. 882.
5. Holloway, S.W.F. *Royal Pharmaceutical Society of Great Britain 1841 to 1991: A Political and Social History* (London, 1991), p. 273.
6. Ibid., p. 186.
7. Matthews, *History of Pharmacy*, p. 149.
8. Ibid., p. 150.
9. Arnold V. *A Brief History of Pharmacy in Canada* (Ottawa, 1967), p. 61.
10. Darwin, John. 'Imperialism and the Victorians: The Dynamics of Territorial Expansion', *English Historical Review*, 112 (1997), p. 629.
11. Breen, Benjamin. *The Age of Intoxication: Origins of the Global Drug Trade* (Philadelphia, 2019).
12. Kroupa, Sebastian. '*Ex epistulis Philippinensibus*: Georg Joseph Kamel SJ (1661–1706) and His Correspondence Network', *Centaurus*, 57 (2015), pp. 229–306.
13. Ibid.
14. Bell, Jacob and Redwood, Theophilus. *Historical Sketch of the Progress of Pharmacy in Great Britain* (London, 1880), p. 227.
15. Ibid., p. 171.
16. Matthews, *History of Pharmacy*, p. 151.
17. Bennell, 'Professionalization', p. 601.
18. Haines, Gregory. *Pharmacy in Australia: The National Experience* (Deakin, 1988), p. 133.
19. Ryan, Mike. *History of Organized Pharmacy in South Africa 1885–1950* (Cape Town, 1986), p. 115.
20. *Chemist and Druggist*, 30 (1887), p. 645.
21. *Chemist and Druggist*, 52 (1898), p. 302.

22. *South African Pharmaceutical Journal* (1945), p. 3.
23. Dyason, 'Medical Profession', p. 195.
24. Dyason, 'Medical Profession', p. 202.
25. Ibid., p. 194.
26. Larkin, G. *Occupational Monopoly and Modern Medicine* (London, 1983).
27. Holloway, *Royal Pharmaceutical Society*, pp. 181–2.
28. Seth, Suman. *Difference and Disease: Medicine, Race and the Eighteenth-Century British Empire* (Cambridge, 2018), p. 6.
29. Bally, C.A. *Imperial Meridian: The British Empire and the World, 1780–1830* (London, 1989).
30. Hall, Catherine. *Civilising Subjects: Colony and Metropole in the English Imagination 1830–1867* (Chicago, 2002).
31. Magee, Gary B. and Thompson, Andrew S. *Empire and Globalisation: Networks of People, Goods and Capital in the British World, c.1850–1914* (Cambridge, 2010), p. 234.
32. Matthews, *History of Pharmacy*, p. 141.
33. *Report of Departmental Committee on Poisons and Pharmacy Acts*, Cmd. 3512 (London, 1930), pp. 4–6.
34. Ibid., p. 6.
35. Linstead, Hugh. *Poisons Law* (London, 1936), pp. 14–8.
36. Holloway, *Royal Pharmaceutical Society*, p. 45.
37. Ibid., p. 110.
38. *Pharmaceutical Journal and Transactions*, 2 (1842–43), p. 741.
39. Earles, M.P. 'The Pharmacy Schools of the Nineteenth Century', in Poynter, F.N.L. *The Evolution of Pharmacy in Britain* (London, 1965), pp. 79–95.
40. Earles, M.P. 'The Development of Pharmaceutical Education,' in Anderson, Stuart (ed.) *Making Medicines: A Brief History of Pharmacy and Pharmaceuticals* (London, 2005), pp. 97–115.
41. Haines, *National Experience*, pp. 164–5.
42. Sonnedecker, G. 'Pharmaceutical Education 1867 and 1967', in *One Hundred Years of Pharmacy in Canada 1867–1967* (Toronto, 1969), pp. 1–10.
43. Ryan, *Organized Pharmacy*, p. 33.
44. Ibid., p. 52.
45. Singh, Harkishan. *Pharmaceutical Education* (Delhi, 1998), pp. 11–22.
46. *Chemist and Druggist*, 23 (1881), p. 467.

47. Ibid., p. 471.
48. Ibid., pp. 490–1.
49. Anderson, Stuart. 'The Quest for Reciprocal Recognition of Colonial Pharmaceutical Qualifications 1896 to 1914', *Pharmaceutical Historian*, 40 (2010), pp. 13–5.
50. *Chemist and Druggist*, 51 (1897), p. 85.
51. *Pharmaceutical Journal and Transactions*, 57 (1896), p. 301.
52. Ibid.
53. Haines, *National Experience*, p. 153.
54. *Chemist and Druggist*, 51 (1897), p. 85.
55. Ibid.
56. Ibid.
57. Ibid., p. 53.
58. Ibid., 52 (1898), p. 71.
59. *Pharmaceutical Journal and Transactions*, 59 (1897), p. 25.
60. Ibid., 52 (1898), p. 71.
61. Ibid., 51 (1897), p. 53.
62. Haines, *National Experience*, p. 152.
63. *Chemist and Druggist*, 51 (1897), p. 53.
64. Ibid., p. 457.
65. Ibid., 53 (1898), p. 279.
66. Ibid., 55 (1899), p. 129.
67. Ibid., 54 (1899), p. 10.
68. Ibid., 55 (1899), p. 169.
69. Ibid., p. 18.
70. *Pharmaceutical Journal and Pharmacist*, 82 (1909), p. 2.
71. *Hansard*, 198 (18 December 1908), pp. 2190–4.
72. *Pharmaceutical Journal and Pharmacist*, 81 (1908), p. 824.
73. Ibid., 86 (1911), p. 223.
74. Lloyd, Alistair. 'Pharmacy in the Australian Colonies: The British Influence, Part 2', *Pharmaceutical Historian*, 19 (1989), p. 6.
75. *Pharmaceutical Journal and Pharmacist*, 86 (1911), p. 223.
76. Crellin, John. *A Social History of Medicines in the Twentieth Century: To Be Taken Three Times a Day* (Binghamton, 2004), p. 283.
77. 'Letter, Newfoundland Pharmaceutical Association to Colonial Secretary', 20 January 1912, Newfoundland Pharmaceutical Association Archives.

78. 'Letter, Newfoundland Pharmaceutical Association to Colonial Secretary', 23 January 1913, Newfoundland Pharmaceutical Association Archives.
79. Lloyd, 'British Influence', pp. 6–8.
80. *Pharmaceutical Journal and Pharmacist*, 121 (1929), p. 76.
81. Ibid., p. 78.
82. Anderson, Stuart. 'The Proposal for an Imperial Pharmaceutical Qualification 1929', *Pharmaceutical Historian*, 41 (2011), pp. 27–30.
83. *Pharmaceutical Journal*, 262 (1999), p. 407.
84. Haines, *National Experience*, p. 141.
85. Monnais, Laurence. *The Colonial Life of Pharmaceuticals: Medicines and Modernity in Vietnam* (Cambridge, 2019).
86. Anagnostou, Sabine. 'Forming, Transfer and Globalization of Medical-Pharmaceutical Knowledge in South East Asian Missions', *Journal of Ethnopharmacology*, 167 (2015), pp. 78–85.
87. Bonnemain, Bruno. 'Colonisation et pharmacie, 1830–1962: une presence diversifiee de 130 ans des pharmaciens francais', *Revue d'histoire de la pharmacie*, 95 (2008), pp. 311–4.
88. Joshi, Madhusudan. 'Pharmacy Profession in Goa During the Portuguese Period, 1842–1961', *Proceedings of World Congress of Pharmacy and Pharmaceutical Sciences* (Lisbon, 2010), p. H4-03.
89. Sowell, D. 'Race and Authorization of Biomedicine in Yucatan, Mexico', in De Barros, J., Palmer, S. and Wright, D. (eds) *Health and Medicine in the Circum-Caribbean, 1800–1968* (Abingdon, 2009), p. 90.
90. Trujillo-Pagan, N. 'The Politics of Professionalization: Puerto Rican Physicians During the Transition from Spanish to US Colonialism', in De Barros, *Health and Medicine*, pp. 152–62.
91. Denzin, N.K. and Mettlin, C.J. 'Incomplete Professionalization: The Case of Pharmacy', *Social Forces*, 46 (1968), pp. 375–81.
92. Goode, W.J. 'Encroachment, Charlatanism and the Emerging Profession: Psychiatry, Sociology and Medicine', *American Sociological Review*, 25 (1960), pp. 902–14.
93. Dingwall, R. and Wilson, E. 'Is Pharmacy Really an Incomplete Profession?' *Perspectives on Social Problems*, 7 (1995), pp. 111–28.
94. Thompson, C.J.S. *The Mystery and Art of the Apothecary* (London, 1927), p. 280.

95. Susskind, Richard and Susskind, Daniel. *The Future of the Professions: How Technology Will Transform the Work of Human Experts* (London, 2020).

Index

S. Anderson, *Pharmacy and Professionalization in the British Empire, 1780–1970*, Medicine and Biomedical Sciences in Modern History,
https://doi.org/10.1007/978-3-030-78980-0

The manufacturer's authorised representative in the EU is Springer Nature Customer Service Centre GmbH, Europaplatz 3, 69115 Heidelberg, Germany. If you have any concerns regarding our products, please contact ProductSafety@springernature.com

Printed and bound by CPI Group (UK) Ltd, Croydon, CR0 4YY

30/09/2026

02225455-0018